The Evolution and Emergence of RNA Viruses

Oxford Series in Ecology and Evolution
Edited by Paul H. Harvey and Robert M. May

The Comparative Method in Evolutionary Biology
Paul H. Harvey and Mark D. Pagel
The Cause of Molecular Evolution
John H. Gillespie
Dunnock Behaviour and Social Evolution
N. B. Davies
Natural Selection: Domains, Levels, and Challenges
George C. Williams
Behaviour and Social Evolution of Wasps: The Communal Aggregation Hypothesis
Yosiaki Itô
Life History Invariants: Some Explorations of Symmetry in Evolutionary Ecology
Eric L. Charnov
Quantitative Ecology and the Brown Trout
J. M. Elliott
Sexual Selection and the Barn Swallow
Anders Pape Møller
Ecology and Evolution in Anoxic Worlds
Tom Fenchel and Bland J. Finlay
Anolis Lizards of the Caribbean: Ecology, Evolution and Plate Tectonics
Jonathan Roughgarden
From Individual Behaviour to Population Ecology
William J. Sutherland
Evolution of Social Insect Colonies: Sex Allocation and Kin Selection
Ross H. Crozier and Pekka Pamilo
Biological Invasions: Theory and Practice
Nanako Shigesada and Kohkichi Kawasaki
Cooperation Among Animals: An Evolutionary Perspective
Lee Alan Dugatkin
Natural Hybridization and Evolution
Michael L. Arnold
Evolution of Sibling Rivalry
Douglas Mock and Geoffrey Parker
Asymmetry, Developmental Stability, and Evolution
Anders Pape Møller and John P. Swaddle
Metapopulation Ecology
Ilkka Hanski
Dynamic State Variable Models in Ecology: Methods and Applications
Colin W. Clark and Marc Mangel
The Origin, Expansion, and Demise of Plant Species
Donald A. Levin
The Spatial and Temporal Dynamics of Host-Parasitoid Interactions
Michael P. Hassell
The Ecology of Adaptive Radiation
Dolph Schluter
Parasites and the Behavior of Animals
Janice Moore
Evolutionary Ecology of Birds
Peter Bennett and Ian Owens
The Role of Chromosomal Change in Plant Evolution
Donald A. Levin
Living in Groups
Jens Krause and Graeme Ruxton
Stochastic Population Dynamics in Ecology and Conservation
Russell Lande, Steiner Engen and Bernt-Erik Sæther
The Structure and Dynamics of Geographic Ranges
Kevin J. Gaston
Animal Signals
John Maynard Smith and David Harper
Evolutionary Ecology: The Trinidadian Guppy
Anne E. Magurran
Infectious Diseases in Primates
Charles L. Nunn and Sonia M. Altizer
Computational Molecular Evolution
Ziheng Yang
The Evolution and Emergence of RNA Viruses
Edward C. Holmes

The Evolution and Emergence of RNA Viruses

EDWARD C. HOLMES

*Center for Infectious Disease Dynamics,
Department of Biology, The Pennsylvania State University,
Pennsylvania, USA and Fogarty International Center,
National Institutes of Health, Bethesda, USA*

OXFORD
UNIVERSITY PRESS

Great Clarendon Street, Oxford OX2 6DP

Oxford University Press is a department of the University of Oxford.
It furthers the University's objective of excellence in research, scholarship,
and education by publishing worldwide in

Oxford New York

Auckland Cape Town Dar es Salaam Hong Kong Karachi
Kuala Lumpur Madrid Melbourne Mexico City Nairobi
New Delhi Shanghai Taipei Toronto

With offices in

Argentina Austria Brazil Chile Czech Republic France Greece
Guatemala Hungary Italy Japan Poland Portugal Singapore
South Korea Switzerland Thailand Turkey Ukraine Vietnam

Oxford is a registered trade mark of Oxford University Press
in the UK and in certain other countries

Published in the United States
by Oxford University Press Inc., New York

© Edward C. Holmes 2009

The moral rights of the author have been asserted
Database right Oxford University Press (maker)

First published 2009

All rights reserved. No part of this publication may be reproduced,
stored in a retrieval system, or transmitted, in any form or by any means,
without the prior permission in writing of Oxford University Press,
or as expressly permitted by law, or under terms agreed with the appropriate
reprographics rights organization. Enquiries concerning reproduction
outside the scope of the above should be sent to the Rights Department,
Oxford University Press, at the address above

You must not circulate this book in any other binding or cover
and you must impose the same condition on any acquirer

British Library Cataloguing in Publication Data

Data available

Library of Congress Cataloging in Publication Data

Data available

Typeset by Newgen Imaging Systems (P) Ltd., Chennai, India
Printed in Great Britain
on acid-free paper by
MPG Books Group, Bodmin and King's Lynn

ISBN 978–0–19–921112–8 (Hbk.)
ISBN 978–0–19–921113–5 (Pbk.)

10 9 8 7 6 5 4 3 2 1

Everyone knows that pestilences have a way of recurring in the world; yet somehow we find it hard to believe in ones that crash down on our heads from a blue sky.

Albert Camus, *La Peste*

For Rachel and Scott.

Preface

Hurricanes are not good for much. This book was conceived in a 'hotel' room in Valladolid, Mexico, during October 2005 where my wife and I had taken shelter from Hurricane Wilma, a category 5 storm responsible for the lowest pressure ever recorded in the Caribbean. With little else to do for 3 days, I set about planning the book that Paul Harvey and Bob May, the series editors, had generously asked me to write. As the good citizens of Cancún, Cozumel, Playa del Carmen, and Tulum will testify, I escaped lightly.

I wish to thank the following people who graciously commented on various chapters: Siobain Duffy, Adrian Gibbs, John McCauley, Andrés Moya, Cadhla Ramsden, and Rafa Sanjuán. The text was greatly improved by their diligent reading, intelligent criticism, and sound ideas. As should go without saying, any errors that remain are entirely my own doing. In addition, I benefited greatly from numerous discussions with John Aaskov, Elodie Ghedin, Bryan Grenfell, Oliver Pybus, Andrew Rambaut, Tony Schmitt, Laura Shackelton, and Paolo Zanotto. I am also grateful to Helen Eaton and Ian Sherman at Oxford University Press for their relaxed encouragement. Finally, I am indebted to my wife Rachel for her patience, support, and willingness to leave me alone on Saturday mornings.

Contents

1 Introduction	1
1.1 Why study RNA virus evolution?	1
1.1.1 Ways to study viral evolution	2
1.1.2 The scope of this book	4
1.2 RNA viruses and evolutionary biology	5
1.2.1 The RNA virus world	6
1.3 The basics of viral biology	8
1.3.1 A cursory history of virology	8
1.3.2 Virology 101	9
1.3.3 Exploring the virosphere	13
2 The origins of RNA viruses	15
2.1 Introduction	15
2.1.1 The perils of deep viral phylogeny	15
2.2 Theories for the origin of RNA viruses	16
2.2.1 The regressive origin theory	17
2.2.2 RNA viruses as escaped genes	18
2.2.3 RNA viruses and the RNA world	20
2.2.4 Eigen's paradox	22
2.2.5 The taxonomic distribution of RNA viruses	24
2.2.6 Conserved protein structures	25
2.3 Deep phylogenetic relationships among RNA viruses	28
2.3.1 The 'higher-order' relationships of RNA viruses	29
2.3.2 Phylogenies based on genome organization	34
2.3.3 Phylogenies based on protein structure	34
2.4 RNA viruses and the evolution of the genetic code	35
3 The mechanisms of RNA virus evolution	37
3.1 The evolutionary dynamics of RNA viruses	37
3.1.1 Mutation rates in RNA viruses and their determinants	37
3.1.2 A comparison of substitution rates in viruses	39
3.1.3 Differences in viral generation time	42

3.1.4 Slowly evolving RNA viruses?	43
3.1.5 Rapidly evolving ssDNA viruses	44
3.1.6 What sets the rate of RNA virus evolution?	45
3.1.7 Trade-offs and the evolution of mutation rates	46
3.1.8 Mutation rates and mutational loads	47
3.1.9 Are RNA viruses trapped by high mutation rates?	48
3.2 Recombination and reassortment in RNA virus evolution	48
3.2.1 Recombination frequency in RNA viruses	50
3.2.2 Detecting recombination in RNA viruses	51
3.2.3 What determines the rate of recombination in RNA viruses?	52
3.2.4 Recombination and deleterious mutation	53
3.3 Natural selection, genetic drift, and the genetics of adaptation	55
3.3.1 Effective population sizes in viral evolution	56
3.3.2 Transmission bottlenecks	58
3.3.3 The dynamics of allele fixation: estimating selection coefficients	59
3.3.4 The importance of hitch-hiking	62
3.3.5 Patterns of synonymous and nonsynonymous evolution	63
3.3.6 Natural selection and transmission mode	63
3.3.7 Escape from intrinsic immunity	65
3.3.8 Strictly neutral evolution in RNA viruses?	66
3.3.9 Determinants of codon bias (and nucleotide composition) in RNA viruses	68
3.4 Deleterious mutation and RNA virus evolution	70
3.4.1 Deleterious mutation and intra-host genetic diversity	73
3.4.2 The importance of defective interfering particles and complementation	74
3.4.3 Complementation may be commonplace in RNA viruses	75
3.5 Epistasis in RNA virus evolution	77
3.5.1 Epistasis and robustness	78
3.5.2 The importance of RNA secondary structure	80
3.5.3 Convergence and pleiotropy	82
3.6 The importance of intra-host viral diversity	83
4 The RNA virus quasispecies	**87**
4.1 What is a quasispecies?	87
4.2 The great quasispecies debate	90
4.2.1 What's in a name: quasispecies or polymorphism?	91
4.2.2 Is quasispecies theory different from 'classical' population genetics?	92
4.2.3 Does genetic drift destroy the quasispecies?	92
4.2.4 The evidence from 'digital organisms'	93
4.2.5 Experimental tests of quasispecies theory	93
4.2.6 Comparative analyses of RNA virus quasispecies	96

	4.2.7 Recombination and the quasispecies	99
	4.2.8 'Memory' in viral quasispecies	99
4.3	Error thresholds, extinction thresholds, and error catastrophes	100
4.4	Concluding remarks	103

5 Comparative genomics and the macroevolution of RNA viruses — 104

5.1 The evolution of genome architecture in RNA viruses — 104
    5.1.1 The evolution of genome size — 104
    5.1.2 The exceptions: coronaviruses and roniviruses — 107
    5.1.3 The evolution of genome organization: an overview — 109
    5.1.4 The evolution of genome segmentation — 111
    5.1.5 The evolution of genome orientation and dsRNA viruses — 113
    5.1.6 The evolution of overlapping reading frames — 114
5.2 The processes of genome evolution — 116
    5.2.1 Gene duplication in RNA virus evolution — 117
    5.2.2 LGT among viruses and hosts — 118
    5.2.3 Modular evolution — 119
5.3 Patterns and processes of macroevolution in RNA viruses — 120
    5.3.1 Speciation in RNA viruses — 121
    5.3.2 A birth-death model of viral evolution — 124
    5.3.3 The birth and death of endogenous retroviruses — 128

6 The molecular epidemiology, phylogeography, and emergence of RNA viruses — 131

6.1 Phylodynamics: linking viral evolution at the phylogenetic and epidemiological scales — 131
    6.1.1 Coalescent approaches to viral epidemiology — 133
6.2 Cross-species transmission, co-divergence, and emergence — 135
    6.2.1 The RNA/DNA divide again — 135
    6.2.2 Inferring co-divergence — 137
    6.2.3 The evolution of persistence in RNA viruses — 138
    6.2.4 Host phylogeny and viral emergence — 139
6.3 The evolutionary genetics of viral emergence — 142
    6.3.1 Adaptation and emergence — 142
    6.3.2 'Off-the-shelf' emergence — 144
    6.3.3 The fitness landscapes of emergence — 146
    6.3.4 Recombination, reassortment, and viral emergence — 147
6.4 The phylogeography of human viruses — 148
    6.4.1 Viruses differ in phylogeographic pattern — 149
6.5 Major transitions in human ecology and viral evolution — 153
    6.5.1 The transitions — 154
    6.5.2 Immunodeficiency and disease emergence — 155

7 Case studies in RNA virus evolution and emergence 156
7.1 The evolutionary biology of influenza virus 156
    7.1.1 The diversity of influenza virus 156
    7.1.2 The evolution of avian influenza virus 158
    7.1.3 Antigenic drift and shift 161
    7.1.4 Antigenic cartography and the punctuated evolution of HA 162
    7.1.5 Genome-wide evolutionary processes 165
7.2 The emergence and evolution of HIV 167
    7.2.1 A brief history of HIV/AIDS 167
    7.2.2 The genetic diversity of HIV 169
    7.2.3 What and why are subtypes? 172
    7.2.4 The origins and spread of HIV 173
    7.2.5 The intra- and inter-host evolutionary dynamics of HIV 176
    7.2.6 The great obsession moves to HIV 177
    7.2.7 Epidemiological scale dynamics 178
7.3 The evolution of dengue virus 180
    7.3.1 The origins of DENV 182
    7.3.2 DENV biodiversity 184
    7.3.3 Lineage birth-death in DENV 186
    7.3.4 DENV fitness 187
    7.3.5 Comparing dengue and yellow fever 188
    7.3.6 Why no yellow fever in Asia? 190
7.4 The phylogeography and evolution of rabies virus 191
    7.4.1 The world of lyssaviruses 192
    7.4.2 The spatiotemporal dynamics of RABV 195

8 Epilogue 198

References 201

Index 249

1

Introduction

1.1 Why study RNA virus evolution?

Viral diseases, particularly the exotic and the fatal, hold a unique fascination to scientists and the general public alike. Because of books like *The Hot Zone* (Preston 1994), which glamorizes outbreaks of highly virulent filoviruses, the public image of RNA viruses is a complex combination of the frightening and the alluring. While this has certainly given them name recognition, the accounts of RNA viruses that are increasingly commonplace in the popular media are also frustratingly inaccurate, as they are often given capabilities that could never have arisen through evolution by natural selection.

The marginalization of viruses also occurs to some extent in evolutionary biology. Although abundant in nature, viruses are sometimes not considered as worthy items for scientific endeavour as the venerable *Drosophila melanogaster* or *Escherichia coli*. A major aim of this book, albeit a rather hidden one, is to show that RNA viruses are as valuable a set of organisms in which to study evolutionary processes as fruit flies or bacteria, and have the added bonus that evolutionary hypotheses can be tested far more rapidly and often with more precision. As a simple case in point, RNA viruses represent one of the few systems in which it is possible to accurately measure the fitness distributions of new mutants (Eyre-Walker and Keightley 2007).

The enormous burden of human mortality and morbidity caused by RNA viruses represents an unfortunate, often highly politicized, but extremely powerful back-drop to discussions of their evolution. This burden is particularly severe for the developing world, where there is little evidence that we winning the war against viral infections. Figures for the year 2002 provided by the World Health Organization show that, globally, almost 28 million people died of HIV/AIDS, and a staggering 18 million of diarrhoeal diseases, a significant proportion of which are due to infection by rotaviruses. Similarly, over 600 000 people died of measles even though an effective vaccine has been available since 1963.

Aside from mortality, the economic costs associated with viral infections of humans, domestic animals, and agricultural plants are staggering. For example, although only 8437 people were known to have been infected with severe acute respiratory syndrome (SARS) coronavirus (SARS-CoV) during the highly publicized outbreak of 2002–2003, with 813 deaths, the global economic bill has been estimated to be in excess of US$50 billion. Similarly, the 2001 epidemic of foot-and-mouth disease in the UK (due to foot-and-mouth disease virus, FMDV) was a major blow to British agriculture, resulting in the death or slaughter of over 3.5 million cattle and a total estimated cost of perhaps $4 billion.

Although it is tempting to over-state the importance of evolutionary ideas as a way of alleviating morbidity and mortality, it is clear that a detailed understanding of the patterns and process of viral evolution can sometimes have major implications for public and animal health. For example, documenting the mechanics of evolutionary change may be critical to the design of future intervention strategies, particularly in the case of antigenically variable pathogens such as the human immunodeficiency virus (HIV), hepatitis C virus, and influenza A virus, where vaccination has proven impossible or only transiently effective. More esoterically, one of the most interesting debates concerning HIV, and which has had a direct impact on strategies for drug treatment, is whether the within-host evolution of this virus is dominated by the stochastic process of genetic drift or the deterministic process of natural selection (Leigh Brown 1997; Rouzine and Coffin 1999; see section 7.2).

1.1.1 Ways to study viral evolution

Broadly speaking, there are three ways in which the study of RNA virus evolution has proceeded, and can do so in the future: the theoretical, the experimental, and the comparative. While all three approaches have their individual costs and benefits, this book is set squarely within the framework of comparative biology, with the evolutionary (and often phylogenetic) analysis of viral gene and genome sequences as the main analytical tool.

The theoretical approach to studying viral evolution has a long history, rooted in ecology and population genetics, and has the power to explore, in exquisite analytical detail, the consequences of specific evolutionary or epidemiological processes. In the context of RNA viruses there is no doubt that good theory has greatly illuminated many areas of study and suggested new avenues for research, including innovative ways to design antiviral agents (Moya *et al*. 2000; Nowak and May 2000; Bull *et al*. 2005; Wilke 2005). Understandably, the potential limitation of all theoretical approaches is that they necessarily construct an idealized model of viral biology, sometimes divorced from the true complexities of the evolutionary process in nature.

The experimental approach has perhaps been the dominant mode of study in RNA virus evolution to date (see Domingo and Holland 1997 for a benchmark review). Such is the success of experimental evolution that it has been used to reveal the mechanisms of evolutionary change in general, and of RNA viruses in particular (see, for example, Turner and Chao 1999; Elena and Lenski 2003; Sanjuán *et al*. 2004b). The power of experimental analyses of RNA viruses is largely a function of their tractability for laboratory study, particularly their rapid rates of mutation, so that evolution can be followed in real time and in a variety of cellular environments. In addition, there is often a direct and simple link between genotype and phenotype, and relative fitness, measured using growth kinetics, is easy to assess. However, evolution *in vitro* is often very different from evolution in nature, particularly for systems other than bacteriophage (where *in vitro* and *in vivo* can be argued to be much the same thing), making it difficult to generalize from individual cell types to whole organisms, and then to the epidemiological scale. For example, regions of the coronavirus genome that appear to have no function *in vitro* or *in vivo* are clearly important in nature (Gorbalenya *et al.* 2006). In broader terms, the desire to use RNA viruses as tools to understand evolutionary processes in general may mean that experimental studies have perhaps revealed rather less about the intricacies of viral evolution than might be hoped.

The power of the comparative approach is that it considers the evolutionary process in nature—genomes are sampled from real populations—and utilizes analytical tools with a strong basis in theory (Harvey and Pagel 1991). Its limitation, as will become evident in this book, is that teasing out the contribution of individual evolutionary processes can be troublesome when the analysis is always 'retrospective' to some extent, and that sample sizes are often both small and biased. Thankfully, the rise of 'pyrosequencing' and related technologies is making the generation of large amounts of complete genome sequence data from a diverse array of RNA viruses increasingly easy, presenting many new opportunities to reveal the nature of viral evolution (Holmes 2007). An important subtext in this book is therefore to demonstrate, perhaps rather shamelessly, the contribution that comparative biology has made to our understanding of RNA virus evolution. In doing so, I will avoid mathematical descriptions of viral evolution as much as possible, in part because these have been undertaken in detail by more able authors (for example, Nowak and May 2000), but more importantly because I wish to highlight the biological limitations in our current understanding of viral evolution. Indeed, many of the major problems in RNA virus evolution are currently limted more by data than theory. Finally, there is also perhaps a lingering mistrust of computational approaches to virology that needs to redressed, although hopefully things have moved on since the *Annual Reviews of Microbiology* published an article entitled 'Coping with computers and computer evangelists'.

Another reason why I take an explicitly comparative approach to the study of RNA virus evolution is that phylogeny is one area where evolutionary ideas have very clearly entered the virological mainstream. The success of phylogeny in virology, the Trojan horse of evolutionary biology, is reflected in the number of papers published in the three top virology journals—the *Journal of Virology*, *Virology*, and the *Journal of General Virology*—that contain phylogenetic trees as part of their data analysis. My own unscientific survey reveals that during 2007 almost 200 papers in these journals described some sort of phylogenetic analysis, and often using the most up-to-date methods. And, of course, where there is evolutionary pattern manifest as phylogeny, there is also evolutionary process. Perhaps of more importance is that although the science of virology has classically been subdivided according to the type of host species a specific virus infects, be it an animal, plant, or bacterium, the *Journal of Virology* now has a more general section devoted to Genetic Diversity and Evolution. In short, the study of viral evolution has come of age.

1.1.2 The scope of this book

Although this book predominantly considers RNA viruses, broadly defined to include both retroviruses and viroids (a simple classification scheme is presented in Table 1.1), this does not mean that I entirely ignore the other classes of viral agent. Indeed, there is mounting evidence, discussed in detail in this book, that single-strand (ss) DNA viruses evolve in much the same manner as RNA viruses. In the

Table 1.1 A simple classification of RNA viruses and viroids according to their replication enzyme (RNA-dependent RNA polymerase (RdRp) or reverse transcriptase (RT)). Classes of viruses are normally shown at the family level, although some unclassified viruses are shown at the genus level. The major type of host is shown in parentheses. Information taken from the 2005 International Committee on Taxonomy of Viruses (ICTV) classification of viruses (Fauquet *et al.* 2005) with a few minor adjustments. The genome sizes of these viruses (excluding viroids) are shown in Fig. 5.1.

		Unsegmented	Segmented
RdRp-utilizing	ssRNA–	*Bornaviridae* (V)	*Arenaviridae* (V)
		Filoviridae (V)	*Bunyaviridae* (IPV)
		Paramyxoviridae (V)	*Ophiovirus* (P)
		Rhabdoviridae (IPV)	*Orthomyxoviridae* (V)
			Tenuivirus (IP)
			Varicosavirus (P)
	ssRNA+	*Arteriviridae* (V)	*Benyvirus** (P)
		Astroviridae (V)	*Bromoviridae** (P)
		Barnaviridae (F)	*Cheravirus** (P)
		Caliciviridae (V)	*Comoviridae** (P)
		Closteroviridae† (P)	*Furovirus** (P)
		Coronaviridae (V)	*Hordeivirus** (P)
		Dicistroviridae (I)	*Idaeovirus** (P)
		Flaviviridae (IV)	*Nodaviridae* (IV)
		Flexiviridae (P)	*Ourmiavirus** (P)
		Hepeviridae (V)	*Pecluvirus** (P)
		Iflavirus (I)	*Pomovirus** (P)
		Leviviridae (B)	*Sadwavirus** (P)
		Luteoviridae (P)	*Tobravirus** (P)
		Marnaviridae (A)	
		Narnaviridae (F)	
		Picornaviridae (V)	
		Potyviridae† (P)	
		Roniviridae (I)	
		Sequiviridae (P)	
		Sobemovirus (P)	
		Tetraviridae† (I)	
		Tobamovirus (P)	
		Togaviridae (IV)	
		Tombusviridae*† (P)	
		Tymoviridae (P)	
		Umbravirus (P)	

Table 1.1 (*Contd.*)

	Unsegmented	Segmented
dsRNA	*Endornavirus* (P) *Hypoviridae* (F) *Totiviridae* (FR)	*Birnaviridae* (IV) *Chrysoviridae* (F) *Cystoviridae* (B) *Partitiviridae** (FP) *Reoviridae* (FIPV)
RT-utilizing	*Caulimoviridae* (P) *Hepadnaviridae* (V) *Metaviridae* (FIVP) *Pseudoviridae* (IVP) *Retroviridae* (V)	
Viroids	*Pospiviroidae* (P) *Avsunviroidae* (P)	

Host groups (in parentheses) are as follows: A, algae; B, bacteria; F, fungi; I, invertebrates; P, plants; R, protists; V, vertebrates. ds, double-strand; ss, single strand; RNA+, positive-sense RNA viruses; RNA−, negative-sense RNA viruses.
* Also multicomponent, as the genome is present in multiple virus particles.
† Sometimes possess multiple segments.

case of large double-strand (ds) DNA viruses, I will tend to use these simply to provide context; to show how their evolution differs from those viruses with RNA genomes, reflecting a fundamental difference in mutational dynamics. My bias towards RNA viruses stems from the large amount of gene and genome sequence data available for these infectious agents, the key role they play as agents of human disease, and, perhaps most important of all, their capacity for remarkably rapid evolutionary change, so that mutations appear and can be fixed within the time frame of human observation. Such rapid evolution has meant that RNA viruses have also been classified as 'measurably evolving populations' (Drummond *et al.* 2003b) (although this does not mean that evolution is only measurable in RNA viruses; Elena and Lenski 2003). I also focus most on those RNA viruses that infect humans, if only because more data are available for human viruses than for any other group.

Finally, because this book considers the evolution of retroviruses—in which the enzyme reverse transcriptase (RT) makes a DNA copy of an RNA genome—and because RT has clear homology to some cellular genes (such as telomerase; see section 2.2), there is an automatic link between the evolution of viruses and that of their hosts.

1.2 RNA viruses and evolutionary biology

Whereas a great deal has been written about the evolutionary biology of RNA viruses, an overview of the patterns and processes of evolution in these fascinating infectious agents is still lacking. Much of the prior work on RNA virus evolution has focused on the largely descriptive molecular epidemiology of specific infections, has only utilized RNA viruses as tools to address broader evolutionary questions, or has not been set within what might be considered the mainstream of modern neo-Darwinism. If there is one over-arching aim of this book it is to place RNA viruses firmly with the mainstream of modern evolutionary biology, rather than being thought of as a quirky off-shoot (see Morse 1994 for a similar argument). To achieve this I will address, in as much detail as space permits, the causes and consequences of evolutionary change in RNA viruses.

1.2.1 The RNA virus world

Although RNA viruses are subject to the same basic evolutionary processes as other organisms, I will argue that they, along with ssDNA viruses, occupy a region of evolutionary parameter space that is very different to that where dsDNA-based organisms reside (Fig. 1.1). Whereas the latter are characterized by a high replication fidelity per nucleotide, large genomes, and often low numbers of offspring (at least in higher eukaryotes), the evolution of RNA viruses is characterized by an extremely low replication fidelity per nucleotide, small genomes, and very high offspring numbers. In particular, population sizes must be enormous in RNA viruses to offset the fact that most progeny contain deleterious mutations that cannot be masked by wild-type

RNA and ssDNA

  High mutation rate (per nt)
   Small genome size (<32 000 nt)
    Population sizes always large
     Little LGT and gene duplication
      Overlapping reading frames common
       Often little recombination

       Often frequent recombination
      Overlapping reading frames uncommon
     Frequent LGT and gene duplication
    Population sizes can be small
   Large genome size
  Low mutation rate (per nt)

dsDNA

Fig. 1.1 The differing evolutionary parameter spaces occupied by RNA- and DNA-based life-forms. Note that the properties shown are general ones, and cannot be applied to every taxon in each category (particularly for DNA-based organisms). Crucially, from an evolutionary perspective, ssDNA viruses behave more like RNA than dsDNA life forms. LGT, lateral gene transfer; nt, nucleotides.

alleles carried in heterozygotes. I will further argue that many of the major aspects of viral biology, including their genome organizations and potential to jump host species, can be set within this fundamental evolutionary dichotomy. Similarly, I will propose that RNA viruses create evolutionary novelty in a way that is profoundly different to that observed in other organisms: while eukaryotes often generate genes with new functions through gene (or genome) duplication, and bacteria through lateral gene transfer (LGT), the constrained genome sizes exhibited by RNA viruses mean that they must again usually rely on a combination of rampant mutation and huge population size to create the genetic diversity they need to adapt to new environments. Whereas the idea that RNA viruses are somehow evolutionarily 'unique' has a long history, and is at the heart of the quasispecies theory of viral evolution that I discuss in detail in Chapter 4, the differences between RNA viruses and other organisms are in reality more quantitative than qualitative.

To extend this theme a little further, perhaps the cornerstone argument of this book is that the key patterns and processes of RNA virus evolution are in large part a function of their intrinsically high rates of mutation, and that it is extremely difficult to evolve the radically higher replication fidelities that characterize most DNA-based organisms. Such highly error-prone replication has a number of direct evolutionary consequences: that the genome sizes of RNA viruses are small, that the burden of deleterious mutation is high, so that major increases in mutation rate (exemplified by the presence of 'mutator' strains) will drastically reduce fitness, that recombination and reassortment are unable to rescue viral populations from this load of deleterious mutation, that complementation is both commonplace and of unappreciated evolutionary importance, and that while rapid evolution ensures that phylogenetic relationships can be resolved with remarkable precision in the short term, the pattern of evolutionary history is soon eroded, thwarting attempts to determine viral origins from gene sequence data alone. I will also argue that the error-prone replication system of RNA viruses most likely reflects their ancient origin in an 'RNA world'.

Another important idea in this book is that despite their rapid rates of mutation, RNA viruses experience more evolutionary constraints than is generally imagined. Although this statement may at first seem rather strange given that the rapidity of RNA virus evolution is something of a mantra for microbiologists, it is supported by a considerable weight of data. Despite frequent suggestions (including a number made by me) that RNA viruses are able to quickly adapt to a myriad of new and changing environments, the reality of the matter is that viruses are often unable to find solutions to major adaptive challenges, or do so relatively slowly. As a particular case in point, although RNA viruses frequently cause transient 'spill-over' infections in new host species (in which only a single individual is infected), most RNA viruses do not evolve sustained transmission cycles in these emergent hosts. Cross-species transmission, the cornerstone of viral emergence, may therefore be more difficult than is often imagined. This topic is discussed in detail in Chapter 6. While HIV is able to rapidly evolve resistance to individual antiviral agents, such as AZT (Zidovudine), it fairs less well against combinations of drugs. Although it is often suggested that a particular RNA virus will eventually evolve a new mode of transmission—for example, a predominantly sexually transmitted virus like HIV will evolve aerosol or vector-borne transmission—in reality such major changes in phenotype usually occur relatively rarely. These examples, and many more, point to same conclusion: that there are major limitations to the adaptability of RNA viruses, particularly when the adaptive pathways in question are complex, and which may in turn have major implications for their control. An important goal of this book is therefore to reveal the exact nature of these constraints and explain why they exist. In particular I will argue that, paradoxically, adaptive constraints are a necessary side effect of rapid mutation (see also Eigen 1996 and Reanney 1982). However, any emphasis on evolutionary constraints should not be taken to mean that RNA viruses are conservative entities: to be sure, their evolutionary rates are still frighteningly quick, they are still able to thwart many of our attempts to control them with vaccines and drugs, and they often jump species boundaries and emerge in new hosts. Rather, my argument is that the cases when evolution is not rapid, when the evolution of drug resistance is unsuccessful, or when adaptation to new hosts does not occur, provide as important an insight into the evolutionary process as those cases of 'fast-forward' change that usually dominate discussions of RNA virus evolution.

As well as considering the nature of evolutionary processes, a considerable chunk of this book (as reflected in its title) is dedicated to the question of viral emergence. Arguably, it is the cross-species transmission of RNA viruses and their establishment in new hosts that has placed them within the sphere of interest of a wider range of biologists, particularly since human populations can be considered as under 'threat' from a number of emerging viruses. While the study of viral emergence is of great importance, I will propose (in Chapter 5) that in a broader sense cross-species transmission is just one of the 'macroevolutionary' processes exhibited by RNA viruses.

1.3 The basics of viral biology

Although this is assuredly not a textbook of virology (and confused readers should stop now) it is important to outline a few of the basic concepts in virus genetics, design, and function, particularly those that will have a major bearing on the evolutionary issues discussed in later chapters. In addition, it is necessary to define some of the virological terms that are used frequently in this book.

1.3.1 A cursory history of virology

The word 'virus' has its roots in the Latin for 'poison', and has been used in the context of disease for almost 400 years. Following a very liberal interpretation, viruses were first described, if only indirectly, by the English doctor Edward Jenner (1749–1823) who is better known as the 'inventor' of vaccination (although a similar approach had been used previously in both China and the Islamic world). Despite the advance made by Jenner, he had no idea that a virus was a 'living' agent, a fact that would not be confirmed for another 100 years, at the junction of the nineteenth and twentieth centuries, when the science of virology truly came of age. Although a number of figures are responsible for the birth of virology, three people working on tobacco mosaic virus (TMV)—the German Adolf Mayer (1843–1942), the Russian Dimitri Ivanovsky (1864–1920), and the Dutchman Martinus Beijerinck (1851–1931)—are generally acknowledged as the discoverers of viruses. In the years that followed the discovery of TMV, a number of 'filterable' agents, which we now know as viruses, were first described, including FMDV (1898; the first animal virus to be discovered), myxomavirus (1898), yellow fever virus (the transmission cycle of which was first described in 1901), rabies virus (1903), and poliovirus (1909). Another important milestone was reached in 1915 when Felix d'Herelle (1873–1949) and Frederick Twort (1877–1950) co-discovered viruses that infect bacteria—the bacteriophages—so presenting the scientific world with some of most important research tools in molecular biology. Bacteriophages have also played a major role in their host's evolution as agents of LGT (Ochman *et al.* 2000). Readers interested in the history of virology should consult the excellent essay on the subject by Arnie Levine (Levine 1996).

1.3.2 Virology 101

In modern parlance, viruses can be defined as obligate parasites that rely on a host cell for replication, be it from an animal, plant, protozoan, or bacterium. In more formal terms, viruses are subcellular organisms whose genomes consist of nucleic acid (RNA and/or DNA), and which replicate inside host cells using host metabolic machinery to differing extents. For example, some viruses carry their own polymerase enzymes, others do not, although all rely on their host's translation apparatus. It is possible that every cellular species of life on Earth experiences at least one RNA virus infection, although it is equally clear that we have only scratched the surface of viral biodiversity, the so-called virosphere. As a dramatic example, to date no RNA virus (nor any ssDNA virus) has been observed in archaebacteria, although there are increasing descriptions of dsDNA viruses from these organisms. Although this may be taken to mean that RNA viruses evolved subsequent to the divergence of the archaebacteria from other taxa, it is more likely that surveys of viral biodiversity in these species have been insufficiently intensive.

Some even debate whether viruses should be classified as 'living' organisms as they must obviously parasitize host cells for key aspects of their life cycle, and a number of viruses, particularly those with ssDNA genomes, do not even encode their own replication enzymes. Indeed, in some respects viruses are no more living than genes are living. Yet, while viruses do not fulfil all the criteria that some may use to classify living organisms (although this is hotly debated; Raoult *et al.* 2004), they do possess perhaps the two most important criteria of all: they reproduce and they evolve by natural selection. Hence the rules of evolutionary change apply to RNA viruses as much as they apply to humans. Therefore, under more liberal definitions of what constitutes a living organism—such as Leslie Orgel's CITREONS (Complex Information-Transforming Reproductive Objects That Evolve by Natural Selection)—viruses are very much alive, and can be studied using the tools of modern evolutionary biology.

There are a number of other infectious agents that have similarities to viruses, but which will only be mentioned briefly in this book, largely because of a lack of available space. These agents are as follows. (i) Viroids: small (246–401 nucleotides, nt) infectious agents of plants that exist as circular, covalently closed ssRNA molecules with complex folding structures that do not encode any proteins (and therefore cannot replicate themselves), and which are becoming increasingly important tools for the study of evolutionary processes (Daròs *et al.* 2006). Viroids will be discussed in more detail with respect to viral origins. (ii) Satellite RNAs: small RNA molecules that require a helper virus for replication upon co-infection but which may encode a coat protein, with human hepatitis delta 'virus' (HDV) an important example. Finally, (iii) virusoids: satellite RNAs of plants that possess a circular, highly base-paired structure like a viroid, but which also differ from viroids in that they require a helper virus.

One reason for the ease of experimental and comparative analysis in RNA viruses is that their genomes are extremely small, ranging from only approximately 2500 nt (with the mitochondrial mitoviruses the smallest) to approximately 31 500 nt (with the mammalian coronaviruses the largest), and with a mean size of approximately 10 000 nt. In contrast, the genome sizes of DNA viruses range from less than 2000 nt for the circoviruses, a group of ssDNA viruses, to an incredible 1 181 404 nt (and over 1000 genes) for the dsDNA mimivirus of amoeba (La Scola *et al.* 2003). Throughout this book I will argue that despite their reliance on DNA-based replication, and even on host DNA polymerases, from an evolutionary perspective ssDNA viruses like circoviruses can effectively be considered as RNA viruses.

The strong constraints on genome size in RNA viruses (which are discussed in more detail in Chapter 5) also mean that they have to adopt a variety of strategies to maximize the amount phenotypic diversity that can be encoded by a small number of nucleotides. This can be thought of as small-genome dynamics. In particular, RNA viruses (as well as a variety of small DNA viruses) commonly express multiple proteins from a single gene sequence. This can be achieved through making a single, large polyprotein and subsequently cleaving this into many smaller proteins, alternative splicing, or via overlapping reading frames, in which multiple proteins are encoded by the same nucleotide sequence. It is also worth noting here that some RNA viruses employ partial genomic copies, known as subgenomic RNAs, as a means of expressing downstream open reading frames (ORFs). The control of gene expression, which I will argue is one of the most fundamental challenges facing RNA viruses and may be responsible for the evolution of major genomic characteristics including segmentation, is explored in Chapter 5.

In many ways viruses do little more than replicate, and differences in how this process occurs have been used as the basis for virus classification (Baltimore 1971). The replication process in RNA viruses involves one of two specific enzymes (Table 1.1). For 'true' RNA viruses, in which an RNA template is used to produce an RNA copy of the genome, the enzyme RNA-dependent RNA polymerase (RdRp; and sometimes known as an RNA replicase) is utilized. This enzyme is of major evolutionary significance because it is notoriously error-prone, lacking any proofreading or repair, thereby generating the extensive genetic variation that serves to characterize RNA viruses. For retroviruses, as well as a number of smaller viral families (notably hepadnaviruses and caulimoviruses which are usually classed as small DNA viruses), a DNA copy of the genome is produced from an RNA template through the use of a RNA-dependent DNA polymerase (RdDp), also known as reverse transcriptase (RT). Like RdRp, RT lacks both proofreading and repair and is therefore subject to abundant mutation. The error rates associated with RdRp and RT are discussed in more detail in section 3.1. Finally, it is also important to recall that because retroviruses integrate into cellular genomes they are also replicated by cellular DNA polymerases. However, because these cellular replication enzymes are of much higher fidelity than either RdRp or RT, with mutation rates some orders of magnitude lower, they do not contribute significantly to viral evolution.

The polymerases are also of note because they represent the only type of protein that all RNA viruses and retroviruses have in common. Indeed, the presence of an RdRp is one of the defining features of RNA viruses and, as such, one of the few characters that can be used to infer their deep phylogenetic relationships (although in Chapter 2 I argue that this exercise has been largely unsuccessful). The only other protein carried by most RNA viruses is the capsid, comprising the protein coat (shell) that surrounds, and so protects, the nucleic acid, although even this is absent from the tiny mitoviruses. The complex of capsid and RNA is often referred to as the nucleocapsid, while the mature viral particle is termed the virion. A number of RNA viruses, including many that are important players in this book (dengue, HIV, influenza A virus), possess virions with an additional outer envelope that contains part of the host cell lipid membrane acquired when they bud from (i.e. exit) the cell, as well as associated sugar-coated glycoproteins. Interestingly, viruses with enveloped virions are far more commonly associated with infections of animals than plants. The virion structure of a fairly typical RNA virus—influenza A virus—is shown in Fig. 1.2.

As discussed in detail in Chapter 5, one of the most remarkable aspects of RNA viruses is their diversity of genome organizations (Table 1.1, Fig. 5.3), which surpasses that seen in any other group of organisms and is still not fully understood. Perhaps the most important division, at least historically, is that between positive-sense (or strand) viruses (ssRNA+), in which the genome is arranged as an mRNA molecule, and negative-sense viruses (ssRNA−), in which genomes are complementary to mRNA molecules. A few RNA viruses, such as the arenaviruses, even possess ambisense genomes, comprising both positive- and negative-sense molecules. A smaller number of viruses, such as the diarrhoea-inducing rotaviruses and the emerging bluetongue virus of cattle, possess dsRNA virus genomes.

Although the replication cycles of ssRNA+ and ssRNA− viruses are similar in many ways, for example that it usually takes place in the cell cytoplasm (for eukaryotic viruses), one key difference between them is that the viral RNA in ssRNA+ viruses functions directly as a mRNA, acting as a template for the production of both proteins and more genomic viral RNA (Fig. 1.3). Hence, the naked RNA extracted from virions is infectious, so that ssRNA+ genomes can be translated upon entry

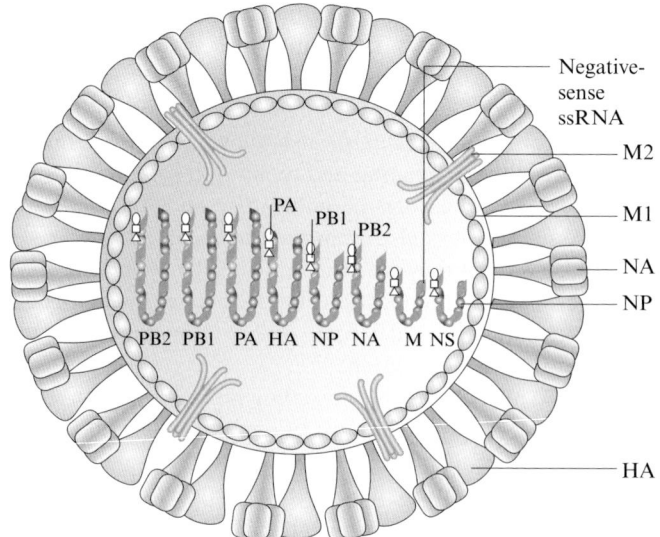

Fig. 1.2 The virion structure of influenza A virus. The genome of influenza A virus is composed of eight segments that are by convention listed from largest to smallest. Three segments encode proteins of the polymerase complex: PB2, PB1, and PA. Two segments encode surface envelope glycoproteins: the haemagglutinin (HA) and the neuraminidase (NA). Segment five encodes a nucleoprotein (NP) that binds to the viral RNA. Segment 7 encodes both the matrix protein (M1), the major component of the viral capsid, and M2, which is an ion channel. Segment 8 encodes two nonstructural proteins: NS1 and NS2. Adapted from Horimoto and Kawaoka (2005) with permission.

into the cell (although sometimes only partially). The replication cycle of a typical ssRNA+ might therefore be considered as comprising three key steps: (i) translation of the original RNA genome (including the RdRp required for replication), (ii) transcription of a negative-sense RNA strand from the positive-sense template (during which a dsRNA intermediate is made), and (iii) transcription of new positive-sense RNA genomes from these negative-sense molecules (see Regoes *et al.* 2005 for an elegant quantitative analysis of the dynamics of this process). In contrast, the genome of ssRNA− viruses must first be transcribed into positive-sense genomic material (mRNA) before the replication cycle can proceed, which then continues in a similar manner to that of ssRNA+ viruses. Consequently, the naked RNA extracted from the virions of ssRNA− viruses is not infectious, and a virion-associated RdRp which initially acts a transcriptase must enter the cell at the same time as the ssRNA− genome. Similarly, transcription takes place before translation in the case of dsRNA viruses, so that a virion-associated transcriptase is also required on cell entry, although mRNA is usually synthesized from one template strand only (conservative replication). The replication of retroviruses, such as HIV, is different again. In this case the RNA genome is replicated using RT to produce dsDNA that is then translocated into the

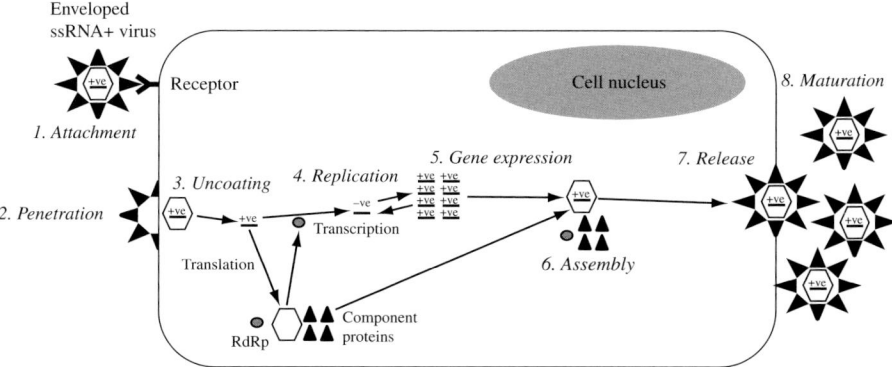

Fig. 1.3 Generalized replication cycle of a eukaryotic enveloped ssRNA+ virus, highlighting the eight major steps in this process. Note that in non-enveloped RNA viruses maturation follows assembly. −ve, negative; +ve, positive.

nucleus, where it is integrated into the host genome and then replicated by host DNA polymerases to produce more RNA molecules. It is this nuclear insertion that makes viruses like HIV so difficult to treat. As might be expected, these generalized replication cycles hide a great deal of virus-specific detail, some of which is touched upon in Chapter 5. Similarly, RNA viruses contain a variety of important sequences and/or secondary structures at their 3′ and 5′ termini and which perform critical roles in the replication process, although there is sadly insufficient space in this book for them to be discussed in any detail.

Another important division is between unsegmented (or monopartite) viruses, in which the viral genome is composed of a single nucleic acid molecule, and segmented (multipartite) viruses, in which the viral genome is composed of multiple nucleic acid molecules referred to as different segments. In ssRNA+ viruses, these different nucleic acid molecules also tend to be contained within different virus particles, so that these viruses are termed 'multicomponent' (strikingly only plant viruses fall into this category). In contrast, in both ssRNA− and dsRNA viruses the different segments are contained within a single virus particle, although arenaviruses (ssRNA−) are notable in that each segment is contained within a separate nucleocapsid, surrounded by a single envelope, while members of the *Partitiviridae* (dsRNA), which infect plants and fungi, possess multicomponent genomes.

1.3.3 Exploring the virosphere

Finally, before continuing, it is essential to remember that many of the ideas presented in this book relate to our current knowledge of the biodiversity of RNA viruses. For example, my discussion on the evolutionary processes that explain why all RNA viruses have genomes of less than 32 000 nt in length obviously assumes that no larger RNA viruses exist in nature. However, given the now intensive efforts to better describe viral biodiversity, particularly using the new techniques of metagenomics to explore marine environments (Angly *et al.* 2006; Culley *et al.* 2003, 2006), human faecal samples (Zhang *et al.* 2006; Finkbeiner *et al.* 2008), or even bee hives (Cox-Foster *et al.* 2007), it is likely that new, and perhaps very different, RNA viruses will be discovered in the near future. As a direct analogy, the discovery of the mimiviruses of algae (Raoult *et al.* 2004), with their huge genomes, and more recently of the small dsDNA viral parasites (so-called virophage) of these giant viruses (La Scola *et al.* 2008), has greatly changed concepts of DNA virus evolution (and even what constitutes a virus) and likely came as a surprise to many working in this area. To make the same point in a more quantitative manner, it has been estimated that there are approximately 10 bacteriophage for each of the approximately 10^{30} bacteria on Earth, such that there are approximately 10^{31} bacteriophage (Hendrix *et al.* 1999). Bacteriophage may therefore constitute the most abundant source of DNA on the planet, although only a small fraction of these organisms have been described to date. In short, we have only just begun to explore the virosphere.

2

The origins of RNA viruses

2.1 Introduction

Of all the issues in RNA virus evolution, and of all those covered in this book, inferring the origins and deep phylogenetic relationships of RNA viruses has proven the most difficult to resolve. This is an extremely frustrating state of affairs as understanding how viruses originated, and how they relate to cellular organisms, is one of the most interesting topics in evolutionary biology, with the potential to shed light on key moments in the early history of life on Earth. Indeed, to some evolutionary biologists viruses have played a central role in the establishment of complex living systems on Earth (although more so for DNA than RNA viruses; Bell 2001; Forterre 2005; Koonin *et al.* 2006; Prangishvili *et al.* 2006).

2.1.1 The perils of deep viral phylogeny

Phylogeny provides an obvious means to get a handle on viral origins. However, inferring the phylogenetic relationships among divergent RNA viruses has been extremely troublesome. There are three particular difficulties here. First, RNA viruses lack any kind of fossil record. At the time of writing, the oldest indisputable sequence of an RNA virus is that of influenza A preserved from the devastating pandemic of 1918 (Taubenberger *et al.* 1997) (the provenance of an earlier sampled avian influenza virus—A/chicken/Brescia/1902—is unclear). Convincing claims of sequences of simian T-cell leukaemia virus (STLV) recovered from samples collected in 1913 have also recently been made (Calvignac *et al.* (2008). Whereas the rapid rate of RNA virus evolution means that the genetic diversity observed over a 90-year time span would translate into roughly 90 million years of eukaryote evolution, this is still a very narrow window of evolutionary time, roughly equivalent to the history of the eutherian mammals. Although there have been assertions that RNA has been isolated and sequenced from older viruses, most notably human T-cell leukaemia virus (HTLV-I) from an Andean mummy (Li *et al.* 1999; Sonoda *et al.* 2000), these have not received widespread acceptance because of the failure to conclusively exclude contamination. Similarly, although it is clear that RNA viruses have been a burden on human health for millennia, the symptoms of individual viruses are often non-diagnostic, making it impossible, in all but a few cases, to determine which pathogen is the agent of a specific infection (although in others, such as rabies or polio, symptoms are more specific so that historical records provide a useful glimpse into evolutionary history).

The lack of an effective fossil record means that it is impossible to determine when RNA viruses have definitely existed in the past, or to use fossil sampling times as calibration points to date viral evolution.

A second major limitation of studies of the deep evolutionary history of RNA viruses is that their rapid rates of nucleotide substitution ensure that phylogenetic signal is eroded very quickly. For example, an average evolutionary rate of approximately 10^{-3} nucleotide substitutions per site, per year (subs/site/year) (section 3.1) obviously means that every nucleotide site will have fixed an average of one mutation every 10^3 years. Although nonsynonymous sites will usually evolve much more slowly, a mean evolutionary rate of, say, approximately 10^{-5} subs/site/year will mean that every nonsynonymous site changes on average once every 10^5 years. Even allowing for the most complex models of nucleotide substitution, as well as major selective constraints on some amino acid sites, it is evident that detailed evolutionary histories that span many millions of years will not be recoverable through an analysis of viral gene sequences, and to expect sequences that perhaps diverged close to life's origin to contain coherent phylogenetic information is fantasy. As such, it cannot be expected that phylogenies inferred from gene sequence data alone will provide many meaningful insights into the origin and early evolution of RNA viruses. As discussed later in this chapter, the most profitable way to infer these deep evolutionary histories may therefore be through the study of protein structure.

The final hurdle facing those wishing to reconstruct the origins of RNA viruses is that the shuffling of genes by recombination and reassortment means that there may not be a single evolutionary history that can be retraced using phylogenetic methods. Although it is still unclear how often such 'lateral' processes have occurred among divergent RNA viruses (see Chapter 5 for a detailed discussion), they are likely to have taken place with at least sufficient frequency that they need to be considered when undertaking any deep phylogenetic study.

2.2 Theories for the origin of RNA viruses

Despite the inherent difficulties in reconstructing the past, a number of important theories for the origins of RNA viruses have been proposed. The difficulty is not so much in generating a theory for the ancestry of RNA viruses, but acquiring sufficient data to test it. As such, all theories for viral origins currently fall into the category of untested hypotheses, and in some cases are little more than educated speculation.

Traditionally, three theories have been proposed for the origins of RNA viruses (Fig. 2.1; see Morse 1994 for an interesting historical review), although sometimes with extensions and modifications (Bândea 1983; Hendrix *et al.* 2000). These theories are: (i) the regressive evolution theory, which postulates that the ancestry of RNA viruses lies with cellular organisms, most probably bacteria, that have so effectively parasitized their hosts that they have been able to gradually off-load their own genes until they become the far 'simpler' organisms we see today as viruses; (ii) the escaped gene theory, that RNA viruses are descendants of escaped host

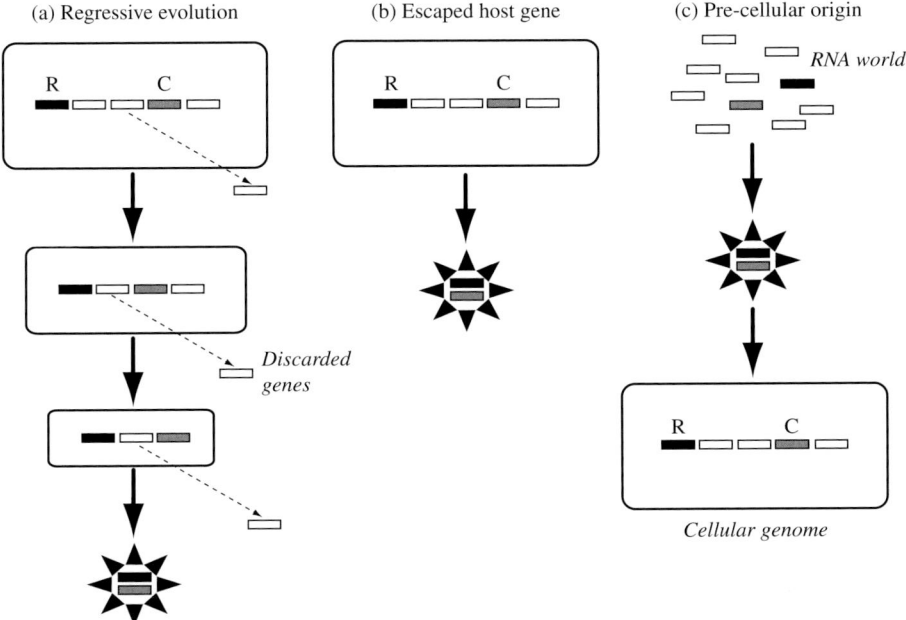

Fig. 2.1 Competing theories for the origins of RNA viruses. (a) The regressive evolution theory, now largely out of favour; (b) the escaped gene theory, and (c) the pre-cellular origin theory, in which RNA viruses evolve from the first RNA or RNA-protein systems. In each case cells are represented by rounded rectangles, and the simplest virus is thought to comprise only replicase (R) and capsid (C) genes.

(cellular) genes which acquired protective protein coats and the ability to replicate autonomously; and (iii) the pre-cellular life theory, that RNA viruses are descendants of pre-cellular RNA life forms, with ancestries dating back billions of years, and so adopted their parasitic lifestyle later in evolutionary time. Although these theories apply equally to DNA viruses, a fact that should not be forgotten in reading what follows (particularly as theories for the origins of DNA viruses are perhaps better formulated; Prangishvili *et al.* 2006), I will necessarily concentrate my discussion on RNA viruses.

2.2.1 The regressive origin theory

Of the theories of viral origins, the idea that RNA viruses are 'regressed' cellular organisms is perhaps the easiest to dismiss. On the one hand, it is clear that some cellular species, particularly endosymbiotic bacteria which have an extremely intimate relationship with their hosts, have been able to streamline their genomes by discarding those genes whose functions can be provided by the host (Moran 2002). A good example is provided by the *Buchnera* species of bacteria that infect various species of aphids, and which may have discarded more than 70% of their genome

(Moran and Mira 2001). In some sense, *Buchnera* have therefore experienced a regressive evolution into a simpler form, although an interesting twist in this tale is that reduced genome size is associated with an increased rate of deleterious mutation (Moran 1996) and the deployment of chaperone systems to correct defective proteins (Fares *et al.* 2002). More importantly, *Buchnera* are still very much bacteria. Another much-quoted example is provided by the Chlamydiae, a group of intracellular bacteria that have no cell wall, and there have been more recent suggestions that the giant mimivirus has regressive origins (Claverie 2006), although this remains to be formally tested.

Despite the regressive evolution observed in some bacteria, there is currently no evidence that this process is responsible for the origin of RNA viruses. In particular, the gene contents of RNA viruses and cellular species differ fundamentally, whereas the regressive theory would predict at least some—if not all—of the genes observed in RNA viruses to have their ancestries in cellular genomes. In addition, it is difficult to imagine how regressive evolution would result in an organism that has to infect new hosts to ensure its persistence (i.e. RNA viruses), rather than those that are inherited vertically along with the host, as is the case with *Buchnera*. Finally, as RNA viruses seem to be genome size limited because of the fitness costs associated with excessive mutational loads (see sections 3.1 and 5.1), it is unclear how they could have previously existed with larger genomes in a non-regressed state and still utilize an error-prone RNA polymerase. It is therefore no surprise that the regressive theory has generally fallen out of favour.

2.2.2 RNA viruses as escaped genes

The second major theory for the origin of RNA viruses is that they are 'escaped' cellular genes—most probably mRNA molecules—that acquired both the ability to self-replicate and the protective protein coats that allowed them to exist independently of cells. Hence, the escaped gene theory proposes that both RNA and DNA viruses existed *after* the first cellular organisms, so that this may be considered a 'post-cellular' theory of viral origins (although it is still likely that at least some of these escape events occurred during the early days of life on Earth). In its earliest formulations this theory was used to suggest that eukaryotic viruses originated in eukaryotic genomes, while bacteria were the progenitors of bacteriophage (reviewed in Prangishivili *et al.* 2006). However, I will take a broader perspective and assign any evidence for a cellular species giving rise to a virus as supportive of the escaped gene hypothesis. Until relatively recently this was also widely regarded as the most likely theory of viral origins (Morse 1994).

Whatever the timescale of gene escape, a fundamental prediction of this theory is that virus genes, including the diagnostic RdRp, should ultimately have their ancestries in cellular genomes. In addition, because such escape events could have occurred multiple times, it is not necessarily the case that RNA viruses, or indeed any class of virus, are monophyletic. Hence, it is possible that each major category of virus—RNA, DNA, retrovirus—has unique ancestries in cellular genes, and/or that different types of RNA virus—ssRNA+, ssRNA−, dsRNA—likewise represent independent escapes from host cells. As a consequence, the demonstration of a deep common ancestry of all viruses would be considered as strong evidence against the escaped gene theory; this theory can only explain the existence of similar traits among divergent viruses, such as the presence of jelly-roll capsid proteins or the palm subdomain of RNA and DNA polymerases (see below), through either extensive convergent evolution or LGT.

The attraction of the escaped gene hypothesis is that it can, in principle, explain the origins of all types of virus, which differ so fundamentally in their genome structures, simply by postulating multiple escape events. For example, ssRNA+ viruses would be descended from escaped cellular mRNA molecules that either possessed or acquired RNA polymerase activity, while retroviruses could be descended from the long terminal repeat (LTR) retrotransposons that are a common component of eukaryote genomes and which encode RT. Similarly, DNA viruses could be descended from DNA transposable elements, or perhaps the genes resident in bacterial plasmids. Indeed, the polymerases of the ssRNA+ and ssRNA− viruses have such different functions that their independent origin seems a reasonable hypothesis (Baltimore 1980).

That mRNA molecules continually pass through the nuclear envelope on their way to the ribosomes indicates that the escaped gene theory is at least mechanistically possible, and a recent and highly compelling analysis of HDV shows that a new 'virus' has been produced by host gene escape at least once in evolutionary history (see below). Two additional pieces of evidence have been put forward in support of the escaped gene theory. The first, and the most simplistic, is that as viruses are currently obligate parasites that rely on host cells for replication, they could never have existed outside of these cells. The second is that despite a number of attempts it has not been possible to use gene sequence data to show, unequivocally, that RNA viruses are monophyletic. Similarly, a recent parsimony-based phylogenetic analysis of tRNAs suggested that although DNA viruses can be considered 'ancient', their origin was subsequent, rather than prior, to that of archaebacteria (Sun and Caetano-Anollés 2008). Unfortunately, the absence of tRNAs in RNA viruses means that they cannot be included in analyses of this type.

The notion that their parasitic nature means that RNA viruses must have evolved after host cells is easy to dismiss. It is now commonly believed that the first replicating molecules on Earth were composed of RNA—in what is known as the RNA world—an idea that first surfaced in the late 1960s (see Joyce 2002 for an excellent review of this concept). It is therefore not a difficult intellectual leap to think that extant RNA viruses could have their origins with these putative ancient RNA replicators. Indeed, this is basis of the pre-cellular theory.

The idea that there is, as yet, no clear monophyletic relationship of all RNA viruses, let alone all viruses, is also rather unconvincing. As is discussed in more detail below, the key issue here is distinguishing firm evidence against a monophyletic origin from a simple lack of phylogenetic signal. While the latter is easy to show, the former is far more challenging. Although it is undoubtedly the case that there is nothing in the phylogenetic analysis of viral gene sequences that unequivocally supports their monophyly, remnants of deep evolutionary relationships may be more apparent in other types of data, most notably protein structures. The potential for protein structure to shed light on the deep phylogenetic relationships of RNA viruses is discussed in section 2.3.

Until very recently there was also strong evidence against the escaped gene hypothesis in that no cellular gene was known to possess RdRp activity, although under this theory the RdRp must have a cellular origin. However, it has now been shown that cellular DNA polymerase (Pol) II, which catalyses the synthesis of RNA from DNA, also has RdRp activity (Lehmann et al. 2007). Pol II had previously been proposed as the agent of replication in viroids, which unlike RNA viruses possess no RNA polymerase (Lai 2005). This remarkable discovery removes one major barrier to the theory that RNA viruses might be escaped cellular genes. However, it is equally likely that the RdRp activity of Pol II evolved before its role in DNA transcription, and there is as yet no evidence that cellular Pol II and viral RdRp are homologous. Shortly, I will argue that a consideration of the mutational loads faced by RNA viruses also provides evidence against the escaped gene theory.

2.2.3 RNA viruses and the RNA world

The final, and perhaps currently the most popular, theory for the origin of RNA viruses is that they represent the modern descendants of an RNA world that is generally thought to have existed before the advent of higher-fidelity DNA-based replication (Gilbert 1986; Szathmáry and Demeter 1987). Similarly, extant DNA viruses could be remnants of the first DNA replicators, while the retroviruses (and other retrotranscribing elements) could be descendants of the first molecules that were able to make the transition between RNA and DNA. Because they are proposed to have arisen in a common environment before the advent of cellular organisms, and may therefore have shared specific genes, a direct phylogenetic link between RNA, DNA and retroviruses would serve as strong evidence for the pre-cellular theory.

There are a number of pieces of evidence for the pre-cellular theory, although as ever in discussions of viral origins the data are scant at best and can often be interpreted in different ways. First, and most obvious, the reliance on RNA for replication automatically connects modern day RNA viruses and the ancient RNA world (although the production of proteins must have also occurred before the origin of the first true RNA virus). Although necessarily hypothetical, the notion of an RNA world is a hugely compelling one, representing a key moment in evolutionary history between the formation of the Earth some 4.5 billion years ago and the appearance of the first fossils at approximately 3.5 billion years ago. As these first fossils are also relatively complex, it is clear that the preceding billion years of Earth's history were the scene of some major evolutionary transitions, the occurrence of which has not been recorded in the fossil record. To some, modern day RNA viruses, and their simpler relatives the viroids, provide a unique insight into these ancient evolutionary events.

The most direct evidence for the existence of the RNA world was the discovery of catalytic RNA molecules known as ribozymes (Kruger et al. 1982; Cech 1987).

2.2 Theories of origin • 21

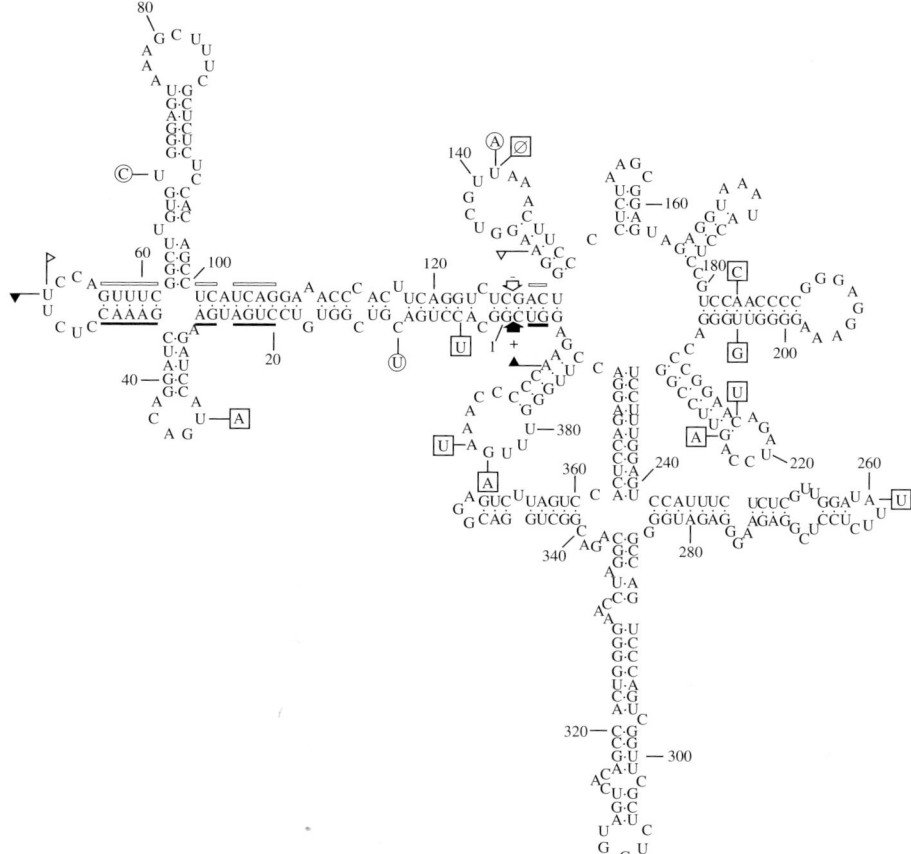

Fig. 2.2 A typical viroid—chrysanthemum chlorotic mottle viroid (CChMVd)—depicting its complex RNA secondary structure. Self-cleavage domains are marked by flags, self-cleavage sites by arrows, and a 13 nt region that is conserved in hammerhead ribozymes by bars. Mutational differences between CChMVd isolates are presented in circles and squares (and do not alter secondary structure). In all cases solid symbols refer to the positive strand, and open symbols to the negative strand. The Ø symbol designates a deletion. Taken from Navarro and Flores (1997) with permission.

These are small, naturally occurring RNA molecules that include the hammerhead ribozyme found in some viroids, the HDV ribozyme, and the *Neurospora* Varkud satellite motif (Fig. 2.2). Ribozymes are of immense biochemical interest because of their involvement in such basic cellular processes as mRNA processing and protein synthesis. It is reasonable to think that the earliest replicators on Earth existed as a form of ribozyme, and that these were also the progenitors of RNA viruses. Indeed, Manfred Eigen's seminal work on early replication systems implies that ribozymes may have set in motion the process of evolution by natural selection, as this can be considered a predictable outcome of self-replication (summarized in Eigen 1992).

Because of their complex secondary structures, lack of protein-coding regions, and most importantly their ribozyme functions, viroids represent potentially good candidates for remnants of the earliest life forms that occupied the RNA world (Fig. 2.2) (Elena *et al.* 1991) and have recently been demonstrated to possess extremely high mutation rates (Gago *et al.* 2009). Yet while it is undoubtedly the case that viroids have a remarkably simple structure, their sequences are also so divergent from those of other organisms to make it very difficult, it not impossible, to accurately infer their origins (Jenkins *et al.* 2000). Again, this lack of phylogenetic signal does not necessary mean that viroids are more recently evolved infectious agents, but rather that it is impossible to use sequence-based phylogenies alone to argue for their antiquity. However, on current data, it seems equally plausible that viroids represent recently escaped host genes or introns which have not yet acquired protein coats, especially as they are only seen in plants, when an origin in the RNA world might predict a far wider taxonomic distribution. Indeed, it is also intriguing that viroids replicate using host cellular DNA Pol II (Lai 2005). Such a reliance on host polymerases for replication strongly suggests that their ultimate origin lies with cellular genomes. A dramatic proof of this principle was the observation that the HDV ribozyme is related to the CPEB3 ribozyme present in a human intron sequence (Salehi-Ashtiani *et al.* 2006). That HDV is only found in humans and requires human hepatitis B virus (HBV) to replicate is powerful evidence that its origins lie in the human transcriptome (Salehi-Ashtiani *et al.* 2006). Although HDV cannot be considered a 'true' virus, its genesis clearly illustrates how host genomic DNA can escape to form an exogenous replicating agent.

2.2.4 Eigen's paradox

Because of the very deep timescales involved and the invocation of the RNA world, understanding the origin of RNA viruses automatically impacts on some of the most fundamental questions in evolution; how to produce the first polymers, how to generate the first replicating molecule, how to make the first cell? Perhaps a more generic problem, and one of special relevance to the study of RNA virus evolution, is how a simple replicating system can evolve increased genotypic, and hence phenotypic, complexity? Although the RNA world is an attractive focus for many of these questions, the reliance on RNA replication comes at a cost: critically, compared to DNA, the copying of RNA is highly error-prone. As discussed in detail below this, in turn, imposes an upper limit on the size of primitive RNA replicators, as overly large RNA molecules will be unable to copy themselves with sufficient fidelity to maintain fitness. It is this phenomenon that also likely explains the highly restricted size of the ribozymes we know today. Although this idea is often discussed in the context of an 'error threshold' (Maynard Smith and Szathmáry 1995), in reality a strict threshold, in the form of a critical value, requires a single-peaked fitness function which may be a poor description of the natural situation (Wiehe 1997). However, irrespective of the precise fitness function evolved, to create greater genetic complexity it is clearly necessary to encode more information in longer genomes by using a replication system with greater fidelity. Unfortunately, there is a serious problem with this idea: to replicate with greater fidelity requires a more accurate and hence complex replication enzyme, but such an enzyme cannot be created because this will itself require a longer genome, and longer genomes will breach any error threshold (Maynard Smith and Szathmáry 1995). This evolutionary chicken-and-egg situation has been called 'Eigen's paradox' and represents one of the most intractable puzzles in the origin of life. It is also fundamental for understanding the origins of viruses, for while current ribozymes are tens of nucleotides in length, the smallest RNA viruses possess genomes of several thousand nucleotides.

How, then, is it possible to increase the genomic complexity of an RNA replicator by an order of magnitude without enduring a mutational meltdown? The answer may lie with RNA secondary structures, and the complex fitness landscapes they enforce, which reduce the impact of deleterious mutations. This results in a certain level of mutational 'robustness': a constancy of phenotype in the face of frequent mutation (see section 3.1 for a more detailed discussion; Sanjuán *et al.* 2006a, 2006b). This results in a 'relaxed error threshold' that buffers genomes against mutation pressure and allows the key evolutionary transition from ribozyme-like to virus-like genomic complexity (Kun *et al.* 2005) (Fig. 2.3). However, the genome sizes of RNA viruses are still very much smaller than either the tiniest cellular life forms

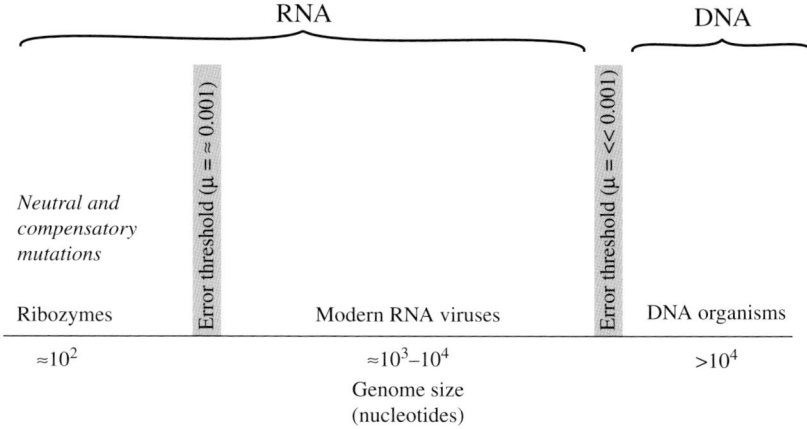

Fig. 2.3 Error thresholds and the evolution of genomic complexity, based on the theory of Kun *et al.* (2005). For evolution to proceed in the RNA world a reduction in the error rate (μ) to approximately 0.001 mutations per nucleotide, per genome replication is required. For more complex genomes to evolve, with sizes greater than 10^4 nucleotides, a second threshold needs to be crossed with $\mu \ll 0.001$. This necessitates the evolution of higher-fidelity DNA replication. Neutral and compensatory mutations tend to dampen the effects of deleterious mutations, allowing a so-called relaxed error threshold (i.e. adding robustness). Adapted from Holmes (2005) with permission.

known today, or that estimated for the last universal common ancestor (Szathmáry 2005). Consequently, although most RNA viruses are larger and more complex than ribozymes, and have greater copying fidelity (Gago *et al.* 2009), they are also at the mercy of excessive mutational loads, which in turn limits their genome size to levels far lower than those seen in most DNA-based organisms. The question then becomes how to move from the genome sizes of RNA viruses to those of more complex life forms? The answer, of course, involves the evolution of DNA replication and its capacity for error correction. Sadly, exploring this major evolutionary transition is beyond the scope of this book. However, the idea that the upper limit on the genome sizes of RNA viruses is determined by their background mutation rate is also generally supportive of the pre-cellular theory of their origins, because if RNA viruses represent escaped transcripts of DNA Pol II we might expect them to be both less error-prone and longer than contemporary RNA viruses.

2.2.5 The taxonomic distribution of RNA viruses

A second piece of evidence cited for the deep antiquity of RNA viruses is that these infectious agents have an expansive taxonomic distribution, and have been described in a wide variety of branches from the tree of cellular life. As noted above, the major exception is the archaebacteria, which could be argued to support an origin of RNA viruses subsequent to the origin of *Archaea*. However, given the evidence for LGT at the base of the tree of life (Brown 2003), it is difficult to believe that an RNA virus would not, at some stage in evolutionary history, have moved between bacteria and/or eukaryotes and archaebacteria. As such, the absence of RNA viruses in *Archaea* more likely reflects the lack of sufficiently intensive sampling in these taxa (although temperature constraints may reduce the frequency of RNA viruses in hyperthermophiles; Zeldovich *et al.* 2007). In a similar manner, the vagaries of sampling, rather than a more recent origin, are also the most likely explanation for other major 'gaps' in the taxonomic distribution of viruses, such as the lack of ssRNA– viruses in bacteria, fungi, and unicellular eukaryotes, and the absence of retroviruses in bacteria. That such major gaps in viral biodiversity exist further highlights the need for more widespread sampling of viruses from a diverse range of habitats.

In some cases it has been possible to use the match between the phylogenies of viruses and their hosts—so-called co-divergence (or co-speciation)—to discuss the timespan of viral evolution in more concrete terms. Because of the rapid erosion of phylogenetic signal in RNA viruses, these studies have proven far more successful in the case of DNA viruses: to date, the oldest example of co-divergence in an RNA virus is that of the retrovirus simian foamy virus (SFV) that may have been associated with non-human primates for 30–40 million years (Switzer *et al.* 2005) (the debatable case of the plant tobamoviruses is discussed in section 3.1). Humans are an interesting exception to the rule of co-divergence in SFV as in this case the virus always represents a spill-over from other animals (Wolfe *et al.* 2004). In contrast, for DNA viruses such as herpesviruses, virus/host co-divergence may extend for up to 400 million years (McGeoch and Gatherer 2005), whereas insect baculoviruses may have established ancient associations with arthropods (Herniou *et al*. 2004). However, although these co-divergence studies provide a useful perspective on the timescale of viral evolution, they are still far too recent to inform on the origin of either RNA or DNA viruses. Although icosahedral dsDNA viruses are found in all three domains of life (Maaty *et al*. 2006; Ortmann *et al*. 2006), which implies that their common ancestor is at least this old, rigorous phylogenetic analyses of these diverse viruses have yet to be undertaken and are likely to represent a major challenge given the necessity to utilize protein secondary structure (section 2.2). An even more remarkable example of deep evolutionary ancestry concerns the giant mimiviruses of amoeba, a discovery that rejuvenated studies of DNA virus origins (Raoult *et al*. 2004; Claverie 2006). In this case phylogenetic analysis placed mimivirus in a position—approximately that of a divergent eukaryote—expected of its amoebae hosts, and relatively close to the major T junction of archaebacteria, bacteria, and eukaryotes. However, the lack of basal eukaryotes in these studies make all phylogenetic trees extremely difficult to interpret and it is impossible to entirely exclude some role for host gene escape. More generally, it is striking that individual RNA virus families rarely contain taxa that infect widely different host species. For example, no family of RNA viruses is known to infect both bacteria and eukaryotes, and few viral families infect hosts as divergent as animals and plants (with the families *Bunyaviridae*, *Rhabdoviridae*, and *Reoviridae* being notable exceptions), although jumps of viruses from plants to animals have been proposed (Gibbs and Weiler 1999). The general separation of plant and animal viruses is likely to reflect the specific adaptations required to infect these very different host types, with the major difference in the structure of the cell wall perhaps prominent among these (Gibbs *et al*. 2008a).

2.2.6 Conserved protein structures

By far the most compelling evidence for the pre-cellular theory of viral origins is that RNA viruses contain a set of genes—or more precisely protein structures—that are not found in cellular species. These have been termed hallmark genes by Eugene Koonin, and their existence provides a powerful argument for the common ancestry of RNA viruses. Even more dramatic are suggestions that some of these hallmark genes are found in both RNA *and* DNA viruses, and are therefore potentially indicative of the common ancestry of all viruses (Bensen *et al*. 2004; Bamford *et al*. 2005; Koonin *et al*. 2006). If upheld, this would represent extremely strong evidence for the pre-cellular theory. Because these highly divergent genes possess almost no primary sequence similarity, all realistic analyses must be based on protein secondary structure. This will be discussed in more detail in section 2.2.

The most important of these genes for RNA viruses is the RdRp, where similarities among highly divergent RNA viruses, and even between RdRp and the RTs found in retroviruses, can be seen in the presence of a small number of conserved sequence motifs and protein structures (Poch *et al*. 1989; Koonin 1991; Goldbach and de Haan 1994). Remarkably, a palm subdomain protein structure, comprising a four-stranded antiparallel β-sheet and two α-helices, is conserved among some

RNA-dependent and DNA-dependent polymerases, and which represents a powerful argument for its great antiquity (Gorbalenya *et al.* 2002). More refined analyses have revealed that two lineages of the palm subdomain exist, representing two distinct protein structures (canonical and non-canonical), and which are also thought to reflect an ancient separation (Gorbalenya *et al.* 2002; Garriga *et al.* 2007). Finally, in the case of RT, it is striking that patterns of sequence (and structural) similarity are also able to link retroviruses, hepadnaviruses, and caulimoviruses with the cellular 'genes' that utilize RT, such as retroelements, group II introns, and telomerase (Eickbush 1994; Malik *et al.* 1999; Chang *et al.* 2008). However, determining the position of the root of the RT tree, which is critical for choosing among competing theories of viral origin, is extremely difficult (Eickbush 1997).

Understandably, most attention has been directed towards those genes that are reportedly shared between RNA and DNA viruses, and which have had a major impact on theories of viral origin (Bamford *et al.* 2005; Jalasvuori and Bamford 2008). Three genes/structures fall into this class: (i) the so-called jelly-roll capsid (JRC) (Fig. 2.4), a tightly structured protein barrel that represents the major capsid subunit of (non-enveloped) virions with an icosahedral structure (Rossmann *et al.* 1985; Coulibaly *et al.* 2005), and which is found in viruses as diverse as picornaviruses (ssRNA+), birnaviruses (dsRNA), herpesviruses (dsDNA), and *Sulfolobus* turreted icosahedral virus (a dsDNA virus of archaebacteria; Maaty *et al.* 2006), (ii) the Superfamily 3 helicase (S3H) which is involved in the initiation and elongation of genome replication (Koonin *et al.* 2006), and (iii) the palm subdomain of RNA and DNA polymerases. Their combined discovery has led Eugene Koonin to propose the existence of an 'ancient virus world': a primordial viral gene pool that marked a key

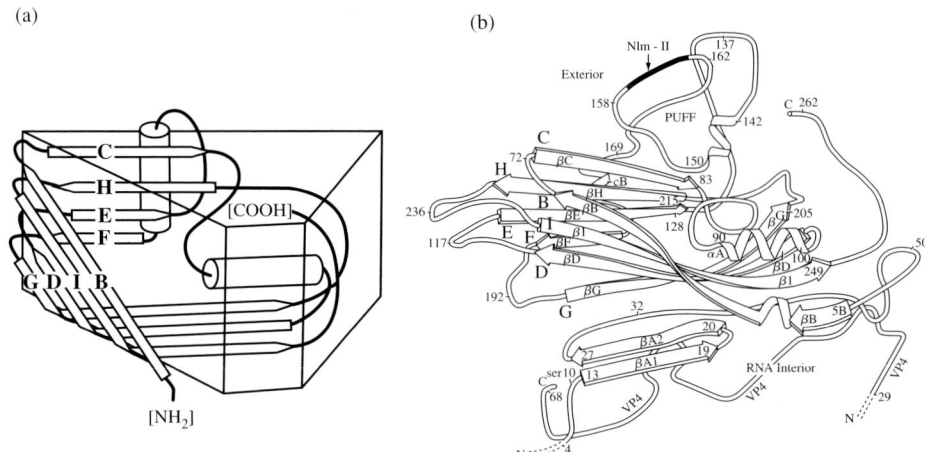

Fig. 2.4 (a) The canonical jelly-roll protein structure contains eight β-sheets (B, C, D, E, F, G, H, I) and two α-helices. Taken from Le Gall (2008) with permission. (b) The VP2 protein of human rhinovirus 14 showing the distinctive jelly-roll. Similar structures are found in the VP1 and VP3 proteins of picornaviruses. Adapted from Rossmann *et al.* (1985).

transition between the RNA world and the evolution of cellular species (Koonin *et al.* 2006) (Fig. 2.5). According to Koonin, the phrase gene pool is a perfect description for the ancient virus world, as there was extensive mixing of genes between viruses of very different types (RNA, DNA, retroviruses), so that we would not expect each type of virus to be strictly monophyletic (Koonin *et al.* 2006).

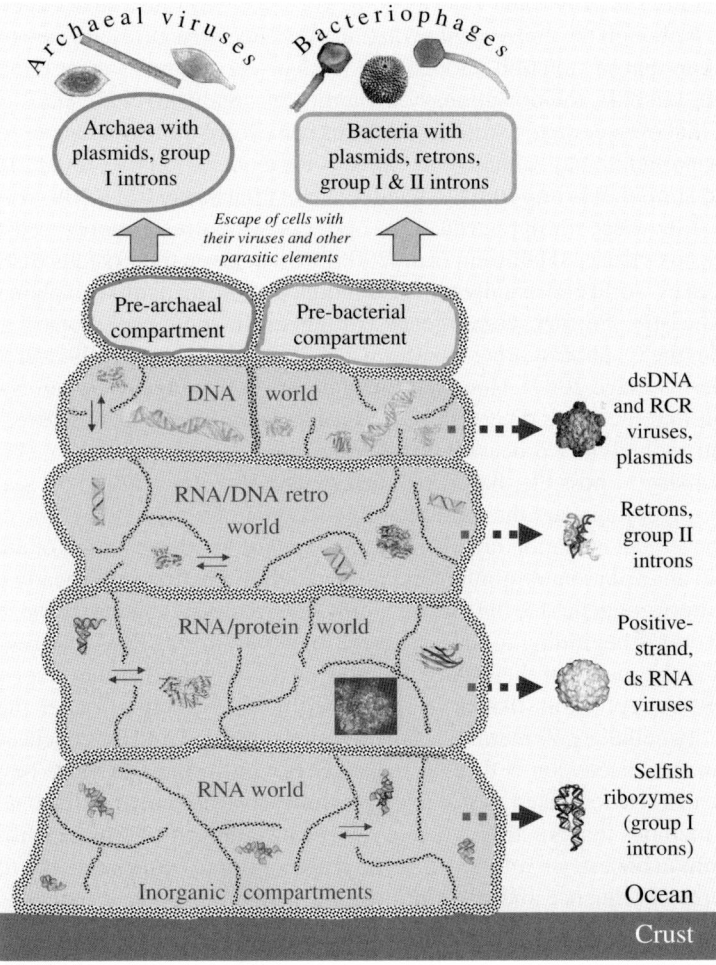

Fig. 2.5 Theory of the 'ancient virus world' developed by Eugene Koonin. The proposed major stages in evolution are shown along with the replicators, including viruses, present during each stage. Under this model RNA viruses have a pre-cellular origin. More details are available in the original publication. RCR denotes viruses that utilize rolling cycle replication. Taken from Koonin *et al.* (2006) with permission.

While the discovery of conserved protein structures in diverse viral species is a hugely important one in discussions of viral origins, it comes with a major caveat: it is currently impossible to rule out that highly similar protein structures, such as JRCs and palm subdomains, could not have arisen through convergent evolution. If true, they would tell us little about evolutionary history and viral origins, although a great deal about selective constraints and adaptation. Convergent (and parallel) evolution is a relatively common occurrence in molecular evolution and may be particularly frequent in viruses because of the major constraint on their genome (and hence) protein sizes (see Bull et al. 1997 and Cuevas et al. 2002 for two illustrative experimental examples). Although the convergent evolution of protein structures appears to be less common than that of function (Gough 2005), clear-cut examples exist (Hamburger et al. 1999). In short, the evolutionary dynamics of organisms with small genomes may mean that even complex protein structures can evolve more than once. Although convergent evolution may seem like an implausible explanation for protein structures that look so similar, it is important to recall that not all icosahedral RNA viruses possess JRCs. Likewise, similar surface glycoprotein structures are observed in some DNA and RNA viruses (Heldwein et al. 2006) which, given their patchy distribution across the virus world and usually rapid evolution, is more difficult to explain through common ancestry. Further, convergence is rampant in some evolutionary systems: for example, the C_4 photosynthetic pathway has evolved independently over 45 times in angiosperms (Sage 2004). However, the more often convergent evolution needs to be invoked, and the more divergent the taxa that carry the same structure, the less plausible an evolutionary process it becomes.

If a summary is possible, based on current data that are admittedly scarce, the pre-cellular theory appears the most plausible account of viral origins. The caveat of possible convergent evolution notwithstanding, Eugene Koonin's notion of an ancient virus world, an evolutionary gumbo that cooked before the first cells, neatly explains many of patterns of gene sharing among highly diverse taxa. Perhaps for the first time the virus world is beginning to look cohesive. In addition, that RNA viruses are still very much at the mercy of their mutation rates further argues that they evolved from primitive RNA replicators that never possessed error-correction, rather than from higher-fidelity cellular polymerases. However, perhaps the strongest conclusion to be drawn from this discussion is that for the study of viral origins to move beyond the speculative to a science based on testable hypotheses, it is critical to develop new computational methods that are able to detect significant similarities—those truly indicative of common ancestry—among highly divergent protein sequences, and then turn these patterns of sequence and structural similarity into robust phylogenetic trees.

2.3 Deep phylogenetic relationships among RNA viruses

The lack of progress in understanding the origins of RNA viruses is mirrored in the difficulties in trying to infer their deep—inter-family—phylogenetic relationships and to devise higher-order classification schemes. Again, the major limitation is that phylogenetic inference based on sequence data alone is severely compromised by the highly divergent nature of the sequences in question. As a consequence, this problem is also more serious for RNA than DNA viruses. In the latter, phylogenetic studies have revealed deep evolutionary relationships among the sequences of DNA polymerases present in both cellular organisms and viruses, although doubts remain over the position of the root and even the monophyly of some groups of DNA virus (reviewed in Shackelton and Holmes 2004).

2.3.1 The 'higher-order' relationships of RNA viruses

There have been a variety of attempts to infer the inter-family phylogenetic relationships of RNA viruses, all necessarily based on the analysis of RdRp sequences. Such phylogenetic studies are prompted by multiple alignments which indicate that there are up to eight very short amino acid motifs that are conserved across all known RdRps (Poch *et al.* 1989; Koonin 1991; Koonin and Dolja 1993; Goldbach and de Haan 1994). The most famous of these motifs is Gly-Asp-Asp (GDD) and its variants, located within the conserved palm subdomain structure (Gorbalenya *et al.* 2002). Phylogenetic trees of sequence alignments centred around these conserved motifs have been used to construct higher-order classification schemes for RNA viruses, involving the delineation of a number of viral 'supergroups'. The earliest studies proposed three such supergroups of ssRNA+ viruses, although each contained viruses with very different genome organizations (Koonin 1991; Koonin and Dolja 1993). For example, the coronaviruses, picornaviruses, and potyviruses were placed together in supergroup I, while the carmoviruses, flaviviruses, and pestiviruses, along with some bacteriophages, were found in supergroup II. Finally, supergroup III contained such diverse infectious agents as alphaviruses and tymoviruses. A later study proposed six supergroups—alpha-like, carmo-like, corona-like, flavi-like, picorna-like, and sobemo-like—each of which contains viruses with broadly similar genome organizations (Goldbach and de Haan 1994) (Fig. 2.6). Other phylogenetic studies have considered the evolutionary relationships between the supergroups of ssRNA+ viruses and the ssRNA− and dsRNA viruses, with a particular focus on trying to determine which group were the first to diverge, although little consensus exists (Bruenn 1991; Koonin and Dolja 1993; Goldbach and de Haan 1994; Vieth *et al.* 2004). Despite this body of work, few higher-order groupings of RNA viruses are officially recognized by the International Committee on Taxonomy of Viruses (ICTV; Fauquet *et al.* 2005), namely the order *Mononegavirales*, a grouping of four families of unsegmented ssRNA− virus—*Bornaviridae, Filoviridae, Paramyxoviridae*, and *Rhabdoviridae*—and the *Nidovirales*, comprising the *Arteriviridae, Coronaviridae* and *Roniviridae* families of ssRNA+ viruses. An order *Picornavirales*, comprising the families *Picornaviridae, Comoviridae, Dicistroviridae, Marnaviridae*, and *Sequiviridae* of ssRNA+ viruses, along with some unassigned viral genera, has also been proposed, largely based on shared patterns of genome and capsid organization (Le Gall *et al.* 2008).

A major reason why the higher-order classifications of RNA viruses have failed to become established, and particularly the notion of viral supergroups, is that their

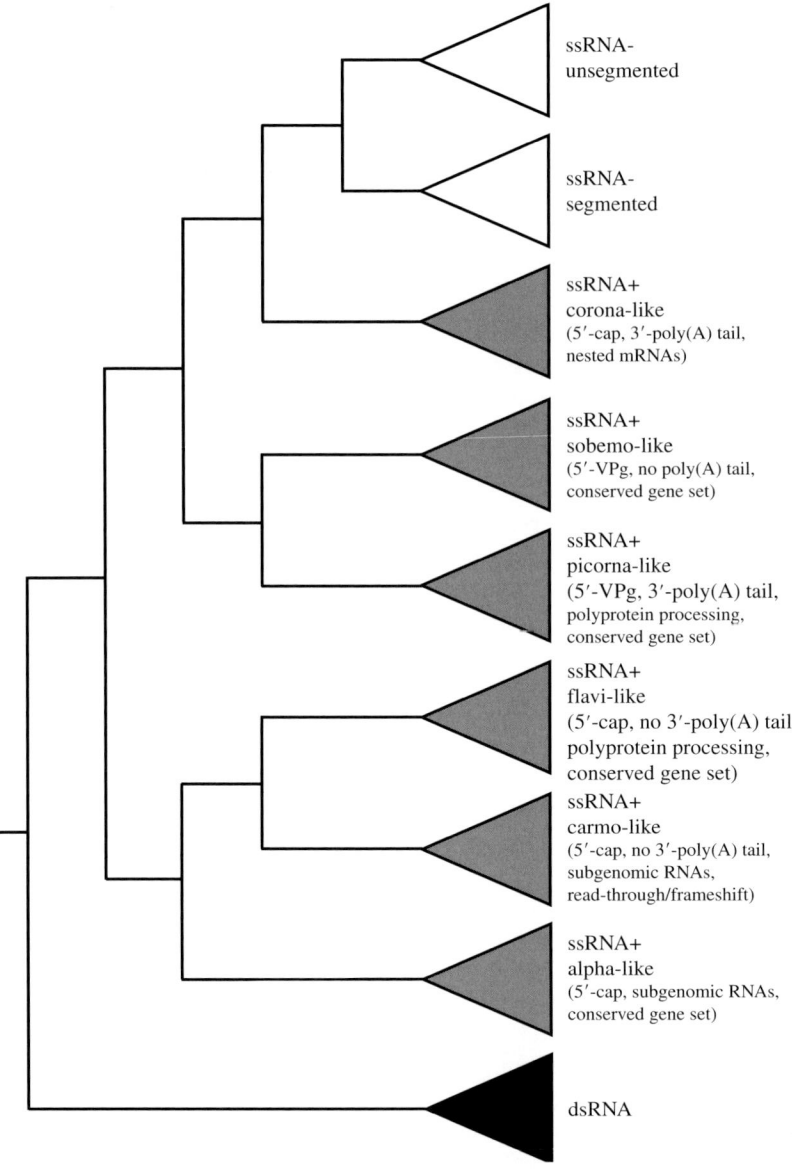

Fig. 2.6 Schematic phylogenetic tree of the proposed 'supergroups' of ssRNA+, ssRNA− and dsRNA viruses based on data provided in Goldbach and de Haan (1994). Some of the major genomic properties of the six supergroups of ssRNA+ viruses are given. Note the divergent position of the dsRNA viruses which is highly debated (as is the common ancestry of the ssRNA− viruses). As can be seen in Fig. 2.8 and discussed in the text, this tree should be regarded as speculatory at best.

phylogenetic support from sequence data is decidedly shaky. It would also be fair to say that most studies of inter-family evolutionary relationships have not utilized the most sophisticated methods for either multiple sequence alignment or phylogenetic analysis. Most importantly, a careful analysis of the RdRp sequences of RNA viruses often found no more phylogenetic signal at the inter-family level than expected by chance alone, and even in cases where there was sufficient evolutionary information to infer a meaningful phylogeny the resultant trees generally lacked any statistical support (Zanotto et al. 1996a). This analysis should strike a cautionary note for all those wanting to undertake deep phylogenetic analyses of RNA viruses. However, and at the risk of repetition, the lack of a well-supported phylogenetic tree of RNA viruses does not necessarily mean that these viruses have independent origins, but rather that there is insufficient information in gene sequence data alone to infer a reliable phylogeny. Likewise, although there are clear similarities in genome organization among the members of some of the proposed supergroups—for example, a conserved gene order, 5' and 3' structures, and polyprotein processing in the case of the 'picorna-like' supergroup (Fig. 2.6), and colinearity among members of the *Mononegavirales*—and which may well be indicative of common ancestry, it is extremely difficult to draw phylogenetic relationships among the supergroups without an explicit model of how genome organization itself evolves (see below).

Detailed phylogenetic analyses of those sequences with RT activity, such as retroviruses, hepadnaviruses, group II introns, LTR retrotransposons, non-LTR retrotransposons, and telomerases, have been far more informative than those undertaken on RdRp, in large part because RT sequences clearly retain more sequence similarity (Xiong and Eickbush 1990; Eickbush 1994; Malik *et al.* 1999; Arkhipova *et al.* 2003; Chang *et al.* 2008). Specifically, phylogenetic analysis has revealed such common patterns as (i) a major division between the LTR and non-LTR retrotransposons, with retroviruses derived from the former, (ii) that hepadnaviruses and caulimoviruses have independent origins (probably from LTR retrotransposons), and (iii) that the RT elements unique to prokaryotes (group II introns, mitochondrial DNA (mtDNA) plasmids, and ms DNAs) form a monophyletic group (Fig. 2.7). However, determining the order of these events has proven extremely difficult (Eickbush 1997), particularly as the 'closest' potential outgroup sequences—the RdRp-utilizing RNA viruses—are so divergent that it is impossible, with current techniques, to unequivocally show that they are related. Indeed, there is so little primary sequence similarity among RdRp and RT sequences—and certainly no more expected than by chance alone—that sequence-based phylogenetic inference becomes a futile exercise at this level (Zanotto *et al.* 1996a). This leaves us in the highly unsatisfactory situation where it is impossible to distinguish evidence for an independent origin of RNA viruses and retroviruses, as expected under the escaped gene hypothesis, from a null case where there is simply insufficient sequence similarity to perform any phylogenetic test of hypotheses of viral origins.

As should be clear by now, 'standard' evolutionary analysis involving multiple sequence alignment followed by phylogenetic reconstruction is clearly insufficient

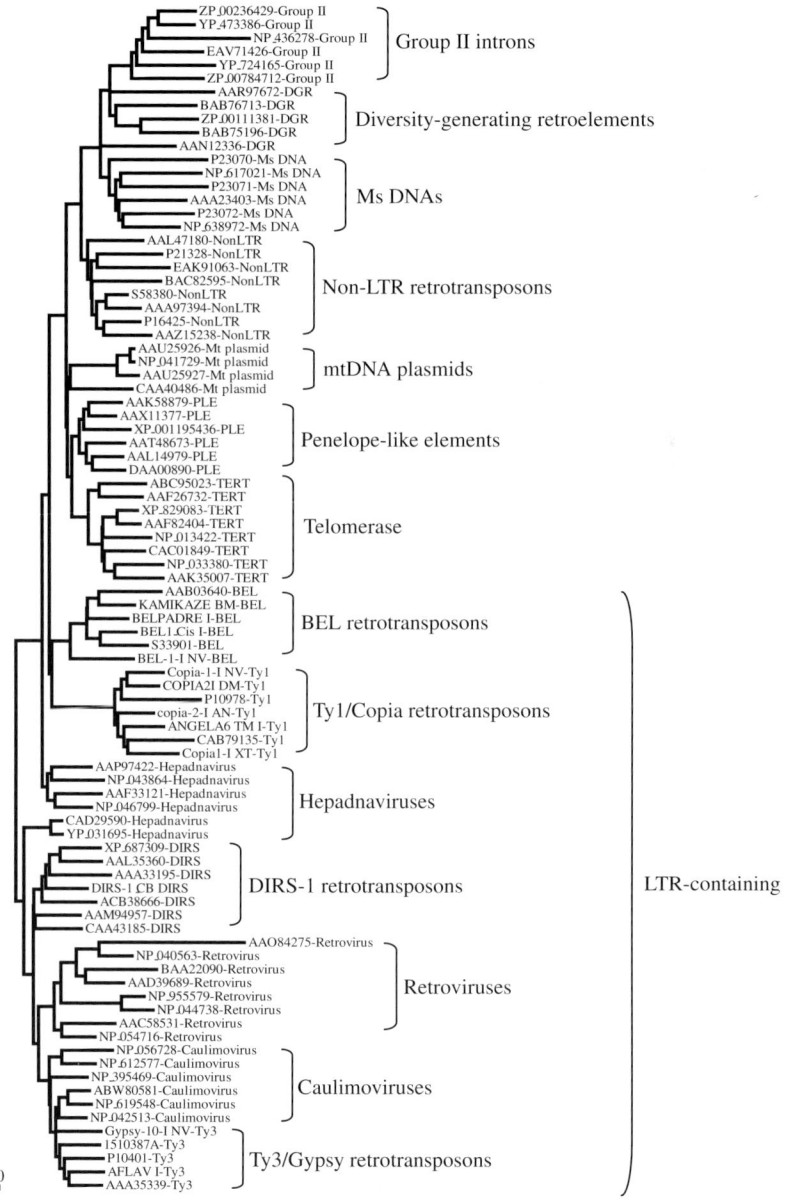

Fig. 2.7 Phylogenetic tree of 88 RT-containing sequences (retroelements). The tree was generated using the minimum-evolution method from pairwise distances estimated under the Gestalt Domain Detection Algorithm-Basic Local Alignment Tool (GDDA-BLAST) of Chang *et al.* (2008). Broadly similar phylogenies have been generated using a variety of methods, particularly the division between the LTR and non-LTR retroelements. Despite its orientation, this phylogeny is unrooted. ms DNA, multicopy single-stranded DNA; mtDNA, mitochondrial DNA. Adapted from Chang *et al.* (2008) with permission.

2.3 Deep phylogenetic relationships · 33

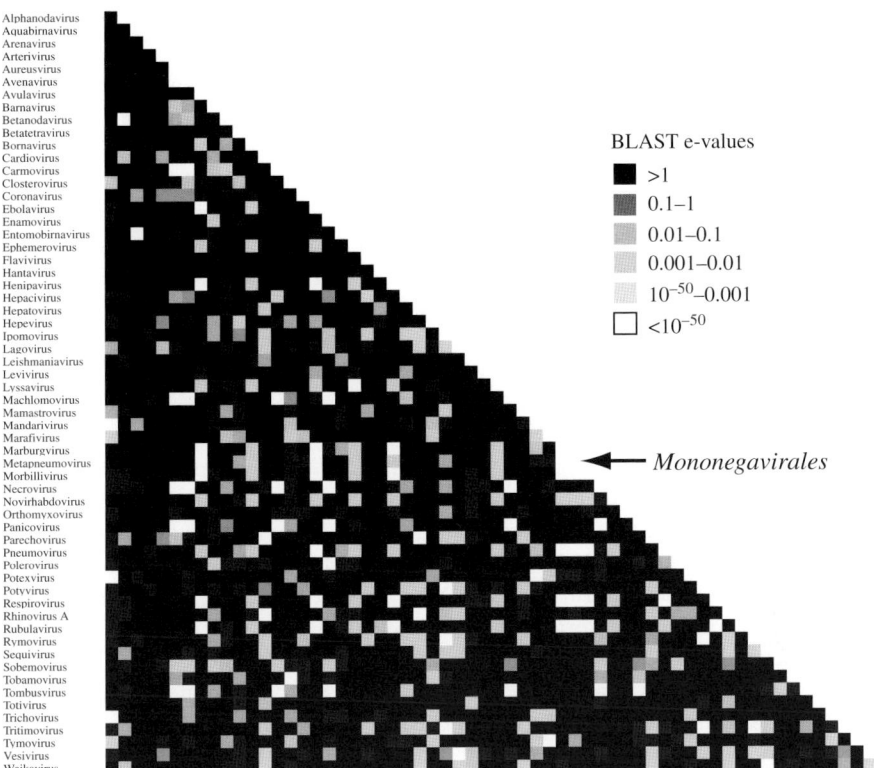

Fig. 2.8 Pairwise similarity matrix of BLAST e-values of the proteomes of 61 genera of RNA virus. The lack of significant sequence similarity across RNA viruses as a whole is striking (black cells), including that between the unsegmented ssRNA− viruses (*Mononegavirales*— some of which are indicated) and the segmented ssRNA− viruses. The clearest cases of significant similarity in amino acid sequence (white cells) are among viruses assigned to the same family and within the *Mononegavirales*. Figure adapted from an original kindly provided by Nathan Deering.

to accurately infer the higher-order relationships of RNA viruses. Despite these inherent difficulties, it is possible to make some evolutionary inferences from amino acid sequence data, albeit very general ones. One simple way this can be done is through a comparison of pairwise BLAST scores (e-values) of a diverse array of viral proteomes (Fig. 2.8). Although only approximate, this analysis does reveal cases where there is clear sequence similarity among viral families (i.e. more than expected by chance alone; very low e-values), and those where no such homology can be discerned. Interestingly, this analysis reveals no significant sequence similarity between the unsegmented ssRNA− viruses (order *Mononegavirales*) and the segmented ssRNA− viruses (for example, *Arenaviridae*, *Bunyaviridae*, *Orthomyxoviridae*). While tentative at best, this suggests that these different types of ssRNA− virus are not closely related, indicative of independent origins, in marked contrast to what is usually assumed in phylogenetic studies. Of course, this begs the bigger question of what forces have acted to favour the evolution of negative-sense genome orientation on more than one occasion? This fascinating question is considered more fully in Chapter 5.

Given the profound limitations in using sequence-based methods to infer the origins and deep phylogenetic relationships of RNA viruses, an urgent requirement for the future are new methods that are able to reveal evolutionary history using other types of biological data. Two data types seem most appropriate here: similarities (and differences) in genome organization (i.e. in gene order and gene content), and protein secondary structure.

2.3.2 Phylogenies based on genome organization

Outside the sphere of viruses, one of the most interesting developments in phylogenetics has been the development of methods that consider gene content and gene order as characters (Sankoff 2003). However, while such approaches may eventually prove useful to the study of large dsDNA viruses, where there is perhaps sufficient genomic patterning for meaningful analysis (McLysaght *et al.* 2003), they are unlikely to have a major impact on inferring the evolutionary history of RNA viruses. There are a number of specific problems associated with the phylogenetic analysis of such secondary genomic data, irrespective of the more generic problem of being able to construct a viable model of genome evolution. First, the genomes of RNA viruses are small, usually containing no more than 10–12 genes. Hence, there are very few phylogenetically informative characters, particularly as all viruses must carry a number of essential genes. Gene content therefore appears to be a weak phylogenetic character. Second, there appears to be little phylogenetic resolution in patterns of gene order. Whereas individual families of RNA viruses tend to be characterized by very similar gene orders and the six proposed supergroups of ssRNA+ viruses tend have similar genome organizations (Goldbach and de Haan 1994; Le Gall *et al.* 2008) (Fig. 2.6), gene orders can vary extensively among these supergroups, such that the inference of deep phylogenetic relationships is inviable. In addition, genome sizes and segment numbers often vary *within* individual viral families. As a consequence, no expansive phylogeny of families of RNA viruses based on genome organization is available at present, and it is difficult to see how one can be determined. Worse, the high levels of sequence divergence exhibited by many viral proteins sometimes make it extremely difficult to accurately determine which proteins are truly homologous. In summary, the study of genome organization does not appear to be a profitable approach to resolve deep phylogenetic structure in RNA viruses.

2.3.3 Phylogenies based on protein structure

Because the study of viral origins necessitates the use of highly divergent sequences, phylogenetic analyses often involve amino acid sequences that fall within the 'twilight zone' of pairwise sequence similarity (≈15–30% identity) (Doolittle 1986). Sadly, accurately recovering evolutionary information in the face of such low levels of sequence similarity is one of the most difficult problems in computational biology, but one that we may need solve if we are to determine the ancient evolutionary history of viruses. A more profitable mode of investigation may therefore be through the analysis of protein structure, such as that of the icosahedral virion, as this seems to be a far more stable character than the underlying primary sequence. While the amino acid sequences of a specific gene may harbour little or no similarity, providing no evidence that they are even homologous, strong similarities can still be observed at the level of protein structure, as the negative fitness costs associated with the disruption of these structures can be profound. This approach, at least in spirit, has already been used in attempts to infer the phylogenetic relationships among negative-sense RNA viruses (Vieth *et al.* 2004). Unfortunately, the fascinating observation of similarities in protein structure among highly divergent viruses, including those with RNA and DNA genomes, has yet to lead to sophisticated methods to infer phylogeny from structural data. The inconvenient truth is that we may not get a good method to infer phylogeny from structure until we can accurately predict structure from sequence. Although progress is being made in the development of viable models of protein evolution, the limitations are all too apparent (Robinson *et al.* 2003; Thorne 2007). To make a final point in passing, it is both interesting and ironic that despite the genomic revolution, a morphological trait—protein structure—may yet provide the most powerful means by which to infer the deep phylogenetic relationships of RNA viruses.

2.4 RNA viruses and the evolution of the genetic code

As a coda to this chapter I will briefly consider one other aspect relating to the origins of RNA viruses that has been ignored to date but which has important implications for their evolutionary history: the evolution of the genetic code. The key observation here is that RNA viruses necessarily employ the same genetic code as their hosts. Hence, for the vast majority of organisms that employ the standard ('universal' or 'canonical') genetic code, so the viruses that infect these organisms also employ the standard genetic code. Similarly, for all those organisms that employ variant genetic codes, the viruses that infect these organisms (or organelles) employ the same variant code. For example, in the fungal mitochondrial code, the 'universal' opal stop codon UGA instead encodes the amino acid tryptophan. In turn, the mitoviruses—ssRNA+ viruses (family *Narnaviridae*) that infect the mitochondria of fungi—contain a number of internal UGA codons and, as RdRp activity has been found in infected mitochondria, these must also express UGA as tryptophan (Cole *et al.* 2000).

Given that RNA viruses are obligate parasites, the strong association between the genetic code in host and virus is entirely predictable: because viruses employ the same apparatus of protein translation of their hosts, it follows that they must also employ the same assignment of codons and amino acids (or stop codons) otherwise non-functional proteins will be produced. But this raises a far larger question: *how* did viruses evolve alternative codes? As the mis-assignment between codons and amino acids is very likely to result in non-functional polypeptides, it is difficult to understand how a virus infecting a host species that employs the standard genetic code could ever jump to successfully infect and replicate in a host species with a variant genetic code. Surely the loss in fitness that such a host jump would entail would be too severe for the establishment of a successful infection? Indeed, it is theoretically possible that preventing viral infections was one of the major selective forces driving the evolution of variant genetic codes: by using an alternative genetic code, a host would have an extremely potent antiviral agent, effectively preventing any new viral infection from establishing itself (Shackelton and Holmes 2008). It is therefore possible to imagine, although currently untested, that the selection pressure imposed by viral infections could result in the reassignment of infrequently used amino acids, therein resulting in the development of an alternative genetic code. Highly circumstantial evidence for this idea is that codon reassignments are particularly common in ciliates, such as *Tetrahymena thermophila*, that inactivate bacteriophage as they filter-feed (Lozupone *et al.* 2001; Pinheiro *et al.* 2007). Such frequent exposure to potentially highly pathogenic viruses may represent exactly the sort of selection pressure required to favour the evolution of variant genetic codes. Finally, another important aspect of this idea, which relates to the overall theme of this chapter, is that those viruses that currently infect organisms with variant genetic codes must have either originated recently as escaped genes from these hosts or, perhaps more likely, entered their hosts before the code reassignment took place. Unfortunately, it is currently impossible to choose among these two theories.

3

The mechanisms of RNA virus evolution

In many ways, this chapter is the heart of the book. In particular, it is arguable that the more 'macroevolutionary' patterns described in Chapters 2, 5, and 6 cannot be understood without a firm grasp of the processes of microevolution in RNA viruses. My specific aim here is therefore to describe the rates, determinants, and consequences of a variety of fundamental evolutionary processes in RNA viruses, particularly mutation, recombination, natural selection, and epistasis. Perhaps to the relief of many working with RNA viruses, debates concerning the respective roles played by natural selection and genetic drift—what John Gillespie has called the 'great obsession of population genetics' (Gillespie 1998)—have not dominated evolutionary arguments. This is probably because students of RNA virus evolution have spent more time considering how frequent mutation may play an even more important role in shaping evolutionary dynamics. However, no discussion of the mechanics and dynamics of viral evolution is complete without at least picking at the great obsession.

3.1 The evolutionary dynamics of RNA viruses

3.1.1 Mutation rates in RNA viruses and their determinants

From an evolutionary perspective, RNA viruses have two uniquely defining features: extremely high mutation rates and extremely small genomes. As I will argue throughout this book, these two features are also inextricably linked. Not only are the mutation rates exhibited by RNA viruses extremely rapid but, with the exception of ssDNA viruses which are discussed in detail below, they are orders of magnitude higher then those seen in DNA-based organisms.

Estimates of mutation rate span several orders of magnitude among RNA and DNA viruses taken together: from up to 1.5×10^{-3} mutations per nucleotide, per replication (mut/nt/rep) in the ssRNA+ bacteriophage Qβ (Drake 1993), to only 1.8×10^{-8} mut/nt/rep in the dsDNA virus herpes simplex virus type 1 (HSV-1) (Drake and Hwang 2005) (Fig. 3.1). For RNA viruses replicating with RdRp, measured mutation rates are usually close to approximately 1 mutation per genome, per replication (mut/genome/rep) (Drake 1993; Drake et al. 1998; Drake and Holland 1999; Schrag et al. 1999; Duffy et al. 2008). Hence, a mutation is made during nearly every round of genome replication. Consequently, it is clear that RNA virus evolution is, to a large extent, dominated by the process of mutation. Rather lower mutation rates are observed in

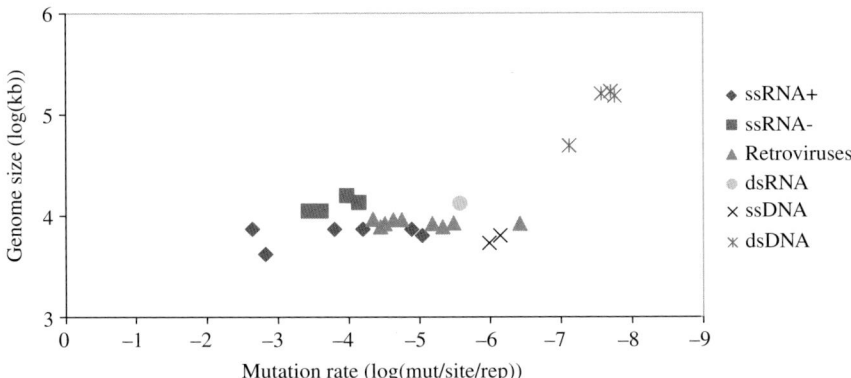

Fig. 3.1 Average rates of mutation per site, per genome replication (mut/site/rep), in different types of virus. Viruses are denoted by different symbols; ssRNA+ are Qβ, poliovirus, tobacco mosaic virus; ssRNA− are vesicular stomatitis virus, influenza A virus, measles virus; retroviruses are spleen necrosis virus, murine leukaemia virus, Rous sarcoma virus, HIV-1, bovine leukaemia virus; dsRNA are bacteriophage φ6; ssDNA are bacteriophages M13, φX174; dsDNA are bacteriophages λ, T2, T4, herpes simplex virus-1. Adapted from Duffy *et al.* (2008) with permission.

retroviruses such as HIV-1, with estimates ranging from 0.1 to 0.3 mut/genome/rep (Drake 1993; Mansky and Temin 1995; Drake *et al.* 1998; Mansky 1998), perhaps five times lower than those observed in most RNA viruses that replicate using RdRp. Although HIV-1 is often cited as the cause célèbre of rapid mutation, in reality it is less error-prone than most other RNA viruses. This has important consequences for the idea that HIV-1 forms a quasispecies (see section 4.2). Finally, Jan Drake proposes that there is a 'universal' mutation rate in DNA viruses and bacteria of 0.0034 mut/genome/rep (Drake *et al.* 1998). Although this is an intriguing idea that fits some of the data, there are also exceptions (Duffy *et al.* 2008). For example, the ssDNA phage φX174 has a mutation rate rather higher than predicted under the universal rate model (Raney *et al.* 2004).

Even though there is a general consensus that replication error rates in RNA viruses are extremely high, it is important to note that the accurate estimation of these rates is challenging under any circumstances. Also, rather lower rates of mutation have been observed in some viruses that replicate using RdRp (Malpica *et al.* 2002; Furió *et al.* 2005), although most fall within the same order of magnitude. The most dramatic exception proposed to the rule of rapid mutation is yellow fever virus (YFV), where a mutation rate of 0.0021–0.0025 mut/genome/rep has been reported (Pugachev *et al.* 2004). Not only is this rate some 400 times lower than that usually associated with RdRp, but it is also lower than Drake's universal mutation rate for DNA microbes. It is therefore difficult to reconcile such a low rate with what is usually seen with RdRp, particularly as estimates of the rate of nucleotide substitution in YFV fall squarely within the normal range associated with RNA viruses (Bryant *et al.* 2007). Indeed, it is notable that this estimate of 'mutation rate' ignores the potentially crucial consequences of natural selection, including that on lethal mutations, such that it cannot be regarded as reliable.

The differences in intrinsic mutation rate among viruses broadly correspond to the fidelity of the various polymerases used in replication: the RdRp used by RNA viruses is more error-prone than the RT used by retroviruses, which in turn has a higher mutation rate than the DNA polymerases used by DNA viruses (Fig. 3.1). Comfortingly, these differences in polymerase fidelity have a solid basis in biochemistry. In particular, DNA polymerases have the ability to correct the errors made during replication, which reduces overall mutation rates by at least an order of magnitude (Garcia-Diaz and Bebenek 2007). In contrast, no RdRp is known to possess this proofreading capability. In addition, DNA-based organisms are able to employ enzymes that perform base-excision repair on mispaired bases, again reducing error rates. In the case of retroviruses, cellular deaminating enzymes, most notably APOBEC3G, add additional transition mutations to those generated during replication (Walsh and Xu 2006), and which also constitutes a potent anti-viral strategy (Mangeat et al. 2003; see section 3.3).

There is also evidence that the mechanism of viral replication affects mutation rate, manifest in the difference between 'stamping-machine' and 'geometric' replication (Duffy et al. 2008). In stamping-machine replication a single virus acts as the template for all progeny genomes, so that mutations accumulate linearly. In geometric replication some of the early progeny genomes are themselves used as templates for further progeny, so that mutations accumulate geometrically as a mutated template propagates the given error to all its replicate copies (Chao et al. 2002), thereby increasing the rate of mutation accumulation (Drake et al. 1998).

However, while it is clear that the error rates associated with RNA polymerase are very high, there are still fundamental gaps in our understanding of the mutational process. Most notably, those estimates of mutation rate undertaken to date usually only consider *mean* mutation rates, and so provide no information on the distribution of error rates within a single replication cycle. Therefore, with very few exceptions (Chao et al. 2002; Malpica et al. 2002), we do not know what fraction of the progeny of replication carry multiple mutations (which may be commonplace; Malpica et al. 2002), or whether this distribution of mutants is Poisson, geometric, or takes another shape (Drake et al. 2005; Drake 2007). However, precisely describing the mutation distribution is critical to understanding adaptation. For example, the occurrence of multiple advantageous mutations in a single replication cycle may be critical for successful cross-species virus transmission (section 6.4; Kuiken et al. 2006).

3.1.2 A comparison of substitution rates in viruses

The division between RNA and DNA viruses in rates of mutation is generally mirrored in their rates of nucleotide substitution, which may differ by some six orders of magnitude. As such, the main factor shaping nucleotide substitution rates in RNA viruses is evidently how frequently mutations are generated. However, there are mounting data to show that although ssDNA viruses obviously possess DNA genomes, and even replicate using host polymerases, their rates of nucleotide substitution are far closer to those of RNA viruses than to those of dsDNA viruses, indicating that the defining issue is not the difference between RNA and DNA (Duffy *et al.* 2008).

Before proceeding, it is important to clear up a long-standing confusion in studies of viral evolution, between levels of antigenic and genetic variation. This confusion is manifest in statements that some viruses, and particularly measles, 'evolve slowly', while others, exemplified by influenza virus, 'evolve rapidly'. In reality, what these statements reflect is that *antigenic variation* is relatively limited in measles virus, such that the vaccines we use to control this infection do not have to be regularly updated, while high levels of antigenic variation are observed in human influenza A virus, so that vaccines need to be updated almost annually and vaccine failure is commonplace. However, although viruses like measles and influenza undoubtedly differ dramatically in levels of antigenic diversity, their underlying mutational and substitutional dynamics fall within the 'standard' RNA virus range which I will define shortly (Schrag *et al.* 1999; Woelk *et al.* 2001, 2002; Kremer *et al.* 2008; Pomeroy *et al.* 2008; Rambaut *et al.* 2008).

Tangible evidence for the rapidity of nucleotide substitution in RNA viruses is that this process can often be observed in real time, simply by analysing the distribution of branch lengths in viruses sampled at different times (Drummond *et al.* 2003a, 2003b). Many RNA viruses therefore evolve on a timescale that can be recorded by human observation. For nearly all RNA viruses examined to date, this translates into overall rates of nucleotide substitution in the range of 10^{-2}–10^{-5} subs/site/year, with most exhibiting rates within an order of magnitude of a value of 1×10^{-3} subs/site/year (Hanada *et al.* 2004; Jenkins *et al.* 2002) (Fig. 3.2). Although there is an absence of reliable estimates of substitution rate in dsRNA viruses, that these viruses also exhibit considerable genetic diversity suggests that their substitution rates fall within the same boundaries (for example, Maan *et al.* 2007; Matthijnssens *et al.* 2008). Finally, whereas rates of nonsynonymous substitution vary more widely among RNA viruses, and among genes within individual viruses, reflecting differences in selective constraint and life history, these differences have only a small impact on overall rates of nucleotide substitution as most substitutions occur at synonymous sites.

In contrast, far lower rates of nucleotide substitution are observed in dsDNA viruses. For large dsDNA viruses of animals, substitution rates have been estimated assuming (probably fairly) that they co-diverged with their hosts over timescales of millions of years. The best documented case is that of the gammaherpesviruses, where related viruses have been obtained from diverse tetrapods (McGeoch and Gatherer 2005). If the assumption of co-divergence is correct this translates into evolutionary rates in the range of approximately 10^{-8} subs/site/year, and so close to the values seen in mammalian mtDNA (Hatwell and Sharp 2000). Similarly, low substitution rates have been estimated in some small dsDNA viruses that appear to have experienced host-virus co-divergence, most notably the papillomaviruses that infect a wide variety of vertebrates (Rector *et al.* 2007) and which are the primary cause of cervical cancer (Bernard 1994).

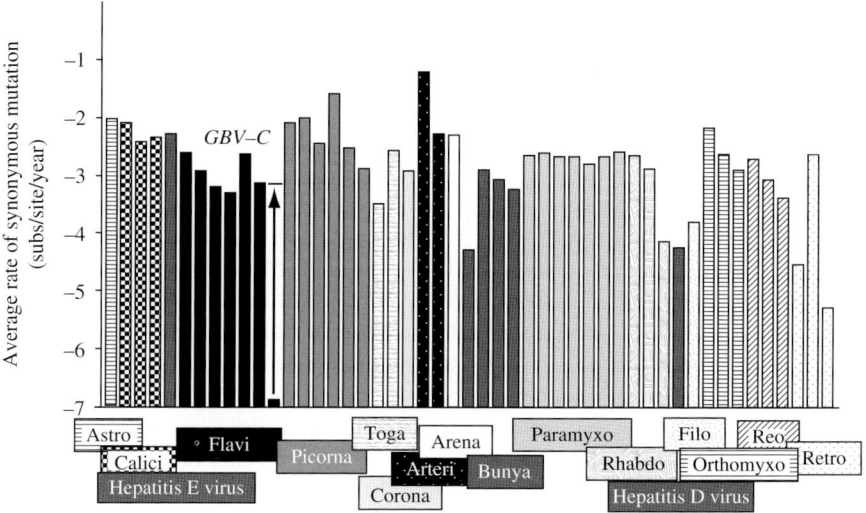

Fig. 3.2 Average rates of synonymous nucleotide substitution (per site, per year on a log scale) in different families of RNA virus (and HDV). Note that although the substitution rate for GBV-C (originally called hepatitis G virus; HGV) appears anomalously low in this figure, more recent analyses (Romano *et al.* 2008) indicate that this virus evolves at approximately the same rate as other RNA viruses so that its rate has been adjusted accordingly here, as shown by the arrow. Adapted from Hanada *et al.* (2004) with permission.

Although there are doubtless some errors in rate estimation, particularly when heavily laboratory-manipulated isolates are included in the analysis, the measurable evolution obvious in many RNA viruses suggests that these rates are, by and large, accurate. To my mind, the difficulties in rate estimation and the inherent sampling errors make it foolish to think of these as measures as anything other than broad-brush indications of evolutionary tempo, so that 'in the range of 10^{-3}–10^{-4} subs/site/year' is probably accurate enough. One important source of such error is that all methods currently used to estimate substitution rate assume that the sampled sequences contain only *fixed* substitutions (as is implicit in the term substitution rate). While nucleotide changes that fall on the deeper branches of phylogenetic trees represent mutations that must have reached high frequency in the population, including true fixation events, a subset of the changes that fall on terminal branches will constitute transient polymorphisms that will ultimately be lost from the population. Given the mutational power and constrained genomes of RNA viruses, it is not surprising that a large proportion of mutations seem to fall into this transient deleterious class (Gao *et al.* 2004; Kosakovsky Pond *et al.* 2006; Pybus *et al.* 2007; section 3.4). Consequently, the evolutionary rates estimated using time-structured data must reflect a composite mutation/substitution rate parameter, which will inflate rate estimates in the short-term (Ho *et al.* 2007). As such, measures of intra-host diversity should never be used to estimate long-term substitution rates. However, as most deleterious mutations may be purged rapidly from RNA virus populations (Holmes 2003a), it is likely that the inclusion of transient polymorphisms will only have a minor effect on rate estimates in most cases.

3.1.3 Differences in viral generation time

Although the substitution rates of RNA viruses are certainly high, they still vary by approximately three orders of magnitude—from roughly 10^{-2} to 10^{-5} subs/site/year—irrespective of estimation method. Why is this so? One possibility is that RNA polymerases differ in intrinsic fidelity. While this is clear from the initial estimates of mutation rate made by Jan Drake and colleagues, and there is good evidence that the fidelity of RNA polymerase can be manipulated experimentally (Vignuzzi *et al.* 2005), and that it may vary according to the host species infected (Pita *et al.* 2007), such localized differences in error rate cannot explain the three order of magnitude variation in substitution rates: for example, the error rate of the 'high-fidelity' strain of poliovirus used by Vignuzzi *et al.* (2005) was still higher than that of DNA viruses. As a consequence, it is likely that differences in viral generation time (replication rate), and specifically the time it takes to go from infected cell to infected cell, explain most of the variance in substitution rates. Unfortunately, accurate estimates of generation time are unavailable for most RNA viruses. At the extremes are viruses that appear to be effectively latent within hosts, replicating only occasionally, which in turn results in greatly reduced rates of nucleotide substitution (see below), compared to those in which replication cycles which can be measured in timescales of hours. As an example of the latter, the intracellular part of the poliovirus replication cycle takes 4–5 h, with cell lysis occurring after 10–12 h (C. Cameron, personal communication). For HIV, where the dynamics of replication have been studied in detail, the cell-to-cell generation time is approximately 2.6 days (Perelson *et al.* 1996).

The idea that latency reduces substitution rates is well developed in the case of the retroviruses HTLV-I and HTLV-II, in which low rates of epidemic transmission may mean that viral evolution is dominated by the occasional clonal expansion of infected cells within individual hosts (Vandamme *et al.* 2000). This results in substitution rates as low as approximately 10^{-7} subs/site/year (Salemi *et al.* 1999; Lemey *et al.* 2005). Another good example is provided by SFV where phylogenetic analysis suggests virus-host co-divergence for periods in excess of 30 million years (Switzer *et al.* 2005; Liu *et al.* 2008). Using the primate fossil record as a calibration point results in substitution rates of only 1.7×10^{-8} subs/site/year, within the range seen in dsDNA viruses but many orders of magnitude lower than those of other RNA viruses (Switzer *et al.* 2005). As some intra-host genetic variation has been observed in SFV (Schweizer *et al.* 1999), suggesting that its RT has not evolved additional mechanisms of error correction, and an extremely low replication rate is the most likely explanation for the low substitution rate in this virus (Meiering and Linial 2001). More generally, because of their ability to integrate into host genomes and so only undergo replication by the much higher-fidelity DNA polymerases, all retroviruses have the capacity to evolve slowly if they can reduce the frequency of RNA-based replication. The ultimate end point of this process are the endogenous retroviruses, which have entirely assimilated the evolutionary dynamics of their host organisms (see section 5.3).

Substitution rates also vary to some extent among persistent RNA viruses that have distinct periods of intra- and inter-host evolution. In the case of HIV-1, for example, there is an apparently inverse relationship between rates of viral transmission and rates of evolutionary change, with the highest substitution rates recorded within individual hosts (Maljkovic Berry *et al.* 2007). The elevated rate of nucleotide substitution at the intra-host level may be because this part of the viral life cycle is dominated by the selective fixation of amino acid changes that enable immune escape (Nielsen and Yang 1998; Williamson 2003). Alternatively, it may be that many of the mutations that occur within hosts are purged ('revert') when the virus is transmitted to new hosts due to a mismatch with cytotoxic T-lymphocyte (CTL) responses, as determined by HLA type (Li *et al.* 2007), which in turn results in strong purifying selection (see section 7.2).

3.1.4 Slowly evolving RNA viruses?

Understandably more controversial are those cases in which RNA viruses replicating with an RdRp purportedly evolve at rates far lower than 10^{-5} subs/site/year, and where nuclear integration does not occur. A handful of RNA viruses fall into this category. Early studies suggested that filoviruses, and most famously Ebola virus (EBOV), might evolve very slowly, largely because of the high levels of sequence similarity observed between viruses sampled from specific EBOV outbreaks (Sanchez *et al.* 1996). However, more recent analyses have revealed that the Zaire strain of Ebola virus (EBOV-Zaire), for which most data are available, evolves at rates similar to those seen in other RNA viruses (Biek *et al.* 2006; Walsh *et al.* 2005). More compelling are suggestions that the flavivirus GBV-C (originally called hepatitis G virus; HGV) evolves at rates as low as 10^{-7} subs/site/year (Suzuki *et al.* 1999; Hanada *et al.* 2004). Indeed, GBV-C has one of the key attributes of a slowly evolving virus—it chronically infects hosts an in asymptomatic manner (see section 6.2)—which has reasonably led to suggestions that it establishes an effectively latent infection (Suzuki *et al.* 1999). Further, it has been proposed that GBV-C, and its close relative GBV-A, have co-diverged with their primate hosts over millions of years (Charrel *et al.* 1999), which again argues for low substitution rates. However, more detailed studies of time-structured sequence data from GBV-C have revealed substitution rates close to those observed in other flaviviruses (Romano *et al.* 2008), and intra-host genetic variation has been observed in this virus (Zampino *et al.* 1999). Similar arguments apply to the rodent-associated hantaviruses (family *Bunyaviridae*), where substitution rates in the range of 10^{-7} subs/site/year have also been inferred, again based on the assumption of host-virus co-divergence (Hughes and Friedman 2000; Plyusnin and Morzunov 2001). However, there are good reasons to doubt such low rate estimates. First, analyses of time-structured data have again unearthed substitution rates that fall within the standard RNA virus range (Ramsden *et al.* 2008), matching new data revealing that intra-host genetic diversity is also extensive in hantaviruses and indicative of a high mutation rate (Sironen *et al.* 2008). Second, newly available hantavirus sequences from shrews (insectivores) are mixed with those sampled from rodents on phylogenetic trees (Arai *et al.* 2008; Ramsden *et al.* 2009), so that neither mammalian order forms a monophyletic group in the virus phylogeny. This suggests a far more complex evolutionary pattern than simple host-virus co-divergence.

The final class of RdRp-replicating RNA viruses proposed to evolve anomalously slowly are some of those that infect plants. For example, both tobamoviruses and closteroviruses exhibit few genetic changes between sequences isolated over long time periods (Fraile *et al.* 1997; Marco and Aranda 2005), with the former group also suggested to have co-diverged with their hosts for perhaps as long as 100 million years (reviewed in Gibbs *et al.* 2008a). The Fraile *et al.* (1997) paper is particularly noteworthy in that limited mutation accumulation was observed among isolates of tobacco mild green mosaic tobamovirus (TMGMV) sampled almost 90 years apart, while one isolate of tobacco mosaic (tobamo)virus (TMV) was sampled as early as 1899. However, as these viruses were not studied with the rigorous techniques associated with 'ancient DNA' that their age merits (Cooper and Poinar 2000), the low substitution rate in TMGMV requires independent verification. Similarly, the co-divergence of tobamoviruses and their principle hosts has not been rigorously tested.

These uncertainties notwithstanding, it has been proposed that the severe population bottlenecks that occur both within and among hosts might act to reduce substitution rates in plant RNA viruses (Li and Roossinck 2004; Ali *et al.* 2006; see section 3.3). However, if viral evolution is in large part neutral (itself the source of much debate), then changes in population size will have no affect on substitution rates, and major population bottlenecks at transmission might equally be expected to occur in rapidly evolving animal RNA viruses. Similarly, it has been suggested that the lack of adaptive immune systems in plants results in weaker immune-mediated positive selection compared to animal viruses and hence lower rates of nonsynonymous substitution (García-Arenal *et al.* 2001). While a reduction in immune selection pressure will undoubtedly reduce nonsynonymous rates, estimates of evolutionary rate in plant RNA viruses using time-structured data are within the range observed in animal RNA viruses (Fargette *et al.* 2008a; Gibbs *et al.* 2008b; Simmons *et al.* 2008), suggesting that they do not evolve anomalously slowly. In short, there is currently no compelling evidence that any virus replicating with an RdRp evolves slower than approximately 10^{-5} subs/site/year.

3.1.5 Rapidly evolving ssDNA viruses

Of the recent developments in understanding the evolutionary dynamics of viruses, perhaps that of most importance is the recognition that ssDNA viruses exhibit rates of nucleotide substitution that approach those of their RNA counterparts. Although

Drake's universal genomic mutation rate requires that ssDNA viruses, all of which have genomes smaller than approximately 11 000 nt, should have high error rates, it was originally thought that these viruses had low mutation rates similar to those in dsDNA viruses as they rely on host DNA polymerases for replication. While high levels of genetic diversity were observed in a number of ssDNA viruses, which is evidently compatible with elevated mutation rates (Isnard *et al.* 1998; Sanz *et al.* 1999; Khudyakov *et al.* 2000; Lopez-Bueno *et al.* 2006; Ge *et al.* 2007), the rapid evolution of ssDNA viruses was not fully apparent until the first phylogenetic analysis of time-structured data from carnivore (Shackelton *et al.* 2005) and human (Shackelton and Holmes 2006; Norja *et al.* 2008) parvoviruses. The case of the carnivore parvoviruses is particularly convincing since the emergence of canine parvovirus (CPV) in dogs from feline panleukopenia virus (FPV) in cats during the 1970s is well documented (Truyen *et al.* 1996). Strikingly, in both viruses the substitution rate was estimated to be approximately 10^{-4} subs/site/year, and hence within the range seen in RNA viruses, but far lower than that of dsDNA viruses. Similarly high rates of nucleotide substitution have now been determined for the plant geminivirus tomato yellow leaf curl virus (Duffy and Holmes 2008) and the anellovirus SEN-V (Umemura *et al.* 2002). Why, mechanistically, ssDNA viruses might evolve (and presumably mutate) rapidly even though they utilize host polymerases is unclear, although it is possible that both proofreading and excision repair are less efficient on ssDNA, perhaps reflecting differences in methylation patterns between host and virus (Sanz *et al.* 1999; Duffy and Holmes 2008). Similarly, deamination, a source of error that is independent of that generated during faulty replication, may also elevate mutation rates in ssDNA viruses (Duffy and Holmes 2008).

The possible exception to the revisionist notion that ssDNA viruses evolve quickly are the circoviruses, such as those responsible for postweaning multisystemic wasting syndrome (PMWS) in pigs. Circoviruses are particularly noteworthy in that they are the smallest of all DNA viruses, with a genome usually comprising only two ORFs and a total length usually no larger than approximately 2000 nt. It has been proposed that these viruses exhibit substitution rates within an order of magnitude of their vertebrate hosts based on the assumption that they co-diverged with mammals and birds over a period of approximately 300 million years (Johne *et al.* 2006). However, while this story clearly merits further study, the evidence for long-term co-divergence was based on the analysis of just seven host species, a number of mis-matches were apparent, and some intra-host genetic variation was observed.

3.1.6 What sets the rate of RNA virus evolution?

As well as demonstrating that RNA (and ssDNA) viruses evolve rapidly, it is also important to document the evolutionary processes responsible for these high rates. The simplest hypothesis is that a high mutation rate is beneficial because it leads to the greater production of advantageous mutations, thereby increasing the rate of adaptive evolution. However, this idea is easily dismissed because the vast majority of the mutations that arise in RNA viruses are deleterious (see section 3.4), so that increased mutation rates will generally reduce fitness. As a consequence, more reasonable hypotheses for the evolution of high mutation rates are based on some sort of evolutionary trade-off, either between replication rate and fidelity, or between the rates of deleterious and advantageous mutation (Sniegowski *et al.* 2000; Duffy *et al.* 2008).

3.1.7 Trade-offs and the evolution of mutation rates

The idea that there is an evolutionary trade-off between replication speed and replication fidelity, such that high mutation rates are simply a consequence of selection for rapid replication, which is also more error-prone (Elena and Sanjuán 2005; Furió *et al.* 2005), is an intriguing one. In direct support of this hypothesis, increased replication fidelity has been observed to result in a fitness cost, associated with a reduced replication rate, in experimental studies of both vesicular stomatitis virus (VSV) (Furió *et al.* 2005) and HIV-1 (Furió *et al.* 2007). Similarly, the higher-fidelity stamping-machine replication produces progeny genomes more slowly than geometric replication (French and Stenger 2003). However, while there is tentative, yet growing, evidence for a trade-off between replication rate and fidelity, whether this can explain mutation rates in their entirety, and whether it operates outside of the laboratory, remains to be established, especially as counter examples exist (Belshaw *et al.* 2008).

The possible evolutionary trade-off between the rate of production of deleterious and advantageous mutations has received rather more attention. Although some quasispecies models predict that high rates of deleterious mutation are advantageous (O'Fallon *et al.* 2007), natural selection should generally favour a reduction in mutation rates in stable environments as this will reduce the load of deleterious mutation, a burden that appears to be particularly severe for RNA viruses (García-Arenal *et al.* 2003; Pybus *et al.* 2007; section 3.4). However, the concept of a 'stable' environment seems alien to most RNA viruses, because of their continual struggle for existence against innate, intrinsic, and adaptive host immunity, as well as their exposure to new hosts and cell types. As a consequence, viruses probably always experience selection for mutation rates that are greater than zero, as is likely to be true of any genetic system (Sniegowski *et al.* 2000). In support of this idea, a selective advantage of lower compared to higher-fidelity RNA polymerases has been observed in experimental systems (Mansky and Cunningham 2000; Furió *et al.* 2005; Vignuzzi *et al.* 2005), although the differences in fidelity are minor compared to the observed range of substitution rates in RNA viruses. Perhaps more importantly, while these studies reveal that there is heritable variation for polymerase fidelity, an obvious pre-requisite for natural selection, RNA viruses are unable to reduce their error rates to the levels associated with DNA polymerases. This implies that there are major adaptive constraints acting against the evolution of very high-fidelity RNA polymerases. As discussed in more detail below, the major implication of this observation is that RNA viruses are in some sense 'stuck' with a highly error-prone replication enzyme, which has profound effects on much of their life history.

Turning the tables, we can also ask what factors determine the upper limit on mutation rates in RNA viruses. The most interesting observation here is that artificially increasing mutation rates results in an excessive mutational load and hence a major loss of fitness (see section 4.3; Crotty *et al.* 2001; Domingo *et al.* 2005). Consequently, there is clearly a selectively determined upper limit on mutation rates. A ceiling on mutation rates also explains why very highly error-prone 'mutator' strains like those seen in bacteria (Taddei *et al.* 1997) have not been observed in RNA viruses (Duffy *et al.* 2008), even though single mutations can increase the fidelity of RNA polymerase (Mansky and Cunningham 2000; Vignuzzi *et al.* 2005) (and selection by viruses may even be responsible for bacterial mutators; Pal *et al.* 2007). Similarly, the reduction in viral fitness that comes as a consequence of increased mutation rates explains why the mutagens 5-fluorouracil and ribavirin have proven useful in reducing viral loads in experimental systems, and may eventually constitute a powerful form of antiviral therapy: so-called lethal mutagenesis (Mansky and Bernard 2000; Sierra *et al.* 2000; Domingo *et al.* 2005; Bull *et al.* 2007; see section 4.3). It is also possible that natural selection favours traits that compensate for the accumulation of deleterious mutations associated with high mutation rates, particularly in the form of mutational robustness (Elena and Sanjuán 2005; Montville *et al.* 2005; Codoñer *et al.* 2006; Lenski *et al.* 2006; Sanjuán *et al.* 2007). This phenomenon is considered in more detail in section 3.5.

3.1.8 Mutation rates and mutational loads

That the artificial elevation of mutation rates in RNA viruses is associated with major fitness costs strongly supports the idea that RNA viruses reside close to their maximum tolerable mutation rates. At higher error rates population structure breaks down because the fittest genotype is lost. This eventual leads to a total loss of information, or the 'melting point' as memorably described by Manfred Eigen (Eigen 1971, 1992; Swetina and Schuster 1982). The mutation rates observed in RNA viruses might therefore represent a trade-off between the generation of sufficient beneficial mutations to rapidly adapt to changing environments, yet not so many as to induce mutational meltdown.

An important consequence of the idea that RNA viruses live at the edge of (i.e. just below) the upper limit on mutation rate is that there must also be an upper limit on their genome sizes: given the same rate of mutation per nucleotide, RNA viruses with larger genomes will suffer more deleterious mutations than those with smaller genomes. It is this relationship that likely explains why the maximum genome sizes of RNA viruses are set at approximately the reciprocal of their mutation rate (Eigen 1992), although it is not necessarily the case that RNA viruses will attain the maximum genome size their error rate allows. Although this theory has traditionally been applied to RNA viruses, that ssDNA viruses also exhibit very high rapid evolutionary rates, and similarly possess very small genome sizes, clearly supports this hypothesis (Duffy *et al.* 2008). A variety of other possible explanations for the small genome sizes of RNA viruses are discussed in section 5.1, although I believe that none are as powerful as the mutation hypothesis outlined here.

3.1.9 Are RNA viruses trapped by high mutation rates?

It is always tempting to construct adaptive explanations for biological phenomena. However, it is possible that a major reason why RNA viruses mutate rapidly is that they are simply unable to do otherwise. This, I should hasten to add, is not the same as saying that RNA viruses are in some respects a 'frozen accident', but rather that they occupy a particular region of evolutionary parameter space where major improvements in polymerase fidelity are extremely difficult. Such a major constraint in evolutionary trajectory again relates to Eigen's paradox (see section 2.2.4). Although this concept is usually applied to debates over the origin of life or the evolution of complexity, Eigen's paradox may also apply to contemporary RNA viruses: their highly error-prone replication dictates that they are unable to evolve long genomes, so to greatly reduce polymerase error rates requires a far higher-fidelity polymerase that can only be attained with a longer genome. An exception—the *Coronaviridae* (and related *Roniviridae*), in which capture of a host exoribonuclease (ExoN) domain may have reduced error rates and elongated genome sizes—is discussed in section 5.1. It is therefore possible that RNA viruses simply lack the requisite genomic flexibility to be able to significantly lower their mutation rates. As a consequence, the life-history strategy of RNA viruses is one in which the cost of abundant deleterious mutation is offset by the production of vast numbers of progeny.

3.2 Recombination and reassortment in RNA virus evolution

While mutation is the ultimate source of genetic variation, there is a growing body of work suggesting that recombination, and its sister process reassortment, can, in some instances, also play a significant role in shaping patterns of genetic diversity in RNA viruses. However, both the frequency with which recombination occurs and the reasons for its occurrence have proven controversial.

Before proceeding it is important to make the essential distinction between recombination, which can in theory occur in all RNA viruses, and reassortment, which only occurs in that subset of RNA viruses that possess segmented (including multi-component) genomes (Fig. 3.3). Although both processes may be regarded as forms of sexual reproduction in the broad sense, and both require two viruses to co-infect a single cell, they are mechanistically very different.

Recombination in RNA viruses, sometimes referred to as 'RNA recombination', is thought to occur when two viruses co-infect a single host cell and a hybrid molecule is produced through a process termed copy-choice replication (Lai 1992) (although other models of recombination have been proposed; Chetverin *et al.* 1997). Under the copy-choice model the RdRp is thought to jump templates during negative strand synthesis,

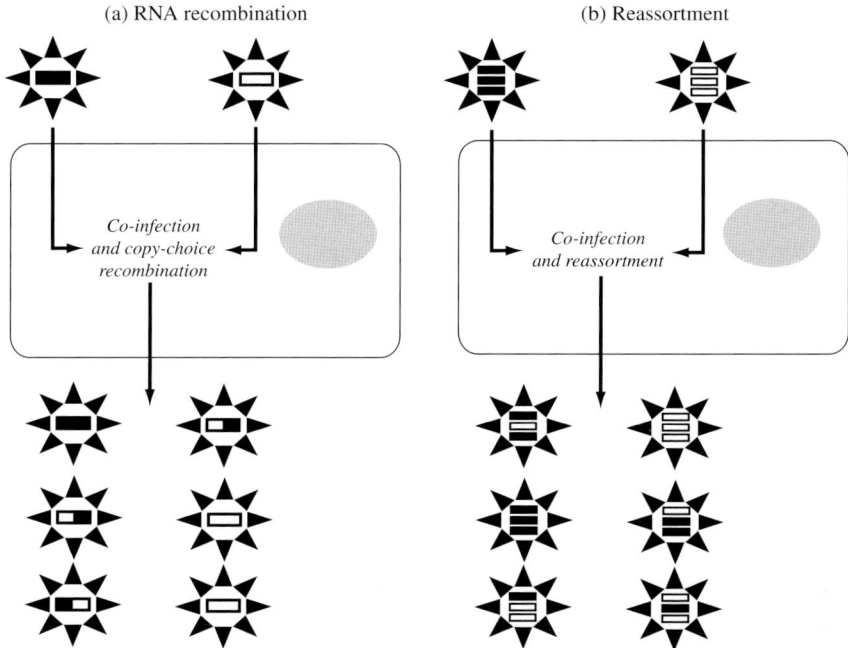

Fig. 3.3 How (a) copy-choice RNA recombination and (b) reassortment create new genetic configurations in RNA viruses. Reassortment only occurs in viruses with segmented genomes.

generating an RNA molecule with mixed ancestry (Aaziz and Tepfer 1999). This form of recombination can be 'homologous', such that the process of template jumping occurs between regions of homologous sequence, or 'non-homologous' (illegitimate), in which genetic material moves between disjunct genomic regions. Both homologous and non-homologous recombination have been described in a variety of RNA viruses, using both experimental and comparative techniques (reviewed in Worobey and Holmes 1999).

The process of reassortment which occurs in segmented RNA viruses is rather different. In this case two viruses co-infect a single cell and reassortants are made when a progeny virus packages segments with different ancestries (see, for example, Borucki *et al.* 1999). As a simple case in point, a '7 + 1' reassortant of influenza A virus occurs when seven genomic segments have their ancestry with one parental lineage, while the remaining segment is derived from a different lineage. Because multicomponent viruses are also segmented, and therefore able to reassort, I will consider them in the same way as other segmented RNA virus (as have others; Chao 1991).

Another form of interaction among RNA viruses stemming from mixed infections that is also analogous to sexual reproduction is phenotypic mixing. In this case, the progeny produced by a mixed infection contain the capsid or envelope protein produced by a genetically *different* parent, leading to an interesting mismatch between genotype and phenotype (Coen and Ramig 1996). This process has been described among closely related viruses, such as the different serotypes of poliovirus (Ledinko and Hirst 1961), as well as among different virus species (Itoh and Melnick 1959). Thus far, however, the evolutionary consequences of phenotypic mixing have not been explored in any detail.

3.2.1 Recombination frequency in RNA viruses

The first question to discuss with respect to viral recombination is the frequency of its occurrence. In what follows, I will concentrate largely on RNA recombination, as the reassortment of segmented RNA viruses is uncontroversial, and has been documented to be very frequent in some cases (Silander *et al.* 2005). For the sake of space, I will bypass discussions of recombination 'hot spots' in viral genomes, although these have been documented (for example, Jetzt *et al.* 2000; Ohshima *et al.* 2007).

At present, the main generality that can be drawn from comparative studies of recombination in RNA viruses is that its frequency varies enormously: it occurs at very high rates in retroviruses (and in other viruses that utilize RT; Mansky 1998; Froissart *et al.* 2005), and also as reassortment in viruses with segmented genomes (including dsRNA viruses), at highly variable frequencies in ssRNA+ viruses (for example, frequently in coronaviruses, enteroviruses, and potyviruses, sporadically in flaviviruses), and is far less common in ssRNA− viruses (Fig. 3.4). Such a broad-brush picture of recombination frequency corresponds with some important biological features of these viral groups. In particular, the virions of retroviruses carry two RNA molecules—so that they can be thought of as 'pseudo-diploid'—which means that viruses with different ancestries that are present in a single cell have a high probability of being packaged together, producing progeny that are effectively heterozygous. Copy-choice recombination may then produce genetically distinct progeny during reverse transcription. For example, in the case of HIV-1, the per-nucleotide rate of (copy-choice) recombination exceeds that of mutation, occurring two to three times per replication cycle (Jetzt *et al.* 2000; Jung *et al.* 2002). Obviously, heterozygous viral progeny cannot be produced when the genomic material is present as a single molecule, which is the case for most RNA viruses, in turn reducing rates of detectable recombination. A far stronger genomic constraint appears to be present in ssRNA− viruses, and explains why recombination rates are likely to be lower in this case. Specifically, the RNA molecules in ssRNA− viruses are very quickly bound to nucleoprotein subunits, and perhaps other proteins, so that RdRp-mediated replication can proceed. However, this tight ribonucleoprotein complex also acts to prevent template switching of the RNA polymerase. Although recombination in ssRNA− viruses is predictably infrequent (Chare *et al.* 2003), a number of interesting cases have been reported in respiratory syncytial virus (Spann *et al.* 2003), arenaviruses (Archer and Rico-Hesse 2002; Charrel *et al.* 2002), and EBOV (Wittmann *et al.* 2007). In sum, there are powerful mechanistic factors, themselves functions of genome architecture,

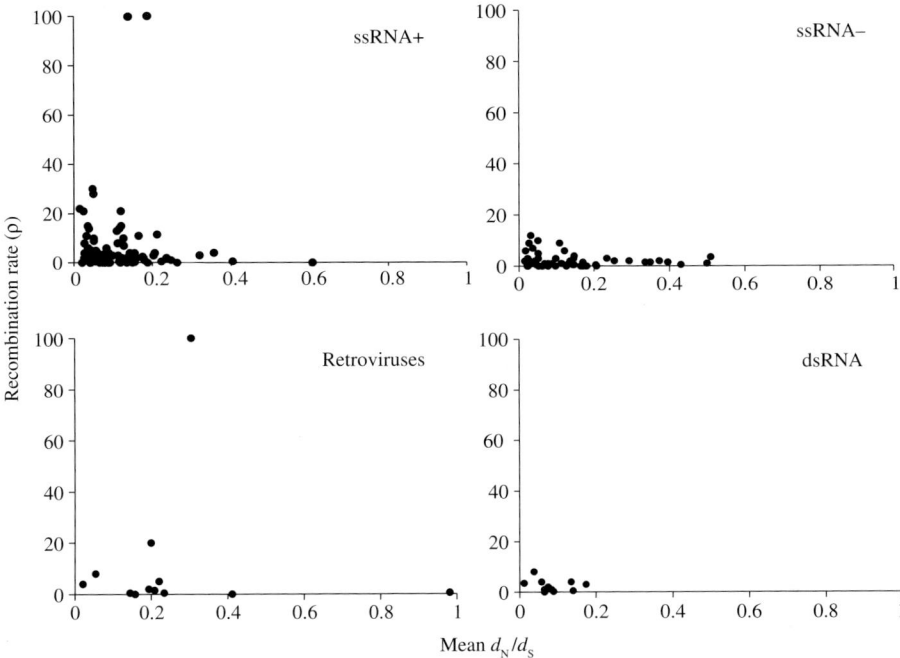

Fig. 3.4 Estimates of recombination rate in the major categories of RNA viruses and retroviruses. The y axis shows a measure of recombination rate (ρ) estimated using the program LDhat (McVean *et al.* 2002), in which $\rho = 2N_e r$ (where r is the recombination rate and N_e the effective population size). The maximum value for ρ was arbitrarily set to 100. The x axis shows a measure of average selection pressure manifest as mean d_N/d_S (the ratio of nonsynonymous (d_N) to synonymous (d_S) substitutions per site). The data used here represent 171 virus genes covering a diverse array of RNA viruses. Note the generally lower values of ρ for the ssRNA–, viruses indicating that they recombine less frequently.

that at least partially determine the rates of recombination and reassortment in RNA viruses.

3.2.2 Detecting recombination in RNA viruses

Although recombination in RNA viruses was traditionally explored using genetic methods (Lai 1992), comparative analyses now probably constitute the main way of determining its frequency. There are, however, three problems associated with the detection of RNA recombination from gene sequence data. First, in some cases it is clear that laboratory error is responsible for the sequences claimed to be recombinant. This phenomenon is well documented in dengue virus (DENV), where a number of the viruses claimed to be recombinant following phylogenetic studies (largely carried about by this author) have later been shown to result from erroneous sequencing

(Goncalvez *et al.* 2002; de Silva and Messer 2004). However, even accounting for this laboratory error there is strong evidence for recombination in DENV (Craig *et al.* 2003; Aaskov *et al.* 2007). Second, in some cases the apparent signal for recombination is due to the use of phylogenetic methods that cannot properly distinguish between this process and rate variation. A widely known case concerns the origin of the strain of influenza A virus responsible for the great global pandemic of 1918–1919. In this case, phylogenetic patterns reasonably thought to indicate a recombination event between human and swine isolates (Gibbs *et al.* 2001) were more probably due to lineage-specific rate variation (Worobey *et al.* 2002). Indeed, the occurrence of homologous RNA recombination in human influenza A virus remains to be demonstrated convincingly (Boni *et al.* 2008).

There is one more major reason why phylogenetic methods sometimes produce erroneous estimates of recombination frequency in RNA viruses: the use of consensus sequences. Most studies of genetic diversity in RNA viruses, particularly those that only cause acute infections in their hosts, have relied on the analysis of consensus sequences, in which each sequence depicts the *most common* nucleotide at any position in the viral sample. As discussed in section 3.6, the use of consensus sequences is greatly limiting when it comes to dissecting intra-host evolutionary dynamics, including that produced by recombination. In particular, with consensus sequences it is difficult to determine whether an isolate with phylogenetic evidence for recombination represents a *bona fide* recombination event, or a mixed infection where PCR has resulted in the amplification of two different viral molecules which then form an artificial hybrid on sequencing. Although mixed infection is a necessary prerequisite for recombination, it is not the same as determining the process itself. Consequently, to accurately determine the frequency of recombination in RNA viruses it will be necessary to obtain the sequences of individual viral particles, either through plaque purification or molecular cloning. In the absence of such data all phylogenetic studies of recombination in RNA viruses should be undertaken with caution, particularly if the putative recombinants are sporadic and/or involve isolates that were sampled many years apart, and which further suggests that they are laboratory artifacts (Boni *et al.* 2008). In contrast, if phylogenetic analyses reveal entire viral lineages to be recombinant, so that the progeny of recombination are 'successful' at the population level, this increases the likelihood that they reflect the true action of recombination in nature (particularly if isolates were derived from different laboratories). Just such a phylogenetic pattern was observed in EBOV (Wittmann *et al.* 2007), thereby constituting some of the best evidence for recombination in a ssRNA– virus obtained to date.

3.2.3 What determines the rate of recombination in RNA viruses?

The second question of evolutionary importance relating to the occurrence of recombination (and reassortment) is what processes dictate its frequency in viral populations? Although a variety of selective hypotheses have been put forward to explain the evolution of recombination in microbial organisms including viruses (Michod and Levin 1988; Michod *et al.* 2008), most discussion has focused on the two major advantages that recombination has over asexual evolution: that it accelerates the rate of production of advantageous genetic combinations, and that it allows the more efficient purging of deleterious mutations, both of which decrease linkage disequilibrium (Felsenstein 1974). Although it has also been suggested that recombination is favoured because it facilitates the repair of genetic damage (Michod *et al.* 2008), the highly variable recombination rates exhibited by RNA viruses with similar burdens of deleterious mutation make this hypothesis unlikely. It is also important to keep in mind that there are costs associated with sex in RNA viruses, such as increasing the degree of intra-host competition (Turner and Chao 1998).

While it is clear that recombination accelerates that rate at which advantageous genetic combinations are produced, it is unlikely that it provides sufficient 'value added' for this to be the reason for its existence. For example, although recombination can clearly assist the development of drug resistance in HIV (Nora *et al.* 2007), most cases of antiviral resistance in this pathogen cannot be assigned to recombination. Similarly, while antigenic escape is commonplace in hepatitis C virus (HCV), recombination has been only sporadically observed in this virus (Kalinina *et al.* 2002; Noppornpanth *et al.* 2006; Sentandreu *et al.* 2008). Hence, it may be that mutation rates are normally so high in RNA viruses, and population sizes often so large, that recombination/reassortment is not required to rapidly generate genotypic variation. In addition, although the competition between advantageous mutations that is a feature of asexual populations—commonly referred to as clonal interference—has been observed in large populations of asexual RNA viruses (Miralles *et al.* 1999; Poon and Chao 2004; Pepin and Wichman 2008), and must act to constrain their adaptability to some extent, that asexual RNA viruses (i.e. at least some ssRNA− viruses) are so successful suggests that this does not constitute a major selection pressure for sexual reproduction.

3.2.4 Recombination and deleterious mutation

The second general theory for the evolution of recombination is that it allows the efficient removal of deleterious mutations: the so-called mutational deterministic hypothesis (Kondrashov 1988; Keightley and Eyre-Walker 2000). This theory has received a great deal of attention from evolutionary geneticists and requires both a high rate of deleterious mutation per genome replication (U), such that U is greater than 1, and that deleterious mutations interact through synergistic (negative) epistasis, so that their combined effect on fitness is greater than expected from their stand-alone effects (Kondrashov 1988).

The idea that the efficient purging of deleterious mutations might be an important aspect of the evolution of recombination in RNA viruses is supported by experimental studies demonstrating the importance of Muller's ratchet—a progressive decrease in fitness due to the monotonous accumulation of deleterious mutations in small, asexual populations where genetic drift is intense (Muller 1964)—in these organisms (Chao 1990, 1994; Chao *et al.* 1992, 1997; Duarte *et al.* 1992; Novella *et al.* 1999; Lázaro *et al.* 2003; Poon and Chao 2004). However, while it is clear that Muller's ratchet can be readily generated *in vitro*, largely because severe population bottlenecks are easy to induce during experimental laboratory passage, how this relates to evolution in nature is an entirely different matter. Indeed, as discussed in more detail in section 3.3, determining the extent of population bottlenecks in natural populations of RNA viruses, as well as their long-term effective population sizes, is one of the most important components in our attempts to understand the evolutionary genetics of RNA viruses (although the power of the mutational deterministic hypothesis is that it does not require finite populations).

There is little doubt that RNA viruses fulfill the criteria of high U, as high rates of deleterious mutation have been both measured experimentally (Elena and Moya 1999; Duffy *et al.* 2008) (although not always; Burch *et al.* 2007) and using comparative approaches (Pybus *et al.* 2007). Remarkably, the estimates of U in some RNA viruses are extremely similar to those estimated for higher eukaryotes, at approximately $U=1$ (Haag-Liautard *et al.* 2007), although in the context of much smaller genomes. In addition, one of the most striking observations from early studies of genetic variation in HIV was how frequently gene sequences carried clearly deleterious mutations (stop codons, deletions). Given the small size of RNA virus genomes and the multiple functions often performed by viral proteins, it is probably reasonable to assume that the majority of mutations occurring within the genome of an RNA virus are deleterious. The issue of epistasis is, as always, more complex. However, as discussed in more detail in section 3.5, those studies undertaken to date suggest that although epistatic interactions are commonplace in RNA viruses, they tend to be positive (antagonistic) rather than negative (Burch *et al.* 2003; Bonhoeffer *et al.* 2004; Sanjuán *et al.* 2004c; Shapiro *et al.* 2006).

More direct evidence against the mutational deterministic theory is that comparative studies suggest that the burden of deleterious mutation is high for *all* RNA viruses studied, irrespective of their genome structure and therefore their propensity for recombination (Pybus *et al.* 2007). Hence, although retroviruses like HIV can evidently recombine rapidly, while segmented viruses like influenza A experience frequent reassortment, they seem to be subject to the same deleterious mutation pressure as unsegmented ssRNA– viruses where recombination is far less common. Although not definitive, this suggests that recombination is insufficient to greatly reduce the load of deleterious mutations experienced by RNA viruses.

Of course, this begs a major question: if deleterious mutation rates are extremely high in RNA viruses, and recombination is insufficient to save them from excessive deleterious mutation loads, what does? As stressed a number of times already, the most likely explanation is that RNA viruses possess very large population sizes for much of their life cycle (although this clearly does not apply to latent viruses). Hence, although deleterious mutations are produced in very large numbers, so are the viral progeny generated during every replication cycle. Large populations also mean that there will be sufficiently frequent reversal and compensatory mutations to off-set the accumulation of deleterious mutations. In short, RNA viruses are able to produce greater numbers of offspring than are removed by purifying selection, and it is this massive reproductive rate that is the key to their evolutionary survival.

While there is currently little evidence that recombination has been selected because of its ability to purge deleterious mutations, it is possible that it has the secondary consequence of disassociating advantageous and deleterious mutations. Again, this effect is largely a function of the extremely high mutation rates experienced by RNA viruses, which means that when an advantageous mutation occurs, there is a chance that it is accompanied by a deleterious mutation (although, as noted above, little is known about the distribution of mutation frequencies in RNA viruses). In these circumstances recombination allows beneficial mutations to be freed from the baggage of deleterious mutations that occur elsewhere in the genome (Rice and Chippindale 2001).

To conclude, at present I believe there is no compelling evidence that recombination and reassortment have evolved in RNA viruses because of their effects on linkage disequilibrium. In particular, if recombination was universally concerned with generating advantageous genotypes or removing deleterious ones, we would expect far higher rates across a diverse array of RNA viruses. Rather, I contend that most available data suggest that the differing rates of recombination and reassortment in RNA viruses reflect the mechanistic constraints associated with particular genome architectures and ecologies (although this does not exclude that natural selection is able to favour specific genetic variants produced by recombination). For example, rates of recombination are low in ssRNA− viruses simply because it is mechanistically impossible to be otherwise. Similarly, if segmentation was a means of facilitating sex through reassortment, why are segmented ssRNA+ viruses so heavily biased towards infecting plants? As such, I propose that recombination and reassortment are by-products of selection for other genomic and/or ecological characteristics, rather than being favoured as particular manifestations of sexual reproduction. Indeed, in Chapter 5 I will argue that genome segmentation, as well as several other important aspects of genome organization in RNA viruses, has evolved as a way of better controlling gene expression.

3.3 Natural selection, genetic drift, and the genetics of adaptation

The theory of evolution by natural selection sits at the heart of the Darwinian revolution. Indeed, to many people evolution is synonymous with natural selection. Given the unique role played by natural selection in evolution, it is essential that we assess its strength and determinants in RNA viruses. Such a goal is not simply for intellectual satisfaction. Because RNA viruses are a major cause of disease, revealing (and quantifying) the dynamics of natural selection is critical if we wish to understand such phenomena as the evolution of drug resistance, vaccine failure, and the ability of viruses to emerge in new host species. Somewhat paradoxically, although a number of population genetic models exist to explain the adaptive process (Fisher 1930; Gillespie 1991; Orr 2002; Rokyta *et al.* 2005), these usually assume low rates of mutation, which clearly rules out their applicability to RNA viruses. Indeed, it is the high rate of mutation that distinguishes the population genetics of RNA viruses from that of DNA-based organisms (and which constitutes the major theme of Chapter 4).

In this section, and those that follow, I will show that RNA viruses constitute some of the best-equipped laboratories to study evolution by natural selection. Rather than giving detailed examples of the adaptive process in RNA viruses—which are numerous—I will confine myself to making more general points which help explain their evolution *in toto* (and some case studies are presented in Chapter 7). In fact, much of what might be called 'classical virology' involves the precise exploration of viral adaptation to different hosts and cell types, although this pursuit is usually phrased in a different way. However, as is probably true of any evolutionary system, it is inadvisable to consider the process of natural selection in isolation. To some researchers, both the substitution dynamics of mutant alleles and larger-scale patterns of genome organization are due to the random-sampling effect associated with genetic drift.

3.3.1 Effective population sizes in viral evolution

The basic principles of population genetics tell us that natural selection should be a potent force in viral evolution. Natural selection is expected to dominate the substitution dynamics of mutant alleles when the product $N_e s \gg 1$, where N_e is the effective population size and s the selection coefficient (i.e. the fitness effect) of the mutant allele in question. To understand the respective roles of natural selection versus genetic drift in RNA virus evolution therefore merely requires us to determine both N_e and s. Unfortunately, estimating these parameters is thwart with difficulties and often relies on highly unrealistic assumptions. As an extremely important case in point, although there are a variety of population genetic measures of N_e, all assume neutrality, which means that they do not constitute an independent means of assessing the respective powers of selection and drift. For instance, relatively low estimates of N_e have been obtained using coalescent-based methods for acute infections like influenza (Rambaut *et al.* 2008) (Fig. 3.5) and measles (Pomeroy *et al.* 2008) that experience population dynamics characterized by distinct epidemic peaks and troughs. However, as these estimates are necessarily based on measures of genetic diversity, it is impossible to disentangle small (neutral) population sizes from successive selective sweeps. In these circumstances it is essential that we turn to our knowledge of viral biology to obtain independent estimates of effective population sizes in nature. Not surprisingly, a major conclusion of this section is therefore that it is impossible to truly understand the dynamics of drift and selection in RNA viruses without estimates of N_e and s in nature, rather than relying on the largely *in vitro* measures obtained to date. Similarly, while I believe there is at least some data to show that long-term values of N_e are by-and-large great enough to allow natural selection to proceed with vigour, it is dangerous, if not foolish, to think that all viruses behave in the same way. Because N_e and s vary among viruses, so selection and drift are necessarily virus-specific.

Although N_e is notoriously difficult to measure in any system, and viruses are no different, there are good *a priori* reasons to think that N_e will be large for at least

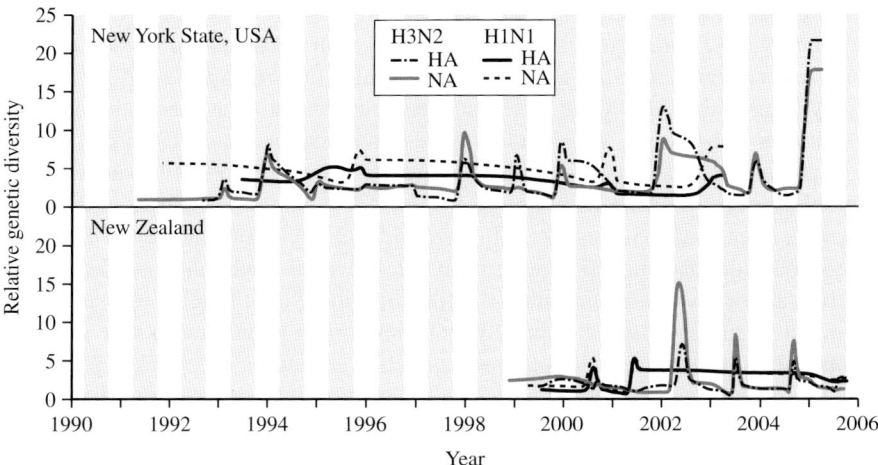

Fig. 3.5 Population dynamics of human influenza A virus showing the changing levels of genetic diversity through time. The figure shows Bayesian skyline plots of the haemagglutinin (HA) and neuraminidase (NA) segments for the H3N2 and H1N1 subtypes sampled in New York State, USA (northern hemisphere) and New Zealand (southern hemisphere). The vertical shaded regions represent the seasons in each locality. The y axes depict relative genetic diversity ($N_e t$; where t is generation time), which can be considered a measure of effective population size under strict neutrality. The shorter timescale of the New Zealand skyline plot is due to the shorter sampling period. Adapted from Rambaut et al. (2008) with permission.

some parts of the life cycle of most RNA viruses. In particular, intra-host population sizes in RNA viruses can be immense, reflecting a large number of infected cells and a high progeny number per cell. For example, in the case of HIV-1 it is estimated that 10^7–10^8 cells may be infected in a typical individual, with some 10^{10} virions produced every day (Perelson et al. 1996), while the equivalent daily production for HCV has been estimated at 10^{12} virions (Neumann et al. 1998). In the case of plant viruses such as TMV, there may be up to 10^{12} virus particles in a single leaf (Gibbs et al. 2008a). Similarly, for poliovirus and perhaps other lytic RNA viruses, the progeny or 'burst size' per single infected cell may be on the order of 10^4 (Kew et al. 2005). Although it is clearly dangerous to extrapolate from census population sizes (N) to N_e, particularly as a large proportion of the viruses sampled contain deleterious mutations and therefore produce few progeny, it is fair to assume that at peak viral loads viral populations are large enough to allow strong natural selection.

It is also critically important to recall that, as well as considering intra-host viral loads, RNA viruses exist at an epidemiological scale, which also contributes to their overall population size. The prevalence of viral infections at the epidemiological scale is often far lower than their (peak) intra-host population sizes. For example, while HIV-1 produces 10^{10} virions per day, its global prevalence—the number of infected hosts—is more like 10^7 (with HCV at 10^8). More dramatically, only a few thousand cases of poliomyelitis are now reported each year, and the virus is endemic in only a few localities. Further, although the number of hosts infected by a virus may be very large (think HIV), the difference between N and N_e at the epidemiological scale may be substantial, and therefore have a major impact on substitution dynamics. In particular, there may be widespread variability in reproductive success at the population level, in part a reflection of transmission mode. For instance, for highly infectious respiratory viruses like influenza there is likely to be relatively little variation in reproductive success among individual viruses: each has a good chance of infecting a new host and continuing the transmission chain. Indeed, influenza virus is thought to infect between 5 and 10% of the world's population at any one time. In contrast, a far greater variation in reproductive success is expected in the case of sexually transmitted viruses, as humans obviously vary enormously in their sexual behaviours, with superspreaders (Lloyd-Smith *et al.* 2005), who acquire multiple sexual partners, playing a key role in transmission dynamics. Such large-scale variation in reproductive success will reduce N_e and hence allow genetic drift to act with more potency.

3.3.2 Transmission bottlenecks

Where N_e might be greatly reduced in RNA viruses, and hence the time when genetic drift is expected to be particularly important, is at inter-host transmission. In particular, there is a long-standing and very reasonable idea that the process of inter-host transmission is accompanied by a large population bottleneck, which in turn introduces a major stochastic element into substitution dynamics. In some cases, such as HIV-1, it is even proposed that inter-host transmission can involve single virus particles (Keele *et al.* 2008). At the low values of N_e associated with inter-host transmission, many slightly deleterious mutations with small selection coefficients that would be purged by selection in large populations will now be subject to the whims of genetic drift, as the product $N_e s$ is expected to be close to zero (Ohta 1992). Transmission bottlenecks may also have a number of secondary effects, including determining some aspects of the evolution of virulence (Bergstrom *et al.* 1999; Elena *et al.* 2001). Although it is obvious that inter-host transmission must, normally, involve some sort of population bottleneck, there are few direct estimates of bottleneck sizes in natural viral populations. In addition, the magnitude of the population bottleneck is expected to vary according to the infecting dose, which may itself be a function of transmission mode.

Whereas little is known about natural populations, major bottlenecks are a common occurrence in experimentally manipulated RNA viruses. In this case, studies of plant viruses have proven particularly informative, in large part because of the relative ease of experimental analysis (de la Iglesia and Elena 2007). Important observations in these systems are: (i) the occurrence of major bottlenecks as the virus moves systemically through the plant (Li and Roossinck 2004; Sacristán *et al.* 2004) and (ii) major bottlenecks during the process of inter-host transmission, particularly as mediated by aphid vectors (Ali *et al.* 2006). Similar population bottlenecks have also been observed during the mosquito stage in the vector-borne RNA viruses that infect animals (Smith *et al.* 2008). However, what effect these bottlenecks will have on long-term evolutionary dynamics is more difficult to determine, and also depends on the associated values of s. As a case in point, studies *in vitro* suggest that population sizes as small as five virions are sufficient to allow natural selection to proceed, although this clearly requires mutant alleles that exhibit very different levels of fitness (reviewed in Bergstrom *et al.* 1999).

Intriguingly, there are a number of indirect studies of viral populations which suggest that transmission bottlenecks in nature may not be as extensive as is often thought. One highly informative case in point concerns DENV. Detailed studies of intra-host genetic diversity revealed that three phylogenetically distinct lineages of DENV-1 were passed successfully among humans and mosquitoes for a number of years (Aaskov *et al.* 2006), suggesting that transmission bottlenecks are not especially severe. There is also a growing body of data to suggest that mixed infections are a common component of intra-host diversity in influenza A virus, including viruses with differing phenotypes, such as resistance and sensitivity to adamantane drugs (Ghedin *et al.* submitted). Such frequent mixed infection, including by strains that should induce strong cross-immunity, suggests that individual hosts can receive multiple viruses on transmission. Similarly, that multicomponent viruses by definition require mixed infections to form a fully functional unit again argues against a substantial transmission bottleneck in the case of some plant RNA viruses.

In sum, although we have no clear idea of long-term values of N_e in RNA viruses, and generalizations are dangerous, there is at least some evidence to suggest that they are sporadically large enough for natural selection to act with great efficiency. This is particularly likely to be the case for RNA viruses whose mode of transmission ensures that they are able to spread through large and well-mixed populations—notably human influenza A virus—and where positive selection is well documented (Fitch *et al.* 1991). Even if transmission bottlenecks are severe, it is possible that their effects on substitution dynamics are off-set to some extent by the ability to infect very large numbers of hosts, or by undergoing sufficient numbers of replication cycles at high population sizes within hosts (Manrubia *et al.* 2005). Similarly, that positive selection is so readily detected in RNA viruses indicates that small effective population sizes do not always succeed in putting a break on adaptive evolution.

3.3.3 The dynamics of allele fixation: estimating selection coefficients

As noted at the start of this section, molecular adaptation is commonly described in RNA viruses, although less often in explicit population genetic terms. This is perhaps unfortunate as RNA viruses are one of the few systems where it is possible to obtain reasonable estimates of selection coefficients, largely because evolution is so rapid and often human-mediated. For example, early work on drug resistance in HIV-1, where the frequency of mutant alleles can be measured with and without drug pressure, resulted in estimates of s of between 0.4 and 2.3% for mutations conferring sensitivity to zidovudine (Goudsmit *et al.* 1996). In the case of mutants that adapt VSV to different cell types, s values of 39% have been estimated (Dutta *et al.* 2008). Similarly, there have been detailed analyses of positive selection relating to such processes as escape from antibody responses (Wei *et al.* 2003), escape from T-cell responses (Gog *et al.* 2003), and the adaptation to new hosts (Anishchenko *et al.* 2006) and vectors (Tsetsarkin *et al.* 2007), sometimes through changes in receptor specificity (Suzuki 2006a), and even for increased virulence (Brault *et al.* 2007). It is also evident that natural selection in RNA viruses can occur at different levels, such as within and between hosts, within individual cells, and at different stages during the viral life cycle (Krakauer and Komarova 2003).

Hence, the rapid pace of RNA virus evolution allows us to visualize stages in the process of allele fixation, a phenomenon most evolutionary biologists can only dream of. This also allows a simple test of natural selection: if mutations are fixed faster than expected by genetic drift, which will take a mean of $2N_e$ generations in a haploid population (although with a large variance), then it must be that they have done so by positive selection, or that they are in linkage disequilibrium with advantageous mutations (Zanotto *et al.* 1999; Williamson 2003; Shih *et al.* 2007). In the hypothetical example in Fig. 3.6, consider an amino acid mutation that is absent in all those sequences sampled from time points t_1 and t_2, but present in those viruses sampled from the later time point t_3, such that it falls on the internal branch linking t_2 and t_3. Clearly, this mutation must have arisen at some point between times t_2 and t_3, placing the *observed* process of allele fixation within a specific time period (although distinguishing an allele that been truly fixed from one that has reached high frequency is difficult). For a typical acute RNA virus with an epidemiological generation time (that is, infected host to infected host) of 4 days, and N_e of 10 000, which may correspond roughly to population sizes during epidemic troughs for infections like measles, the mean *expected* time to fixation under genetic drift is then $2 \times 10000 \times 4$ days, or approximately 200 years. Consequently, if mutations are fixed very much faster than this then it seems reasonable to conclude that positive selection has been involved. Given these parameter estimates it is also possible to obtain approximate values of s using standard population genetic theory (Zanotto *et al.* 1999).

Such an analysis of fixation times has provided important insights into the process of CTL escape in HIV. One of the puzzles of intra-host HIV evolution is that although the virus has a remarkable mutational power—such that every individual mutation is generated every single day within each infected patient—the process of CTL escape at single amino acid sites can often take several years. Such a delay is particularly well characterized in the case of individuals who carry the HLA-B27 allele (Kelleher *et al.* 2001), and raises the question of how a mutation that is obviously of enormous benefit to the virus, and which is made on a regular basis, can take so long to spread? The answer is that CTL epitopes fall in genes that undertake a wide array of functions, including those that are normally subject to strong selective constraints such as those encoding the RT or the capsid. This is clearly true of some of those mutations that occur in the HLA-B27 epitope in the HIV *gag* gene, notably at amino acid residue 264: although these mutations facilitate CTL escape, they also disrupt

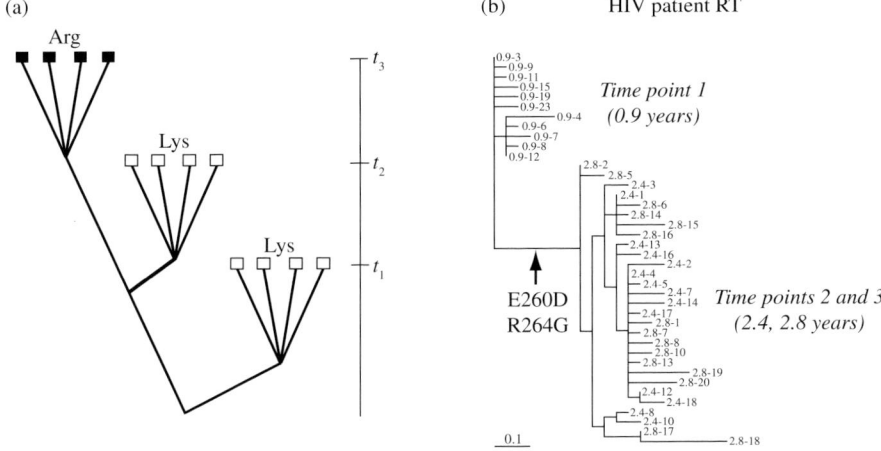

Fig. 3.6 Using fixation times to estimate selection coefficients in RNA viruses in which time-structured sequence data are available. (a) An idealized time-structured data set where all those viral sequences collected from each of three time points (t_1, t_2, t_3) form nested phylogenetic groups. A Lys→Arg amino acid change occurs between time points t_2 and t_3. (b) Real data example of the fixation of the *gag* R264G CTL escape mutation in HIV-1 and the E260D compensatory mutation that occurs simultaneously (based on Kelleher *et al.* 2001).

key aspects of capsid structure, thereby reducing viral fitness. As a consequence, mutations at this residue have a net deleterious effect. To overcome this fitness cost it is necessary for the virus to fix, effectively simultaneously, compensatory mutations that allow the virus to form a viable capsid structure, in this case at residues 260 and 268, a process clearly visible through phylogenetic analysis (Kelleher *et al.* 2001) (Fig. 3.6). It is this requirement for compensatory mutations that delays CTL escape, as multiple mutations are now required to occur in the same molecule and without the baggage of deleterious mutations. Indeed, CTL escape mutations with only weak functional constraints generally appear to occur more frequently in HIV (Liu *et al.* 2007). A similar process has been invoked to explain the constraints to CTL escape in simian immunodeficiency virus (SIV; Friedrich *et al.* 2004b) and influenza A virus (Berkhoff *et al.* 2005).

In principle, looking at the process of allele fixation as a way of quantifying adaptive evolution can be applied to a whole range of RNA viruses where the times of sampling are known, although this form of analysis needs to be reformulated in more rigorous mathematical terms. The largest compounding factor is genetic linkage. For clonally evolving RNA viruses, or individual viral segments, it will often be difficult, if not impossible, to determine which of those amino acid changes that appear on a specific branch are positively selected as all share a single evolutionary pattern. In these circumstances additional information is required to determine evolutionary processes, such as whether the mutations fall in known epitopes or antigenic sites (Wolf *et al.* 2006; Shih *et al.* 2007).

3.3.4 The importance of hitch-hiking

An interesting example of how linkage—hitch-hiking—can confound the analyses of site-specific selection pressures was recently documented in human influenza A virus (Simonsen *et al*. 2007). Here, isolates of H3N2 influenza virus have, since 2005, acquired resistance to adamantanes (amantadine and rimantadine), a class of first-generation antiviral drugs often deployed against influenza. The pace with which this resistance has evolved is dramatic: in the 2005–2006 influenza season in the USA more than 90% of influenza A viruses sampled were adamantine-resistant, a rise from only approximately 15% in the 2004–2005 season (Bright *et al*. 2005, 2006; Hayden 2006) (Fig. 3.7). Adamantane resistance is due to a single Ser (S)→Asn (N) amino acid replacement at position 31 in the M2 protein of influenza A virus, which functions as an ion channel. Although the S31N mutation could confer sufficient selective advantage to facilitate its spread in parts of South-east Asia where over-the-counter drug use is relatively common (Bright *et al*. 2005), such direct selection pressure cannot apply to most nations, including the USA, where the use of adamantanes is limited. The explanation for the dramatic rise of adamantane resistance on a global scale in the absence of drug pressure is hitch-hiking. A phylogenetic analysis revealed that the viral lineage conferring adamantane resistance was generated by a complex reassortment event (Simonsen *et al*. 2007), during which the S31N mutation most likely became linked, by chance, to advantageous mutations located elsewhere in the viral genome that were selected for a different reason, such as immune escape.

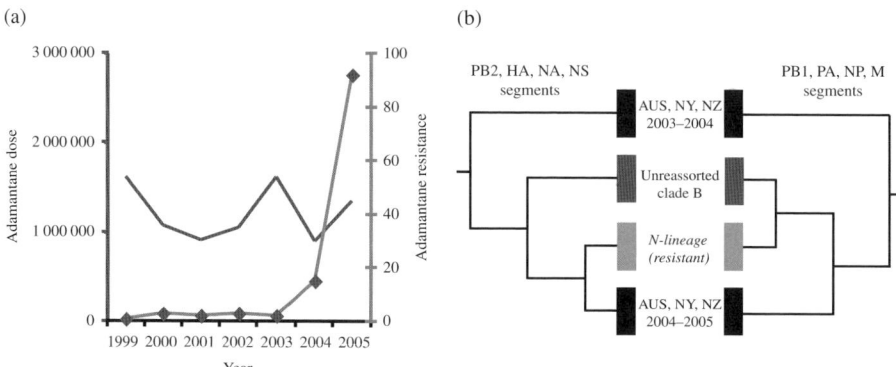

Fig. 3.7 The evolution of adamantane resistance in H3N2 human influenza A virus. (a) Rise in prevalence of adamantane resistance (diamonds) in comparison to the number of drug doses used in the USA. (b) Schematic representation of the pattern of phylogenetic incongruence suggestive of a 4+4 segment reassortment event. The reassortant adamantane-resistant viruses, characterized by the S31N mutation in the M2 protein, are denoted the N-lineage. Adapted from Simonsen *et al*. (2007) with permission.

3.3.5 Patterns of synonymous and nonsynonymous evolution

In many respects, measures of selective pressure based on fixation times are more informative in the peculiar case of RNA viruses than d_N/d_S ratios (the ratio of non-synonymous (d_N) to synonymous (d_S) substitutions per site), even though the latter have commonly (and often successfully) been used to study adaptive evolution in these infectious agents. It is sometimes forgotten that d_N/d_S methods were originally designed for inter-species sequence comparisons in which the nucleotide changes observed between species reflect fixed substitutions (Nei and Gojobori 1986). However, in the case of RNA viruses there is mounting evidence that d_N/d_S values are often elevated towards the tips of phylogenetic trees, representing transient deleterious mutations that will eventually be removed by purifying selection (Pybus *et al.* 2007; section 3.4). In addition, although analysis of d_N/d_S provides a useful overview of the selection pressures that act on gene sequences, and which explains their popularity, it is also the case that attempts to use d_N/d_S to infer positive selection on individual amino acid sites or branches are inherently difficult (Kosakovsky Pond and Frost 2005), have a potential for false-positive results if not used with care (Suzuki and Nei 2002a), and are confounded by such factors as extensive variation in nucleotide composition and recombination, a non-independence of sites, and a lack of selective neutrality at synonymous sites (Anisimova *et al.* 2003; Novella 2003; Novella *et al.* 2004b; Kryazhimskiy *et al.* 2008). Measures of d_N/d_S are also inherently conservative in that that they generally require the repeated fixation of nonsynonymous changes at individual sites to infer the action of positive selection. The occurrence of single beneficial amino acid changes on individual lineages are far harder to detect, even though many of the adaptive mutations that underpin cross-species transmission are likely to fall into this category (Anishchenko *et al.* 2006).

Yet, irrespective of how selection pressures are measured, and even accounting for the complex effects of genetic linkage, it is clear that RNA viruses are perhaps the class of organism where adaptive evolution has been most readily identified (Yang and Bielawski 2000). Indeed, the interaction between RNA viruses and their hosts constitutes a classic evolutionary 'arms race'. In addition, viruses like HIV have proven important testing grounds for new methods to measure selection pressures (Kosakovsky Pond *et al.* 2006), while the occurrence of positive selection in host genomes constitutes a simple and powerful way to detect genes with antiviral functions (Sawyer *et al.* 2004; Obbard *et al.* 2006).

3.3.6 Natural selection and transmission mode

One of the most consistent observations in studies of the adaptive process in RNA viruses is that selection pressures differ according to whether the virus is vector-borne or transmitted by some other means. Specifically, those viruses that are transmitted by aphid, mosquito, or tick vectors are subject to stronger purifying selection (and weaker positive selection) than those viruses that are transmitted without the assistance of arthropod vectors, and therein evolve more slowly at nonsynonymous

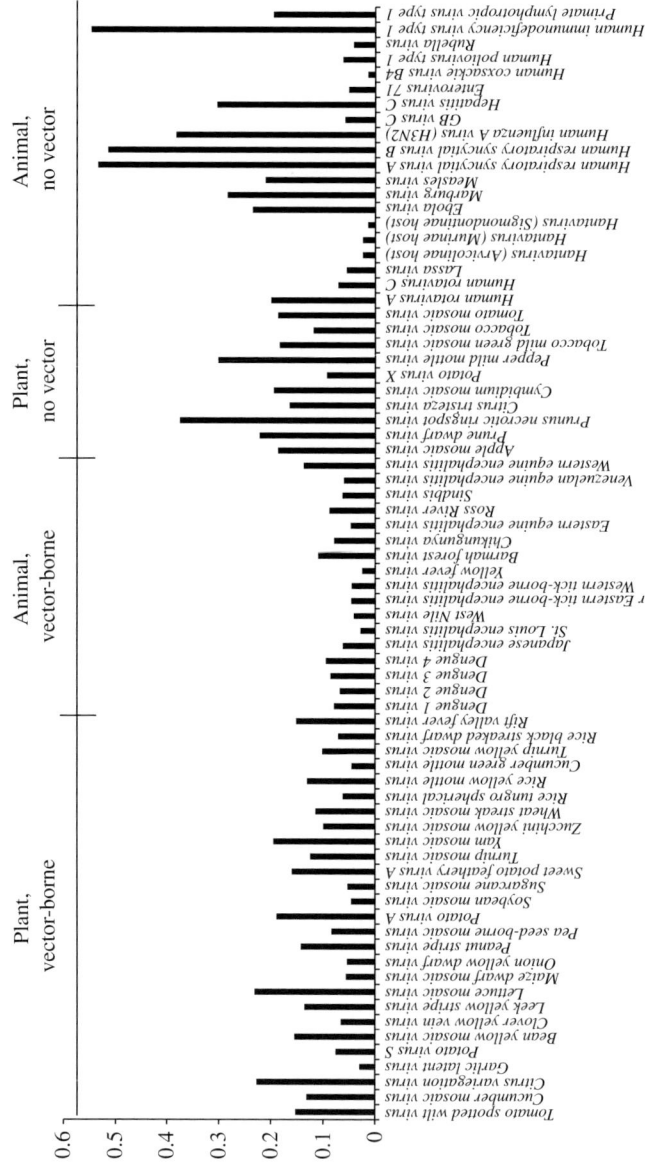

Fig. 3.8 Increased purifying selection on vector-borne RNA viruses compared to those RNA viruses transmitted by other routes. Selection pressure is measured as the mean value of the number of synonymous (d_S) and nonsynonymous (d_N) substitutions per site (ratio d_N/d_S). Data taken from Chare and Holmes (2004) and Woelk and Holmes (2002).

sites (Fig. 3.8). Importantly, such 'constrained' evolution of vector-borne RNA viruses holds true for those that infect both animals and plants (Chare and Holmes 2004), and has been demonstrated *in vitro* (Scott *et al.* 1994; Weaver *et al.* 1999; Greene *et al.* 2005), *in vivo* (Jerzak *et al.* 2005; Coffey *et al.* 2008), and using comparative analyses (Woelk and Holmes 2002; Chare and Holmes 2004). This important evolutionary pattern undoubtedly reflects the inherent difficulties in infecting hosts and cell types that are so phylogenetically divergent: mutations that increase fitness in one environment, such as mammalian cells, decrease fitness in another environment, such as insect cells. The streamlined genomes possessed by RNA viruses mean that such antagonistic pleiotropy, and fitness trade-offs in general, are expected to be a relatively common occurrence (Elena 2002).

As well as influencing the rate at which nonsynonymous mutations are fixed in RNA viruses transmitted by vectors, antagonistic pleiotropy might also have a major bearing on the process of host switching, the main mechanism of viral emergence, as different hosts may represent very different selective environments (Duffy *et al.* 2006). As discussed in detail in Chapter 6, understanding the evolutionary processes that underpin viral emergence is a major theme in current work on infectious diseases. Generalities are few and far between in this area. However, one of the strongest is that although a frequent cause of spill-over infections, vector-borne RNA viruses are statistically less likely to evolve sustained transmission cycles in new host species (M. Woolhouse, personal communication). Antagonistic pleiotropy provides a powerful explanation: vector-borne RNA viruses inhabit a world dominated by evolutionary trade-offs, so that most (if not all) mutations will have a major bearing on fitness. Hence, those mutations that are needed to confer adaptation to a new mammalian host are likely to have a detrimental effect on some other component of fitness, resulting in a major adaptive constraint. As a consequence, vector-borne viruses may only be able to spread successfully in new host species if these hosts are phylogenetically similar, thereby minimizing the number of advantageous mutations required.

3.3.7 Escape from intrinsic immunity

Some of the most interesting examples of positive selection acting on RNA viruses described in recent years involve what might be called the 'intrinsic' host immune response (or, alternatively, intracellular 'restriction factors'). Of these, that involving the *APOBEC3G* (apolipoprotein B-editing catalytic polypeptide) gene—a remarkable anti-retroviral response of primates (Mangeat *et al.* 2003)—has generated the most interest. *APOBEC3G* (and its anti-retroviral relative *APOBEC3F*) is a member of a family of genes that are involved in the editing of RNA and/or DNA through the deamination of cytosine. When directed to the reverse-transcription step of the HIV life cycle, *APOBEC3G* induces monotonous G→A mutations, a phenomenon known as G→A hypermutation, and which had been observed by HIV researchers for many years without a clear understanding of its basis (Vartanian *et al.* 1991). Such G→A hypermutation results in the generation and incorporation of multiple deleterious mutations that have a fatal impact on key viral functions. As these distinctive mutational signatures have also been observed in human endogenous retroviruses (HERVs), it is possible that *APOBEC3G* has been functioning as an anti-retroviral agent for millions of years (Armitage *et al.* 2008). In an analogous manner, the *TRIM5α* gene is able to induce to a cytoplasmic barrier to some retroviral infections, restricting HIV infection in Old World monkeys (Stremlau *et al.* 2004).

Not only do the genes involved in intrinsic immunity constitute interesting and informative examples of natural selection in their own right but, more broadly, they show that the innate immune response can be subject to as strong positive selection as that which occurs on antibodies and T cells (Sawyer *et al.* 2004, 2007). They also represent beautiful examples of the intensity of the virus/host arms race. As a case in point, despite the potent affect of *APOBEC3G*, HIV-1 has evolved an anti-*APOBEC3G* response controlled by the *vif* gene (Sheehy *et al.* 2002). A strong prediction for the future is that more intrinsic immunity genes will be discovered.

The arms race between RNA viruses and host intrinsic immunity is not restricted to mammals. A similar evolutionary phenomenon is observed in the genes involved in RNA interference (RNAi), and which appear to represent a front-line defence against viral infections in plants and invertebrates (Wilkins *et al.* 2005; Wang *et al.* 2006; Marques and Carthew 2007; Gibbs *et al.* 2008a). That RNAi has such an important antiviral role in these organisms, but has not been clearly observed (at the time of writing) in vertebrates, suggests that in some sense it compensates for the lack of a truly adaptive immune system. Further, in an analogy with *APOBEC3G*, RNAi represents a simple way in which a method to control post-transcriptional gene expression has been co-opted as an antiviral strategy. In the case of RNAi, the targets are the dsRNA molecules that are an obligatory by-product of RNA virus replication (excluding retroviruses). The virus/RNAi arms race is best described in *Drosophila*, in which the genes involved are among the most rapidly evolving in these species (Obbard *et al.* 2006).

3.3.8 Strictly neutral evolution in RNA viruses?

Early analyses of rates of nucleotide substitution in a limited number of viruses suggested that a significant component of viral evolution was likely to be neutral, because of the apparently clock-like increase in genetic diversity (Gojobori *et al.* 1990; Sala and Wain-Hobson 2000). However, clock-like evolution as quantified using simple regression measures can also be observed under selective regimes (Fitch *et al.* 1991), and more recent analyses utilizing far larger numbers of sequences have found widespread rate variation among viral lineages, such that a strict molecular clock is often rejected (Jenkins *et al.* 2002). More fundamentally, whether or not viral evolution proceeds in a clock-like manner in reality constitutes a poor test of the neutral theory, as it both deals with long-term averages and fails to account for the differences in generation times among viruses.

At the amino acid level, it is likely that few strictly neutral mutations ($s=0$) occur in RNA viruses, particularly given the extensive epistasis and pleiotropy that characterize their evolution (see section 3.5). Indeed, biophysical studies suggest that little protein evolution in general can be considered strictly neutral (DePristo *et al.* 2005). Further, the low rates of recombination observed in many RNA viruses mean that the evolutionary fate of any neutral mutation is inextricably linked to those mutations that are either removed by strong purifying (background) selection or which are swept to fixation. In those RNA viruses, or genome segments, that evolve asexually, the classification of individual mutations as selectively neutral may therefore have little meaning.

As in any genetic system, the most likely class of neutral sites in viral genomes are those that do not code for protein. In eukaryotes, a number of different types of DNA sequence may fall into this category, including introns, inter-genic DNA, pseudogenes, and synonymous sites, although selection in some of these classes has been observed (Eyre-Walker 1999; Chamary and Hurst 2004; Plotkin *et al.* 2004; Andolfatto 2005; Eyre-Walker and Keightley 2007). For RNA viruses, introns are very rare (see Cubitt *et al.* 2001 for one example), as are truly functionless regions of non-coding RNA and pseudogenes, which in itself argues against a major role for genetic drift (see Chapter 5) (suggestions of a pseudogene in rabies virus have not withstood more detailed analysis; Ravkov *et al.* 1995). As a consequence, the only major class of sites that might fall into the neutral category are synonymous ones. Although there is obviously extensive inter-virus and inter-genic variation, a general picture seems to be that, on average, levels of synonymous variation are far higher than those observed at nonsynonymous sites (Jenkins *et al.* 2001b, 2002; Woelk and Holmes 2002), as expected if the average selection coefficient is less than that at nonsynonymous sites (Fig. 3.9). Of course, this does not necessarily mean that synonymous sites are strictly neutral, although they may behave as such when N_e is small and, as I have argued above and will below, there is growing evidence that synonymous changes may have a major impact on fitness.

In theory, synonymous sites can possess a number of important functions, including containing signals for promotion, transcription, and encapsidation. If a generality can be made, it is that there is increasing evidence that many synonymous sites evolve in a blatantly non-neutral manner (Novella 2003; Novella *et al.* 2004b). A powerful example was recently documented in influenza A virus in which synonymous mutations in the PB2 polymerase gene were shown to have a major impact on viral packaging, such that these sites were normally highly conserved (Marsh *et al.* 2008). These special functional properties notwithstanding, the two most obvious types of natural selection that might act on synonymous sites are those caused by RNA secondary structure and by codon usage bias. As the role of RNA secondary structure is discussed in more detail in section 3.5, a simple summary statement—that there is also mounting evidence for its importance in viral evolution—will suffice now. In what follows, I consider codon usage bias in rather more detail.

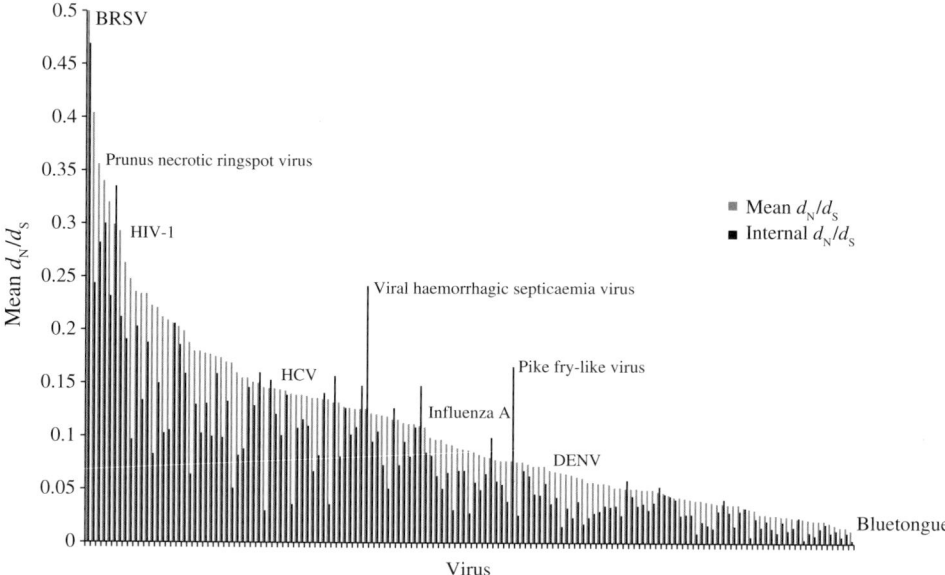

Fig. 3.9 Mean d_N/d_S ratios in the structural genes of 143 RNA viruses. Viruses are ranked from highest mean d_N/d_S (bovine respiratory syncytial virus; BRSV) to the lowest (Bluetongue virus) (grey bars). Note that no mean d_N/d_S value exceeds 1.0. For comparison, the d_N/d_S on internal branches of phylogenetic trees of these viruses are also depicted (black bars). The positions of some other interesting and outlier viruses are shown. Note the general similarity of the mean d_N/d_S and internal d_N/d_S distributions. Data taken from Pybus et al. (2007).

3.3.9 Determinants of codon bias (and nucleotide composition) in RNA viruses

By far the most common way to explore the nature of selection pressures acting on synonymous sites, and of assessing the relative strengths of genetic drift and natural selection more generally, is to determine the extent and causes of biases in codon usage. Under neutral models codon usage bias simply reflects the background mutational bias. Although a variety of selective models for codon choice have been proposed (Qin et al. 2004), the most popular is that optimizing the match between codon and anti-codon will enhance the accuracy and/or efficiency of the translational machinery, particularly in highly expressed genes (Bulmer 1987). If a general conclusion can be drawn from the analysis of codon bias in other organisms it is that natural selection is most able to shape the substitution dynamics of synonymous codons, which are likely to be characterized by small selection coefficients, when effective population sizes are sufficiently large, as in bacterial populations (Ikemura 1985; Sharp and Matassi 1994), their DNA bacteriophages (Lucks et al. 2008), and *Drosophila* (Akashi 1994).

Traditionally, analyses of codon usage bias have played only a minor role in studies of RNA virus evolution. This is in part a reflection of the fact that viruses

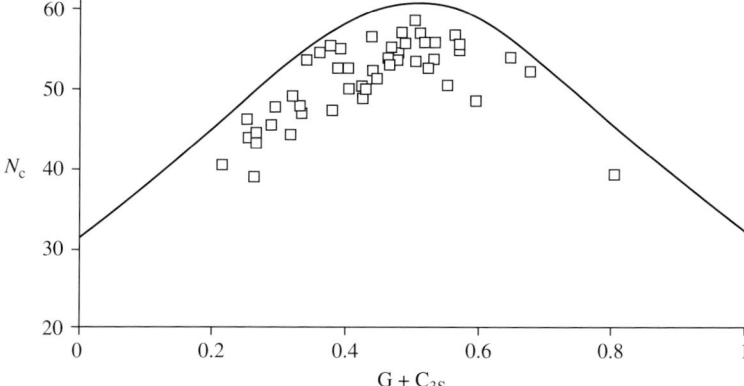

Fig. 3.10 The extent of codon usage bias in selected human RNA viruses measured using the effective number of codons (N_C) statistic and compared with the genome-wide G+C content at synonymous third-codon positions (G+C_{3S}). Each square represents a different virus and the curved line indicates the codon usage bias expected under the constraints of nucleotide composition alone. Taken from Jenkins and Holmes (2003) with permission.

must utilize the host translation machinery, so codon usage bias in viruses should, at least to some respect, match that of the host organism. However, because RNA viruses only infect particular cell types within a host, and host codon biases—in mammals at least—are often tissue-specific (Plotkin *et al.* 2004), knowing the overall codon bias in host species is insufficient for determining the cell-specific selection pressures that might act on viruses. Similarly, host-induced mutational biases caused by *APOBEC3G*-like mechanisms will also complicate the analysis of codon usage.

The small number of studies of codon usage bias in RNA viruses undertaken to date suggest that codon choice is mainly determined by mutation pressure (Adams and Antoniw 2003; Jenkins and Holmes 2003). In particular, while RNA viruses often exhibit strong biases in codon usage, the nucleotides utilized as synonymous codons tend to match the nucleotide biases across the viral genome as a whole (Fig. 3.10). Although this suggests a generally weak role for natural selection in determining the choice of individual codons, it is possible that the overall nucleotide composition in RNA viruses is itself selectively determined, and that codon usage is just one manifestation of this larger-scale selection. Indeed, experimentally altering synonymous codon usage in poliovirus had a major effect on viral fitness, even resulting in attenuation (Coleman *et al.* 2008).

In further support of a role for natural selection in determining codon choice are observations of major host-specific differences in nucleotide composition among RNA viruses (Greenbaum *et al.* 2008), which are particularly apparent when viruses jump species (Rabadan *et al.* 2006). For example, a measurable change in nucleotide composition occurs coincident with the cross-species transmission of influenza

A virus from birds to mammals, and which may reflect differences in the core body temperatures between these species that in turn affect thermodynamic stability, or selection for escape from innate immune responses (Greenbaum *et al.* 2008). Further, nucleotide compositions can vary extensively within individual viral families, such as the *Flaviviridae*, perhaps reflecting differences in the host and/or vector species utilized (Jenkins *et al.* 2001a). For example, in the case of the flaviviruses a significantly higher GC content was observed in non-vector-borne compared to tick-borne flaviviruses, with intermediate values in those viruses transmitted by mosquitoes (Jenkins *et al.* 2001a). More directly, a recent analysis of insect RNA viruses revealed that unrelated viruses infecting honey bees converged on the same pattern of codon usage, indicative of a strong host-species effect (Chantawannakul and Cutler 2008). Although a host-specific pattern of nucleotide composition could equally reflect large-scale mutational bias in the absence of natural selection, especially as RNA viruses must utilize host nucleoside triphosphate (NTP) pools as they replicate, the possibility that the overall nucleotide compositions of RNA viruses have been selectively optimized should clearly be assessed, particularly as more host transcriptomes become available for study.

3.4 Deleterious mutation and RNA virus evolution

Although I have spent a good deal of time considering the ins and outs of adaptation, arguably the most profound observation stemming from experimental and comparative studies of RNA virus evolution is that most mutations arising during replication—and perhaps the vast majority at nonsynonymous sites—are deleterious, and so will only persist for short time periods before being removed by purifying selection. As a simple example, it is estimated that only about 1% of poliovirus virions released from a cell are able to complete a full replication cycle (Krakauer and Komarova 2003), a proportion that may be common to many RNA viruses. Consequently, it is not simply the mutational power of RNA viruses that determines whether they are able to adapt to new environments, but the percentage of these mutations that increase fitness relative to those that decrease it.

A high rate of deleterious mutation in RNA viruses can be inferred from a number of observations. Most directly, there have been several experimental measurements of the deleterious mutation rate per generation (U). The rates inferred from these studies are usually on the order of $U=1$ (Elena and Moya 1999; Gao *et al.* 2004) and hence similar to overall mutation rates (section 3.1). This further supports the notion that the vast majority of mutations produced during viral replication are deleterious. Similarly, analyses of the fitness spectrum of random mutations arising in experimental populations of VSV reveal that approximately 40% can be considered lethal in single cell types, with most (≈30%) of the others falling into the deleterious class, although often with fitness values close to the neutral expectation (so that the fitness distribution of deleterious mutations is bimodal) (Sanjuán *et al.* 2004b) (Fig. 3.11). Interestingly, over 25% of mutations could be considered neutral, with less than 5% as beneficial under

3.4 Deleterious mutation

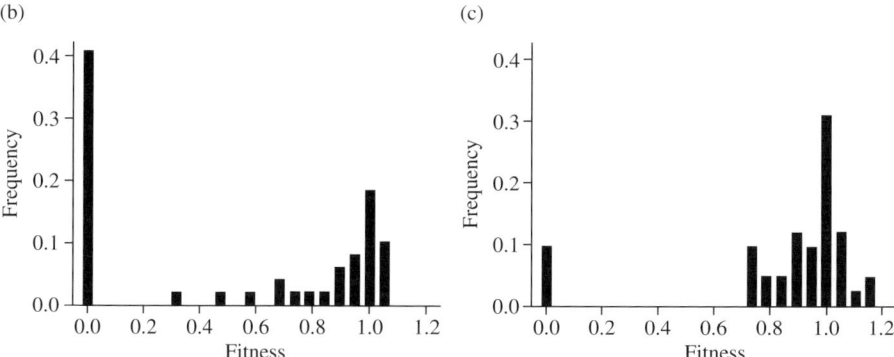

(a)	Random		Pre-observed		Total	
	Proportion, %	Effect, %	Proportion, %	Effect, %	Proportion, %	Effect, %
Lethal	39.6 (19)	−100	11.6 (5)	−100	26.4 (24)	−100
Deleterious	29.2 (14)	−24.4	41.9 (18)	−16.4	35.2 (32)	−19.9
Neutral	27.1 (13)	−3.8	32.6 (14)	−0.9	29.7 (27)	−2.3
Beneficial	4.2 (2)	4.2	14.0 (6)	7.9	8.8 (8)	7.0
Total	100 (48)	−47.6	100 (43)	−17.7	100 (91)	−33.4

Fig. 3.11 The fitness spectrum of single point mutations in experimental populations of VSV. (a) Table showing the proportion (and number in parenthesis) of mutations falling into different fitness categories for both random and previously described single point mutations in VSV. The mean fitness is shown in each case. The figures below show the frequency of fitness values associated with (b) random single point mutations, and (c) previously described single point mutations. Adapted from Sanjuán et al. (2004b) with permission.

the particular assay system used. A similarly high frequency of deleterious mutations was observed in experimental analysis of plant viruses (Malpica et al. 2002). An extensive literature review undertaken by Santi Elena and colleagues revealed that the median selection coefficient against single mutations in RNA viruses was 10.8%, and hence far greater than that (1.7%) seen in DNA-based organisms, where more mutations are classed as selectively neutral (Elena et al. 2006). Such a burden of deleterious mutations argues against the existence of expansive 'neutral spaces' in RNA virus evolution. A high rate of deleterious mutation can also be inferred through various comparative analyses. In particular, a survey of the structural genes of 140 RNA viruses revealed a large excess of nonsynonymous changes on the tips of phylogenetic trees (Pybus et al. 2007) (Fig. 3.12), as expected for transient deleterious mutations, and independent of genome architecture.

In sum, it is possible to imagine RNA viruses as bacterial mutators gone mad: whereas their extremely high mutation rates allow them to rapidly generate the genetic variation needed to adapt to changing environments, and natural selection can be extremely effective in achieving this, it comes at the cost of producing a multitude

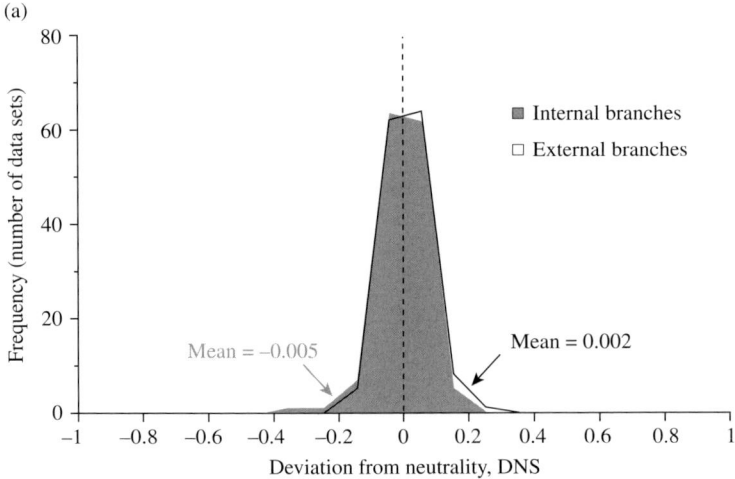
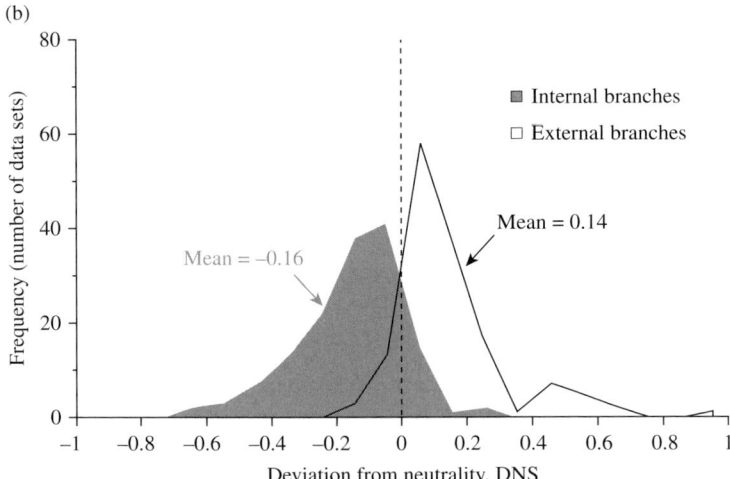

Fig. 3.12 Frequency of transient deleterious mutations in RNA virus evolution. The DNS is the deviation from neutrality statistic which reflects the difference in d_N/d_S values on internal and external branches of phylogenetic trees. (a) The frequency distribution of DNS values under simulated neutrally evolving data. (b) The frequency distribution of DNS values in the structural genes of 143 RNA viruses (the same data as used in Fig. 3.9). The positive DNS values on external branches reflect an increase in the number of nonsynonymous mutations, indicative of transient deleterious mutations. Adapted from Pybus *et al.* (2007) with permission.

of defective progeny. To my mind, the only way that RNA viruses can survive with such a mutational burden is to have extremely large population sizes in the long term, reflected in high rates of viral production per cell, many infected cells within an individual host, and a multitude of infected hosts. Crucially, and in clear contrast to the situation seen in bacteria, viruses with high deleterious rates do not experience competition with viruses that have greatly increased fidelity and which are expected to dominate in stable environments. As I have already noted, as substantial improvements in polymerase fidelity seem extremely difficult to achieve, it is likely that all RNA viruses are equally weighed down by a high deleterious mutation rate.

3.4.1 Deleterious mutation and intra-host genetic diversity

The heavy burden of deleterious mutations in the short term, manifest as high d_N/d_S values on the tips of phylogenetic trees, is also apparent in the few studies of mutational patterns within and among hosts that have been undertaken to date. DENV is particularly informative in this respect, with abundant nonsynonymous mutations observed at the intra-host level: in many cases intra-host d_N/d_S is close to 1.0 (Holmes 2003a; Chao *et al.* 2005; Table 3.1), which is exactly as expected if we were to observe the mutational spectrum prior to the action of natural selection, either positive or negative. In addition, the amino acid changes observed are generally singletons, falling once in the alignment, and occur at sites that are normally invariant at the population level, as expected if they represent transient deleterious mutations. In contrast, far lower d_N/d_S ratios—in the region of 0.1—are observed at the inter-host level (Holmes 2003a). Hence, there is a massive purging of nonsynonymous mutations as the virus moves among hosts, again as predicted if the majority of these mutations are deleterious. Indeed, approximately 90% of the nonsynonymous mutations in DENV that occur within hosts may be deleterious (Holmes 2003a). A similar pattern was observed in West Nile virus (WNV), in which d_N/d_S ratios were 5-fold greater within than among hosts, although in all cases d_N/d_S was less than 1.0, indicating that some purifying selection had occurred within hosts (Jerzak *et al.* 2005). The same story can also be told for *in vivo* (inter-host) versus *in vitro* comparisons of FMDV (Carrillo *et al.* 2007), and in some plant RNA viruses (Teycheney *et al.* 2005). More generally, that rates of nucleotide substitution tend to be higher in the short term than the long term, as appears to be the case in a number of RNA viruses including WNV (Bertolotti *et al.* 2008), GBV-C (Romano *et al.* 2008), and the rodent hantaviruses (Ramsden *et al.* 2008), is compatible with the idea that intra-host viral populations still contain a significant number of transient deleterious mutations that have yet to be purged by purifying selection.

The transient deleterious mutations observed in RNA viruses not only involve point mutations, stop codons, and indels. It is also possible that novel recombinants represent at least some of the mutations in the deleterious class, in part explaining why this process has often been difficult to observe in viral populations where only consensus sequences are generated. For example, in the case of DENV, the in-depth analysis of genetic diversity within single patients not only uncovered viruses of differing genotypes, but multiple recombinants among them (Aaskov *et al.* 2007).

Table 3.1 Measures of intra-host genetic diversity in the E gene of DENV.

Data set	No. of clones	Genetic diversity (%)	Mean d_N/d_S	Reference
D1.Myanmar.Mos059/01	20	1.0	0.52	Aaskov et al. (2006)*
D1.Myanmar.Mos194/01	20	1.5	0.69	
D1.Myanmar.Mos206/01	10	2.3	0.12	
D1.Myanmar.Mos305/01	20	0.9	0.86	
D1.Myanmar.37045/00	20	0.4	0.46	
D1.Myanmar.38518/01	20	1.3	0.26	
D1.Myanmar.40530/01	20	1.2	0.23	
D1.Myanmar.40906/01	20	0.8	0.56	
D1.Myanmar.43549/01	20	0.4	0.85	
D1.Myanmar.43826/01	20	0.4	0.63	
D1.Myanmar.47185/02	20	0.9	0.63	
D1.Myanmar.47662/02	20	0.7	0.34	
D1.Myanmar.48572/02	20	0.6	1.11	
D1.Myanmar.49440/02	20	0.9	0.62	
D1.Myanmar.50457/02	20	1.2	0.58	
D2.Myanmar.32309/98	14	0.1	0.85	Craig et al. (2003)
D2.Myanmar.35004/99	12	0.1	2.00	
D2.Myanmar.36273/00	11	0.2	1.00	
D2.Myanmar.Mos196/00	20	0.3	0.58	
D2.Myanmar.Mos025/00	20	0.3	0.74	
D2.Myanmar.Mos376/00	17	0.2	0.69	
D3.Taiwan.ID3/98	17	0.1	0.42[†]	Wang et al. (2002)
D3.Taiwan.ID7/98	10	0.1	0.67[†]	
D3.Taiwan.ID8/98	11	0.4	0.88[†]	
D3.Taiwan.ID17/98	10	0.8	3.36[†]	
D3.Taiwan.ID19/98	10	0.1	0.50[†]	
D3.Taiwan.ID20/98	11	0.3	0.73[†]	
D3.Taiwan.1F/98	26	0.6	2.04	Chao et al. (2005)
D3.Taiwan.1H/98	21	0.8	1.11	
D3.Taiwan.2F/98	18	0.7	1.38	
D3.Taiwan.2H/98	25	1.2	1.06	
D3.Taiwan.3F/98	23	0.7	0.81	
D3.Taiwan.3H/98	23	0.9	1.46	
D3.Thailand.D92.431/92	20	0.2	0.52	Wittke et al. (2002)
Overall mean		0.7	0.86	

*Some of the viruses from the Aaskov et al. (2006) study are especially diverse because they harbour a defective stop codon lineage.
[†] The estimate of mean d_N/d_S in this case is only approximate.

3.4.2 The importance of defective interfering particles and complementation

The importance of deleterious mutation in RNA virus evolution can also be inferred from one of the most intriguing observations stemming from 'classical' experimental virology: that deleterious mutation produces defective viral genomes that can 'interfere' with the growth of fully functional viruses by competing with them during replication. These are referred to as defective interfering (DI) particles, and they have been observed in cell culture (Huang and Baltimore 1970) and sometimes *in vivo* (Holland and Villarreal 1975), and especially at high multiplicity of infection (MOI) when a single cell is infected by more than one virus. Indeed, DI particles are likely to be a common occurrence in many RNA viruses. Despite their name, the success of DI particles seems to be built more on out-growth than true interference. As a particular case in point, the greater the proportion of DI particles, the lower the total yield of poliovirus (Cole and Baltimore 1973). While DI particles are usually thought to contain long deletions (perhaps up to 90% of the genome), which gives them a replication advantage, in theory any deleterious mutation can fall into this class.

The classic explanation for the maintenance of DI particles is complementation, in which defective viruses effectively parasitize fully functional viruses that co-infect the same cell, utilizing their complete complement of proteins to complete their own life cycle and continue their existence (which is why they are most common at high MOI values). For viruses that undergo frequent recombination or reassortment, a second possible cost of DI particles (aside from reducing viral yield) is that a fully functional virus could reduce its fitness by undergoing recombination with a DI particle, in so doing acquiring a deleterious mutation in a manner other than through simple point mutation alone. Indeed, an important evolutionary cost of complementation as a whole is that it allows deleterious mutations to persist for extended periods (Froissart *et al.* 2004), although given the very high deleterious mutation rate in RNA viruses this additional cost may not be excessive.

Although their name implies that DI particles are usually a hindrance to co-infecting functional viruses, it is theoretically possible that they are on occasion beneficial. In particular, if a DI particle is able to initiate infection of a new host cell, and if this cell is then subject to an immune response, the host will have deployed some of its finite immune resources against a virus that could never produce functional progeny. In these circumstances DI particles can be considered as inadvertent immune 'decoys', although it is difficult to see how this phenomenon could have evolved by natural selection.

3.4.3 Complementation may be commonplace in RNA viruses

Complementation is an increasingly common observation in RNA viruses (see, for example, Geigenmüller-Gnirke *et al.* 1991; Mansky *et al.* 1995; Moreno *et al.* 1997; Tzeng and Frey 2003; Wilke and Novella 2003; García-Arriaza *et al.* 2004; Smallwood *et al.* 2002; Appel *et al.* 2005). Indeed, at high MOI values complementation is a predictable and important genetic process, although its true evolutionary significance has yet to be fully appreciated (García-Arenal *et al.* 2003). As noted above, complementation allows defective viruses to persist by 'stealing' the proteins of co-infecting functional viruses, an ability that seems to decline with the extent of genetic divergence among them (Simon *et al.* 1995). Although there is evidence that some viruses have evolved mechanisms to prevent the superinfection of individual cells, including HIV (Geleziunas *et al.* 1994) and influenza A virus (Huang *et al.* 2008), that recombination/reassortment occurs frequently in these viruses indicates that these barriers are not absolute. Complementation, in the broad sense, is also the mechanism by which the individual particles of multicomponent RNA viruses are able to reproduce successfully (Chao 1991).

Until recently, complementation had only ever been observed *in vitro*, and DI particles were only thought to survive a few generations through complementation. This is expected as DI particles are obviously defective, and their continued survival is critically dependent on a high MOI of functional viruses. However, more recent studies, including some comparative analyses, suggest that complementation is likely to be frequent in nature (Froissart *et al.* 2002) and may extend over very long time periods (Aaskov *et al.* 2006). More importantly, as complementation can also be thought of as a type of buffer against deleterious mutation, a high frequency of complementation will reduce selection for mutational robustness (Montville *et al.* 2005).

Simple, yet unappreciated, evidence for the action of complementation in RNA viruses are the occurrence of premature stop codons within consensus sequences, and which are readily available on GenBank. A stop codon, or any other clearly defective mutation, will only be present in a consensus sequence when it represents the *majority* form at any nucleotide site. By default, misplaced stop codons in consensus sequences signify defective mutations at high frequency, which in turn implies complementation, although some may also be read through by the RNA polymerase (a process which is commonplace in some alphaviruses).

It is this same logic that indicates that complementation may allow defective mutations to survive for extended time periods, including multiple rounds of inter-host transmission. The most compelling example is a lineage of DENV-1 collected over a period of 18 months in Myanmar (formerly Burma) that contained a stop codon in the E (envelope) gene and a variety of other apparently deleterious nonsynonymous mutations (Aaskov *et al.* 2006) (Fig. 3.13). As additional population genetic and biochemical analyses confirmed that this stop codon lineage was indeed defective, the only viable explanation for its survival over multiple cycles of human–mosquito transmission is complementation. Specifically, in conditions of frequent viral transmission, as might characterize regions of South-east Asia where DENV is at very high prevalence, the multiple infection of individual hosts and cells is so commonplace that complementation becomes an important evolutionary process. The outstanding question is whether this defective lineage of DENV is a 'hyper-parasite', praying on the abundance of functional viruses that happen to co-infect a single cell, or whether it is in some way beneficial to the co-infecting functional virus, perhaps by a producing a subgenomic RNA that provides an additional function (García-Arriaza *et al.* 2004). Finally, that this stop codon lineage was effectively 'hidden' by the consensus sequence also highlights the importance to undertaking in-depth analyses of intra-host genetic variation. This is the theme of section 3.6. Indeed, the transmission of multiple lineages hidden by a dominant consensus sequence has also been observed in WNV (Jerzak *et al.* 2005), one of the few other viruses for which extensive intra-host sequence data are available.

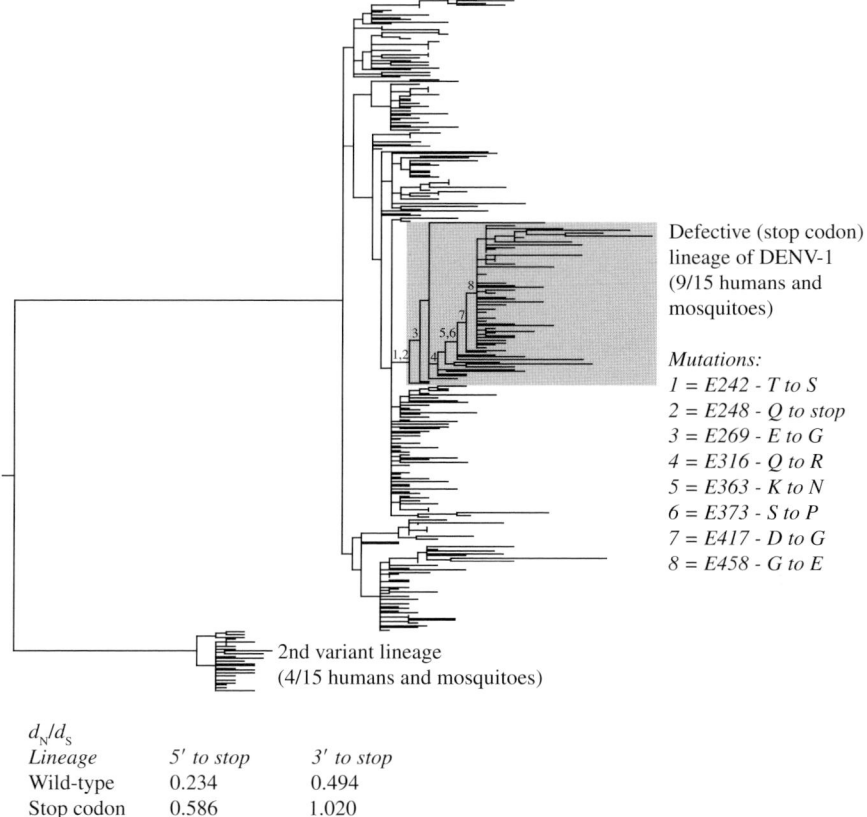

Fig. 3.13 Evolution of the defective (stop codon) lineage of DENV-1 sampled in Myanmar (Burma) between 2000 and 2002. The phylogenetic distribution of the eight amino acid mutations in the E gene associated with the stop codon lineage are shown, including the termination codon at E248, as are the selection pressures (d_N/d_S) of this compared to the functional 'wild-type' lineage of DENV-1. The phylogenetic position of a second variant lineage, 'hidden' by the consensus sequence, is also shown. Adapted from Aaskov et al. (2006) with permission.

3.5 Epistasis in RNA virus evolution

Most phylogenetic studies of RNA virus evolution have assumed that each site, be it nucleotide or amino acid, evolves independently. Indeed, the assumption that nucleotide sites are independent random variables is a common component of models of molecular evolution. However, there is now abundant evidence, both experimental and comparative, that the nucleotides and amino acids of RNA viruses interact in a variety of complex plays. As a simple case in point, both protein and RNA secondary structures may be ubiquitous in the genomes of RNA viruses, generating frequent compensatory mutations (Sanjuán et al. 2005). These properties, along with the highly compact nature of viral genomes, the use of overlapping reading frames, and frequent pleiotropy, also mean that epistatic interactions are predicted to be commonplace in RNA virus evolution (Holmes 2003b). Such epistatic effects can be either antagonistic or synergistic, depending on the direction of deviation from multiplicative fitness effects: antagonistic epistasis reduces the effect of combined mutations on fitness, whereas synergistic epistasis increases this effect (see Michalakis and Roze 2004 for an informative discussion in the context of RNA viruses). Understandably, determining the rules of epistasis is one of the most important problems in contemporary evolutionary genetics (Azevedo *et al.* 2006; Martin *et al.* 2007).

3.5.1 Epistasis and robustness

Although measuring the extent and sign (positive or negative) of epistasis is challenging, the emerging consensus from studies of this process in RNA viruses, particularly through experimental studies, is that most epistatic interactions are antagonistic (Bonhoeffer *et al.* 2004; Burch and Chao 2004; Sanjuán *et al.* 2004c, 2006b; Sanjuán 2006; de la Iglesia and Elena 2007; Pepin and Wichman 2007). Importantly, the general lack of evidence for synergistic epistasis provides a strong argument against the mutational deterministic hypothesis for the evolution of recombination in RNA viruses (Bonhoeffer *et al.* 2004).

The seminal work on epistasis in RNA viruses has been undertaken by Santi Elena and Rafa Sanjuán (and colleagues) from Valencia, Spain, who in a beautiful series of experiments have measured not only the fitness of individual mutations (Sanjuán *et al.* 2004b, 2006a), but also how they interact, with antagonistic epistasis again the most common observation (Sanjuán *et al.* 2004c, 2006b; Sanjuán and Elena 2006; de la Iglesia and Elena 2007). However, the importance of this work is not simply in the demonstration and impact of epistasis, but in how it is linked to mutational robustness. Robust genomes are protected against deleterious mutation through the creation of neutral spaces, genetic redundancies (such as duplicated genes), modularity, or through frequent complementation (Krakauer and Plotkin 2002; de Visser *et al.* 2003; Montville *et al.* 2005). Perhaps the key observation in this context is that antagonistic epistasis and robustness are inversely correlated: when antagonistic epistasis is commonplace, as appears to be true for RNA viruses, robustness is weak (Elena *et al.* 2006; Sanjuán and Elena 2006). In contrast, larger DNA-based organisms are characterized by synergistic epistasis and greater robustness (Fig. 3.14). In this sense there is also a strong correlation between epistasis and genomic complexity (Sanjuán and Elena 2006). The explanation for the dominance of antagonistic epistasis in RNA viruses is most likely that genome sizes are small, with many overlapping functions, and few mechanisms of robustness (for example, duplicated genes), such that mutations will repeatedly damage the same functions (Wilke and Adami 2001; Wilke *et al.* 2003; Sanjuán 2006). In addition, antagonistic epistasis in RNA viruses, and more particularly in viroids, may also be due in part to an abundance of secondary structures (Sanjuán 2006; Sanjuán *et al.* 2006b). In contrast, more complex DNA-based genomes have evolved buffering mechanisms

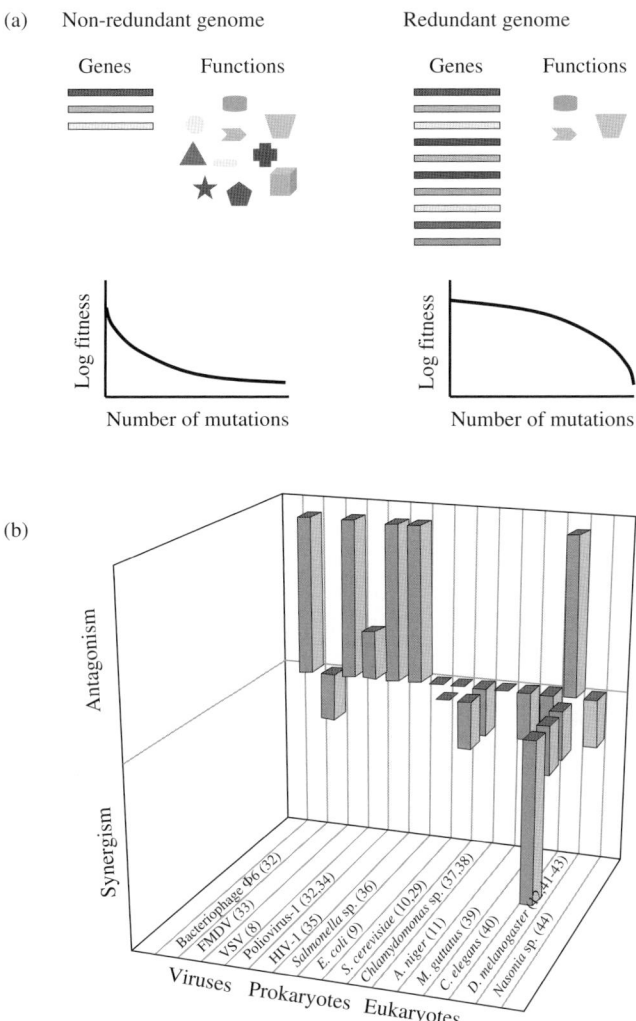

Fig. 3.14 Genetic robustness and epistasis. (a) Schematic representation of the differing properties of robust (redundant) and non-robust (non-redundant) genetic systems. In non-robust systems, such as RNA viruses, where there are few genes, single mutations have strong fitness effects and antagonistic epistasis, whereas in robust genetic systems with many genes, individual mutations have weak fitness effects and synergistic epistasis. Taken from Elena *et al*. (2006) with permission. (b) Evidence for generalities depicted in (a). A compilation of the sign of epistasis in a variety of organisms: long bars show significant evidence for epistasis; short bars show non-significant evidence for epistasis; flat bars show no evidence for epistasis. *A. niger*, *Aspergillus niger*; *C. elegans*; *Caenorhabditis elegans*; *M. guttatus*, *Mimulus guttatus*; *S. cerevisiae*, *Saccharomyces cerevisiae*; Taken from Sanjuán and Elena (2006) with permission.

that allow them to tolerate a certain number of deleterious mutations. When these mechanisms are exhausted by too many mutations such cryptic deleterious effects become visible, resulting in synergistic epistasis.

Although robustness is generally thought to be weak in RNA viruses, there are two other ways that it might be manifest, first through the action of cellular chaperones that increase protein stability (Elena *et al.* 2006), and which been shown to operate with dramatic effect in endosymbiotic bacteria that are also characterized by high mutational loads (Fares *et al.* 2002), and second in quasispecies models in which very high mutation rates push populations towards flat fitness landscapes (i.e. where more mutations are neutral; Wilke *et al.* 2001). The discussion of this latter model, often called 'survival of flattest,' lies at the heart of Chapter 4 on the RNA virus quasispecies. However, it is sufficient to say here that RNA viruses in nature are unlikely to regularly attain the very high mutation rates required to experience quasispecies dynamics. Rather, it is more likely that RNA viruses are engendered with a kind a 'population robustness' derived from their large population sizes (Elena *et al.* 2006). This ensures that a sufficient number of unmutated progeny are available for the next generation, facilitates the efficient removal of deleterious mutations, and will result in an increased frequency of compensatory mutations. As should be clear by now, a common theme in this book is that it is the large population sizes of RNA viruses that saves them from their deleterious mutation rates. Conversely, in large multicellular genomes, genetic redundancy through gene duplication provides a perfect route to genetic robustness.

Those comparative analyses undertaken to date also suggest an important role for antagonistic epistasis in RNA virus evolution. One simple manifestation of epistasis that can be detected in sequence analyses is the occurrence of co-varying nucleotide or amino acid changes, such that pairs (or more) of mutations co-occur across the branches of phylogenetic trees more frequently than expected by chance (Fig. 3.15). Although this test is highly conservative, it does give some indication of the frequency of one form of epistasis in nature. Such an analysis found that 55 of 177 RNA data sets showed significantly more co-varying changes than expected by chance alone and irrespective of viral type (ssRNA− or ssRNA+), and with a particularly strong effect at synonymous sites and in short sequence regions (generally less than 15 amino acid residues) (Shapiro *et al.* 2006) (Fig. 3.15).

3.5.2 The importance of RNA secondary structure

That epistatic effects are often localized suggests that they may be in part determined by RNA or protein secondary structure. Indeed, there is growing evidence for the importance of RNA secondary structure as a general cause of epistasis (Higgs 1998). Although RNA viruses are often viewed as a linear series of nucleotides, this is a gross simplification as they may form a complex set of RNA secondary structures. The functional importance of some of these structural elements means that they have been studied in great detail, such as the internal ribosome entry site (IRES), that enhances viral translation in a variety of viruses (Pelletier and Sonenberg 1988),

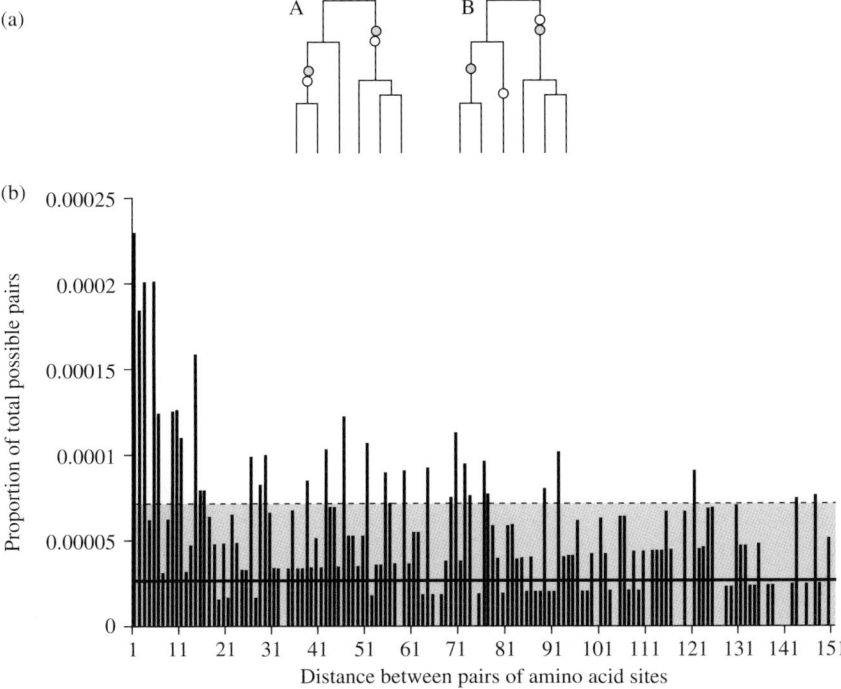

Fig. 3.15 Compensatory mutations as a measure of epistasis. (a) Detecting compensatory mutations using phylogenetic trees. Each circle represents a mutation. Those in tree A are compensatory because they co-occur on independent lineages, while those in tree B do not. (b) The spatial distribution of pairs of compensatory amino acid changes in 177 RNA genes (vertical bars). The mean (solid horizontal line) and 95% highest value obtained from simulation (shaded area) are shown. The bias toward compensatory mutations that are close in primary sequence is evident. Adapted from Shapiro *et al.* (2006) with permission.

or the 3′ untranslated region (3′ UTR) of flaviviruses, which is fundamental to a number of aspects of the viral life cycle (Alvarez *et al.* 2005a), and perhaps virulence (Proutski *et al.* 1997; Leitmeyer *et al.* 1999). These structural interactions can also be very long range, sometimes causing viral genomes to circularize (Alvarez *et al.* 2005b). Importantly, functional RNA secondary structures are also present within the coding regions of viral genes (Thurner *et al.* 2004; Robertson *et al.* 2005; McMullan *et al.* 2007; Yang *et al.* 2008), and doubtless many more examples will be discovered. Whatever their cause, the observation of widespread RNA secondary structures again argues against the selective neutrality of synonymous sites (Simmonds and Smith 1999).

A more dramatic proposal is that genome-scale secondary structures are also commonplace in RNA viruses (Palmemberg and Sgro 1997; Simmonds *et al.* 2004; Thurner *et al.* 2004). For example, extensive work by Peter Simmonds (University of Edinburgh, UK) suggests that ssRNA+ viruses frequently contain 'genome-scale ordered RNA structures' and which are associated with the ability of these viruses to generate persistent infections, perhaps through the evasion of innate immunity (Simmonds *et al.* 2004). Although a fascinating proposal, it is important to note that large-scale elements of structure are notoriously difficult to determine, particularly through bioinformatic approaches. For example, most of the computational methods used to predict RNA secondary structure only consider energetic criteria, whereas the kinetics of RNA folding may be equally important.

3.5.3 Convergence and pleiotropy

Before finishing this section, it is important to briefly mention two other evolutionary phenomena which similarly reflect the evolutionary constraints that result from highly constrained genome sizes: convergence (and parallelism) and pleiotropy (Holmes 2003b). A variety of experimental studies have now revealed both convergent and parallel evolution to be particularly commonplace in both RNA viruses and small DNA viruses (Wichman *et al.* 1999; Cuevas *et al.* 2002; Novella and Ebendick-Corpus 2004; Novella *et al.* 2004b; Greene *et al.* 2005; Duffy *et al.* 2006; Remold *et al.* 2008). For example, Cuevas *et al.* (2002) found widespread parallel evolution in 21 independently replicating lines of VSV, while Remold *et al.* (2008) similarly showed that homogeneity in environment (cell type) led to frequent parallelism (Fig. 3.16). Notably, both studies also showed that synonymous mutations can

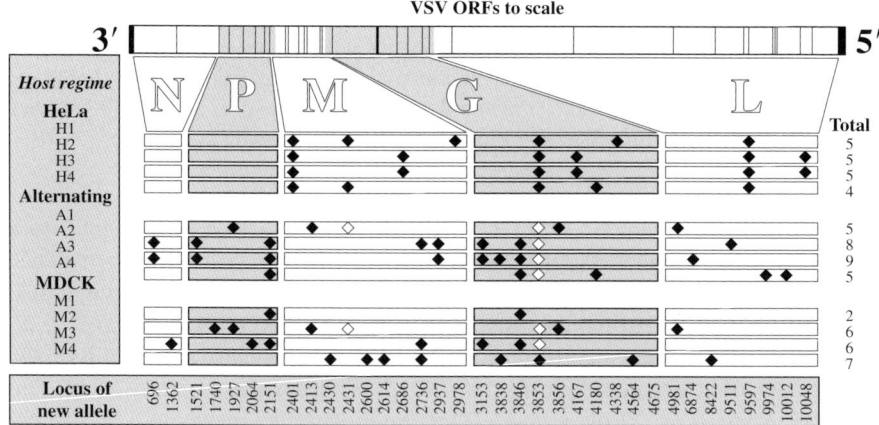

Fig. 3.16 Frequent parallel evolution in an RNA virus (VSV). N, P, M, G, and L denote individual genes. Each diamond represents a mutation relative to the ancestral sequence in 12 independently evolved populations cultured on different cell types (HeLa cells or Madin–Darby canine kidney (MDCK) cells, or both (Alternating)). Filled and open diamonds represent different mutations at the same site. The single dash represents a deletion. Taken from Remold *et al.* (2008) with permission.

be selectively advantageous, casting another vote against their universal neutrality. That nearly all studies of convergence and parallelism have involved experimental, as opposed to comparative, analyses highlights the problems of obtaining a phylogeny that is both sufficiently well known that homology and homoplasy can be clearly distinguished (Holmes *et al.* 1992), and accounting for the occurrence of multiple substitutions. Finally, although less often studied in an evolutionary context, pleiotropy also appears to be common in RNA viruses (and small DNA viruses), particularly given the frequent use of overlapping reading frames, and more directly from the observation that single proteins often possess multiple functions (Pepin *et al.* 2006; Remold *et al.* 2008). For example, whereas influenza A virus possesses separate haemagglutinin and neuraminidase genes, a single *NH* gene performs both functions in paramyxoviruses. Further, important pleiotropic mutations have been described in VSV (Frey and Youngner 1984), influenza A virus (Kilbourne *et al.* 1998), rotavirus (Au *et al.* 1993), and HIV (Shin *et al.* 1994).

3.6 The importance of intra-host viral diversity

Despite the remarkable capacity for RNA viruses to rapidly generate genetic variation it is striking that the overwhelming majority of comparative studies of their evolution have utilized consensus sequences. Rather than isolating and sequencing individual viral genomes, and which is often referred to as 'clonal' sequence data (although it most certainly does not mean asexual!), most work has utilized sequences that represent something of an average of the entire viral population within a single patient. The exceptions to this consensus versus clone rule are HIV and HCV, both of which generate persistent infections in their hosts and where the importance of examining the full extent of genetic diversity within single hosts was realized very early on. In stark contrast, clonal studies of intra-host genetic diversity within those RNA viruses that cause transient, acute infections are notable for their absence.

In some cases, a reliance on consensus sequences for evolutionary studies seems justified. In particular, for broad-scale investigations in molecular epidemiology, where the goal is to reveal the origins or phylogeographic structure of a specific virus, the use of consensus sequences is both sensible and far cheaper than obtaining clonal data. Further, if all the genetic diversity observed within a single host was generated *de novo* in that host, then the use of consensus sequences is also phylogenetically appropriate. The difficulty, of course, comes with more detailed studies of evolutionary dynamics where the reliance on consensus sequencing does not provide sufficient resolution. In these cases it is important to explore what Bryan Grenfell has termed 'beyond the consensus' evolution.

Before describing the results of analyses of clonal diversity in RNA viruses it is important to make an extremely important caveat: that laboratory error, most notably involving faulty PCR or sequencing, and where polymerase accuracy is an extremely important consideration, is likely to have contributed to at least some of the apparent sequence diversity recorded within hosts (Bracho et al. 1998). Annoyingly, the structure of genetic diversity expected under laboratory error—in which there is a roughly even distribution of mutations at nonsynonymous and synonymous sites—is also that expected under a legitimate process of random mutation prior to the imposition of natural selection. The burden of proof therefore lies with those who generate or analyse intra-host sequence data to show that their conclusions are robust to laboratory error. In particular, it is obviously important to use the highest-fidelity enzymes for PCR amplification.

To date, studies of natural intra-host genetic diversity in acute RNA viruses and involving a reasonable number of sequences are only available for a small number of viruses. However, even these allow a number of important generalities to be made. First, most studies in this area have found relatively abundant intra-host genetic diversity, testament to the rapidity of both mutation and replication in RNA viruses. For example, in the case of DENV, average levels of intra-host pairwise genetic diversity are usually between 0.1 and 1% (Wang et al. 2002; Holmes 2003a; Lin et al. 2004; Chao et al. 2005; Aaskov et al. 2006) (Table 3.1), whereas in the case of Banana mild mosaic virus, an unclassified member of the *Flexiviridae* (ssRNA+), equivalent values are generally less than 2% (Teycheney et al. 2005). However, rather lower levels of intra-host diversity have been documented in some other cases, such as the flavivirus GBV-B (McGarvey et al. 2008), while mean pairwise diversity in WNV was 0.016%, with ~20% of clones differing from the consensus (Jerzak et al. 2005). Second, and of more importance, the nature of the mutations that are observed within hosts is often very different to that observed at the population level, with a predominance of putative deleterious nonsynonymous mutations that are later purged by purifying selection (see section 3.4 for more details).

Admittedly, it is possible that DENV is unrepresentative because the difficulties of replicating in hosts as divergent as humans and mosquitoes mean that purifying selection is especially strong in this virus. However, as argued throughout this book, deleterious mutation appears to be pervasive in RNA viruses. Further, a notable result from the WNV studies is that levels of genetic diversity are greater in mosquito compared to vertebrate (bird) hosts, indicative of different selection pressures in these different host environments (Jerzak et al. 2005). A major focus for future studies of RNA virus evolution in nature should therefore be to measure the extent and structure of intra-host genetic variation in a far wider array of RNA viruses, covering a range of ecologies and genome structures. It is also the case that most studies of deleterious mutation load in RNA viruses are highly conservative in that they do not directly consider synonymous mutations, even though I have documented the mounting evidence that many synonymous changes are not neutral. Measuring the fitness of the silent changes that occur within hosts, perhaps initially by determining how they affect RNA secondary structure, therefore represents another important research goal for the future.

Finally, it is also evident that intra-host genetic diversity is not simply the product of *de novo* mutation, as another of the key insights provided by studies of clonal

3.6 Intra-host viral diversity • 85

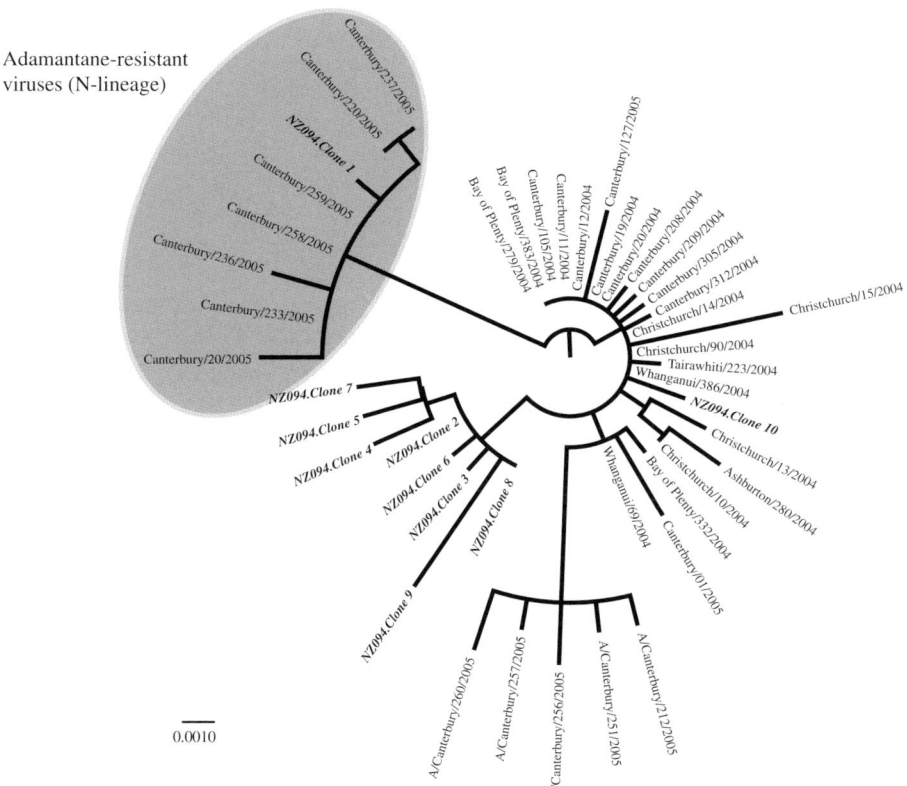

Fig. 3.17 Phylogenetic evidence for mixed infection in human influenza A virus. Clones of the matrix (M) segment for isolate NZ094 (from New Zealand), shown in italics, fall into three groups, one of which (grey box) clusters with adamantane-resistant viruses that possess the S31N mutation (i.e. they are members of the N-lineage; see Fig. 3.7). The other viruses in this tree were also sampled from New Zealand. The tree is drawn to a scale of substitutions per site.
Data taken from Ghedin *et al.* submitted

genetic diversity is that the mixed infection of individual hosts may be commonplace. This is beautifully illustrated in the case of influenza A virus described in section 3.3, in which adamantane-resistant and -sensitive lineages co-circulate in individual patients (Fig. 3.17). As at least 3% of all (≈3000) viral isolates generated by the Influenza Genome Sequencing Project (Ghedin *et al.* 2005) have some evidence of the large-scale sequence polymorphism indicative of mixed infection (Ghedin *et al.* submitted), this process clearly occurs at relatively high frequency. Such a high level of mixed infection also provides the potential for both frequent reassortment and complementation.

4

The RNA virus quasispecies

Since its formulation in the 1970s, the notion that RNA viruses form complex population structures known as quasispecies has dominated discussions on the mechanisms of RNA virus evolution. At the time of writing over 1000 articles on the PubMed online database contain the word quasispecies in the title or abstract, and a whole volume has been written on quasispecies in virology (Domingo et al. 2001). Although there is no doubt that quasispecies theory is a valuable evolutionary model, has been instrumental in introducing evolutionary ideas into virology, and can shed new light on evolutionary dynamics when mutation rates are extremely high, in this chapter I will argue that it is still highly debatable whether it applies to RNA viruses in nature. However, this should not be seen as an all-out attack on the quasispecies concept. Indeed, it is clear that for all its rights and wrongs quasispecies theory has had the major positive effect of making both virologists and evolutionary biologists think more carefully about the consequences of rapid mutation, and has perhaps inspired a new form of antiviral therapy. Rather, my gripes are that (i) quasispecies theory is often misunderstood, and I will be honest and say that I have erred here at times myself, (ii) that it is often described in quasi-scientific, almost mystical terms, and (iii) that much of the evidence said to support the quasispecies over other evolutionary models does nothing of the kind. In essence, my problem is not with the quasispecies *per se*, which is a rigorous and important theory, but rather with the loose manner in which it has sometimes been applied to RNA virus evolution. I believe that a critical, yet constructive, discussion of the value of quasispecies theory to understanding the evolutionary biology of RNA viruses in nature is of great importance to the field.

4.1 What is a quasispecies?

The theory underlying the quasispecies was originally developed by Manfred Eigen as a mathematical model of the self-replicating macromolecules that likely characterized the early evolution of life on Earth (Eigen 1971), although the term itself was not coined until a few years later (Eigen and Schuster 1977). Although these early papers by Eigen and colleagues are seminal in many ways, it has been argued that their innovation from a population genetics perspective was as important extensions of mutation-selection balance models (Wilke 2005). The quasispecies concept was first applied to viral populations by Esteban Domingo in the late 1970s, following the observation of widespread genetic variation in the RNA bacteriophage Qβ

(Domingo et al. 1978), and who has since published an impressive number of papers on this subject. As increasing amounts of gene sequence data were generated from RNA viruses (see, for example, Domingo et al. 1985 and Steinhauer et al. 1989), so the term became increasingly synonymous for genetic variation, eventually dominating discussions of viral evolution. Such is the undeniable success of the quasispecies concept that it has now become the dominant model of RNA virus evolution.

My first task is to develop a more precise definition of the quasispecies as applied to RNA viruses. In keeping with the rest of this book I shall do this in biological rather than mathematical terms, although mathematically it has been defined simply as the 'distribution of mutants that belongs to the maximum eigenvalue of the system' (Eigen 1996). Readers interested in a more quantitative definition should consult the original formulation by Eigen and Schuster (1977), or more recent considerations by Martin Nowak and Bob May (Nowak and May 2000), Rafa Sanjuán (Sanjuán 2008), and Claus Wilke (Wilke 2005). Interested parties might also consider a number of the theoretical updates that have been made to quasispecies theory in recent years (for example, Wilke 2003).

In simple terms, the quasispecies is a particular form of mutation-selection balance in which a distribution of variant genomes is ordered around the fittest, or 'master', sequence. This distribution of mutants is sometimes referred to as a cloud or a swarm. However, rather than simply providing a description of genetic diversity, the key element in quasispecies theory is that the mutation rate in the system is so high that the frequency of any variant in the population is not only a function of its own replication rate—and hence individual fitness—but also the probability that it is produced by the erroneous replication of other variants in the population that are linked to it in sequence space (i.e. that differ from each other by only a small number of mutational changes). As a consequence of this 'mutational coupling' viral genomes are not independent entities, but rather form a distribution of evolutionarily interlinked genomes, so that the entire mutant distribution behaves as if it were a single unit (Eigen 1996) (Fig. 4.1). To rephrase this slightly, high mutation rates ensure that

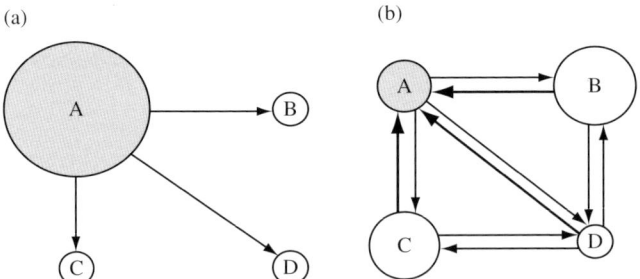

Fig. 4.1 The quasispecies. (a) A viral population without quasispecies structure in which natural selection favours variant A which, by mutation, produces variants B, C, and D at a low rate. (b) A quasispecies in which mutational coupling among variants (denoted by arrows in both directions) ensures that the viral population evolves as a single unit, although variant A still possesses the highest individual fitness (but is not at the highest frequency). The frequency of each variant in the population is reflected in the size of each circle.

the evolutionary dynamics of individual variants depend on the fitness of others in the population. Crucially, this particular population structure also means that even variants with a low individual fitness can reach a relatively high frequency in the quasispecies if they have mutational links to variants with higher fitness (Biebricher and Eigen 2005; Wilke 2005). Similarly, quasispecies dynamics mean that the most common genotype is not necessarily the fittest, and that the 'wild-type' may only comprise a small proportion of the total population.

In original models of quasispecies dynamics there was no room for the random diffusion of mutants through the population by genetic drift. As population sizes were assumed to be effectively infinite, the sequence space—the distribution of all possible mutants—was thought to be fully explored (Eigen 1971, 1996). An important consequence of this lack of genetic drift is that although the master sequence continually generates mutants, it maintains a stable frequency in the population. However, more recent manifestations of quasispecies theory have relaxed this limiting assumption so that it can apply in finite populations given particular parameter values (Wilke *et al.* 2005; see below).

By far the most important evolutionary consequence of quasispecies dynamics—indeed, their essence—is that the mutational linkage among genomes means that natural selection acts on the mutant distribution as a whole, rather than on individual variants as in the population genetic models normally applied to cellular organisms. As such, the quasispecies as a whole evolves to maximize its *average fitness*, rather than that of individual variants. This, in turn, leads to one of the most interesting, and controversial, implications of quasispecies theory: that under particular mutant distributions low-fitness (i.e. slow-replicating) variants can sometimes outcompete variants of higher fitness if they are surrounded by beneficial mutational neighbours. It is this effect that has been proposed to explain classic experimental observations in VSV that high-fitness mutants can be 'suppressed' by their low-fitness neighbours (de la Torre and Holland 1990).

As an example of this key aspect of quasispecies dynamics imagine two hypothetical viral populations; population A in which there is an individual variant of highest fitness, but which has low-fitness mutational neighbours, and population B, in which all mutants have a similar, average fitness that is higher than the average fitness of population A (i.e. population B is more robust) (Fig. 4.2). Under classic 'survival of the fittest' models of Darwinian evolution population A is superior to population B as it contains the individual variant of highest fitness. However, under quasispecies dynamics, population B can in some instances gain superiority: high mutation rates mean that the variant of highest fitness in population A is connected, by mutational coupling, to its low-fitness neighbours, reducing the average fitness of the population as a whole to potentially below that of population B. This effect has cleverly been referred to as the 'survival of the flattest' (Wilke *et al.* 2001), although it is perhaps more correctly thought of as increased mutational robustness. In fact, it is interesting to note that in recent years the debate over the existence of quasispecies in RNA viruses has turned into a debate over the extent of mutational robustness (Montville *et al.* 2005; Codoñer *et al.* 2006; Sanjuán *et al.* 2007; Belshaw *et al.* 2008).

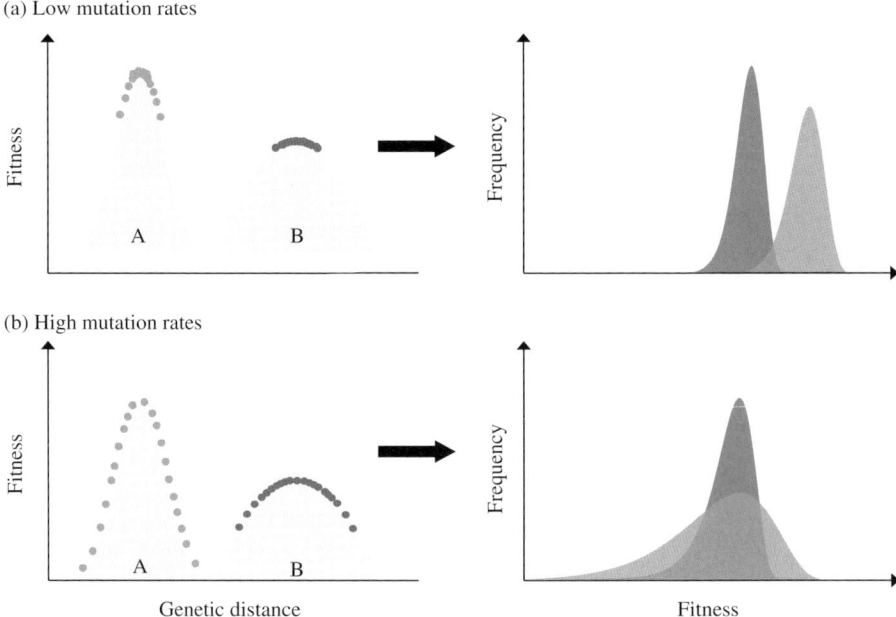

Fig. 4.2 The survival of the flattest. The figure depicts two populations—A and B—that are located in different regions of sequence space. The light-grey population (A) is characterized by a high replication rate but low mutational robustness. In contrast, the dark-grey population (B) has a lower replication rate but greater mutational robustness. Dots depict individuals located on each peak at low and high mutation rates. The expected distribution of individual fitness values for the two populations are shown on the right of the figure. At low mutation rates, population A, which possesses the variant of highest individual fitness, will always out-compete the flatter population B. However, the situation can be reversed at high mutation rates, in which case selection favours the population (B) with greater mutational robustness.
Adapted from Sanjuán et al. (2007) with permission.

4.2 The great quasispecies debate

It is easy to see the power of quasispecies theory, why it differs from the individual-based models more commonly used in population genetics, and its potential applicability to rapidly mutating RNA viruses. However, over the last decade there have been a number of major debates relating to the validity of quasispecies theory in virology, generating considerable controversy (Domingo 1992, 2002; Domingo and Holland 1997; Smith et al. 1997; Domingo et al. 1999; Holmes and Moya 2002; Novella 2003; Moya et al. 2004; Biebricher and Eigen 2005; Sanjuán 2008). Importantly, the main bone of contention is whether the quasispecies represents a viable model to describe the evolution of RNA viruses in nature, and not whether the quasispecies is a viable model, period, because this is clearly so. The validity of the quasispecies for RNA virology has been addressed using three types of data: (i) *in silico* experiments involving so-called digital organisms, (ii) classic experimentation *in vitro*, and (iii) comparative, sequence-based studies. As I will show below, the quasispecies debate spins on whether the mutation rates experienced by RNA viruses in nature are of sufficient magnitude to allow the onset of quasispecies dynamics through mutational coupling. However, before I turn my attention to this central talking point, it is important to consider a number of other aspects of the quasispecies debate.

4.2.1 What's in a name: quasispecies or polymorphism?

The first source of contention regarding the viral quasispecies, and the easiest to resolve because it is largely semantic, concerns the use of the term as a valid way to describe genetic variation within viral populations. Remarkably, despite the frequency with which the word 'quasispecies' is used in virology, it is clear that most people use the term simply to describe the occurrence of intra-host (or intra-cell in culture systems) genetic variation in RNA viruses. This is an unquestionable misuse of the word, as the identification of genetic variation in no way fulfils all the defining criteria of a quasispecies (an error that was pointed out by Eigen himself some years ago; Eigen 1996). This being said, this mistake is now such a common occurrence that it will probably never be corrected. Indeed, the term quasispecies has become so synonymous with the observation of genetic variation that it has even been applied to bacterial populations (Kuipers *et al.* 2000), even though they can never experience the error rates required to establish mutational coupling. Confusingly, some early studies even applied the term to viral isolates sampled from different geographical localities (Dopazo *et al.* 1988). This is also an incorrect usage because the required mutational coupling among variants cannot occur unless a viral population is sampled from an individual infected host. In only a minority of cases have those investigators who have used the term quasispecies to describe a viral population considered its defining feature: that natural selection acts on the whole population. Similarly, much of the terminology associated with the use of quasispecies theory in everyday virology is divorced from the rigour of its underlying mathematical model. Words such as swarm and cloud are good examples: although these may generate useful images of genetic diversity, they are some distance from the evolutionary model that underlies quasispecies theory.

More importantly, given the rapid mutational dynamics that seem to characterize RNA viruses as a whole, intra-host genetic variation is an expected and predictable occurrence, even in the case of acute viral infections that only infect their hosts for a few days. In short, high levels of genetic variation—sequence polymorphism—do not equate to the existence of viral quasispecies. Ironically, a greater challenge to the orthodoxy of RNA virus evolution are cases where *no* genetic variation is observed within viral populations despite repeated sampling.

4.2.2 Is quasispecies theory different from 'classical' population genetics?

One of the most contentious, yet unnecessary, aspects of the quasispecies debate is that this theory is somehow qualitatively different from the models used in what might be regarded as the 'classical' population genetics edifice built by Fisher, Haldane, and Wright. For example, even Eigen himself states that 'The new [quasispecies] formulation of the concept of selection and its application to molecular systems *differ sharply* from the original Darwinian approach and from its later reformulation in population genetics' (Eigen 1992, p. 27; my emphasis). However, the truth of the matter, as neatly demonstrated by Claus Wilke (Wilke 2005), is that quasispecies theory can be framed within the mainstream of modern population genetics, although its intellectual history is rather different. In short, the quasispecies is no more than a form of mutation-selection balance that applies to genetic systems characterized by very high mutation rates. As natural selection acts on the population as a whole in the quasispecies, it can also be considered a form of group selection. Indeed, scenarios involving very high mutation rates represent one of the few cases in which group selection—usually an anathema to evolutionary biologists—is considered to be viable. More generally, there is no doubt that the mathematical theory underlying the quasispecies correctly describes the dynamical behaviour of genomes given the underlying assumptions of the model. Hence, there is nothing inherently alien or heretic about quasispecies theory. The only issue worthy of serious debate is whether it accurately describes populations of RNA viruses in nature.

4.2.3 Does genetic drift destroy the quasispecies?

As noted above, in original formulations of quasispecies theory the sequence space of possible mutations was thought to be fully occupied, thereby preventing genetic drift. It was this lack of random diffusion that enabled the mutational coupling essential for quasispecies formation. Indeed, computer simulations show that simple models of quasispecies structure break down in the face of widespread genetic drift (Jenkins *et al.* 2001a). However, the uncertainty over the role played by genetic drift in viral evolution notwithstanding (discussed in detail in section 3.3), more modern descriptions of the quasispecies allow its applicability to finite populations. In particular, if the product of the effective population size (N_e) and mutation rate (μ) is significantly greater than 1 (i.e. $N_e \mu \gg 1$), then a finite population effectively behaves as an infinite one (Bull *et al.* 2005; Wilke 2005). In addition, it is also possible, and perhaps more informative, to think of the quasispecies only with respect to mutations that directly impact on fitness, rather than to genome sequences as a whole. This is sometimes called the 'phenotypic quasispecies' (Schuster and Stadler 1999).

While the potential for genetic drift to disrupt mutational coupling is no longer a major argument against the validity of the RNA virus quasispecies, the impact of random sampling in small populations may in part explain why beneficial mutations sometimes appear to be 'suppressed' in viral populations as claimed in early studies *in vitro* (de la Torre and Holland 1990). As the probability that a mutation reaches fixation is partially dependent on its initial frequency, most advantageous mutations will lost by genetic drift in small populations. Similarly, in large populations clonal interference will also give the impression that beneficial mutations are somehow suppressed.

4.2.4 The evidence from 'digital organisms'

One interesting way to explore the importance of quasispecies dynamics is through the use of so-called digital organisms, in reality a sophisticated form of computer simulation involving self-replicating entities that complete for the resources provided by CPU cycles (Wilke and Adami 2002). The most important observation stemming from these studies *in silico* is that there are situations when genetic systems characterized by high mutation rates—which can be considered as analogous to RNA viruses—do indeed exhibit quasispecies dynamics (Wilke *et al.* 2001; Comas *et al.* 2005). While there is little doubt that these results, and their derived conclusions, are correct given the parameter values used, it is less clear that evolution *in silico* can be equated to viral evolution in nature. The main issue here is how often the high mutation rates required to drive systems into quasispecies dynamics occur in natural systems. In the Wilke *et al.* (2001) study the mutation rates required to achieve quasispecies dynamics were usually greater than 1 mut/genome/rep (range of 1.13–3.5 mut/genome/rep), while in that of Comas *et al.* (2005) mutation rates of more than 2 mut/genome/rep were necessary. At lower mutation rates evolution conformed to standard survival of the fittest models. As noted in section 3.1, these mutation rates *in silico* are usually greater than the mean mutation rates observed in natural RNA viruses (and far higher than those seen in retroviruses). As a consequence, if the mutation rates estimated in the studies *in silico* are accurate, then these computer experiments paradoxically represent a blow to quasispecies theory as applied to RNA viruses in nature: while the theory represents a powerful description of evolutionary dynamics at high mutation rates, these high mutation rates are unlikely to be a common occurrence in 'real' RNA viruses.

4.2.5 Experimental tests of quasispecies theory

The use of experimental virology to explore aspects of quasispecies dynamics has a long history following the pioneering work of Esteban Domingo in Madrid, Spain, and John Holland in San Diego, USA, and based on even earlier work using the phage Qβ (Mills *et al.* 1967). A full description of the results of these studies is sadly beyond the scope of this book, although interested readers should consult their major review article (Domingo and Holland 1997) or book (Domingo *et al.* 2001) on this subject. As a necessity, I will focus only on those results that are most pertinent to determining whether the quasispecies represents a viable model of RNA virus evolution.

Despite a long history of experimental study, the most important work on quasispecies dynamics has probably occurred in the last 10 years. Arguably, the first true demonstration of what might be considered quasispecies behaviour *in vitro*, and still one of the most important papers on the subject, was the observation by Christina Burch and Lin Chao that 'evolvability' in the RNA phage $\phi 6$ was critically dependent on its mutational spectrum; in this case the 'accessibility of advantageous genotypes' (Burch and Chao 2000). The same study also showed that a high-fitness clone evolved to *lower* mean fitness because its mutational neighbours were of low fitness, exactly as expected under quasispecies theory. In short, although it did not demonstrate the key quasispecies effect—the survival of the flattest—this paper did show that, *in vitro*, fitness can be thought of as an average property of the viral population, concordant with the predictions of quasispecies theory.

This key result has been extended by more recent experimental analyses of mutational robustness, using VSV (Sanjuán *et al.* 2007) and viroids (Codoñer *et al.* 2006) as model systems. Importantly, both these studies recapitulated the major findings of the work *in silico* described above: that quasispecies dynamics do occur at high mutation rates, but to achieve these mutation rates the error frequencies in RNA viruses have to be elevated artificially, and perhaps to levels that are unsustainable in nature. For example, in the study of VSV undertaken by Sanjuán *et al.* (2007), the detailed characterization of both fitness distributions and genetic variability revealed that a viral population with a lower replication rate was able to outcompete one characterized by a higher replication rate in the presence of chemical mutagens, indicating that the former was more robust to mutation (Sanjuán *et al.* 2007) (Fig. 4.3). In comparison, the faster-replicating population was fitter in the absence of elevated mutational pressure and may therefore better reflect natural populations of RNA viruses. Essentially identical conclusions can be drawn from the experimental study using viroids (Codoñer *et al.* 2006). In this case, viroids were subjected to treatment with ultraviolet C light as a way of elevating their mutation rate. Under these conditions the system also favoured the more robust viroid, again compatible with quasispecies dynamics (Codoñer *et al.* 2006), whereas traditional survival of the fittest behaviour was observed at spontaneous (and hence 'normal') mutation rates. In these circumstances it is possible that complementation represents a more powerful buffer against the effects of deleterious mutation than robustness, particularly as high rates of co-infection are associated with weaker selection for robustness (Montville *et al.* 2005).

Experimental studies have also suggested that quasispecies dynamics are critical to viral pathogenesis (Vignuzzi *et al.* 2005). In this case a mutant of poliovirus (denoted G64S) that produced an RNA polymerase with 6-fold-higher fidelity than the wild type was unable to infect the full range of tissues that are associated with severe disease, most notably the brain. Such widespread tissue diffusion only occurred when a more diverse viral population was used (G64S^{eQS}), as generated by chemical mutagenesis, suggesting that increased genetic diversity is somehow central to pathogenesis. However, while this paper clearly demonstrates that the fidelity

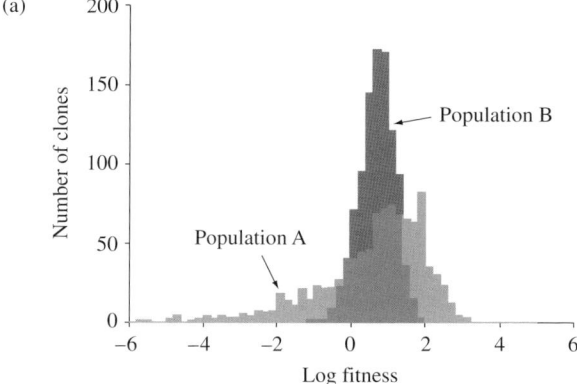

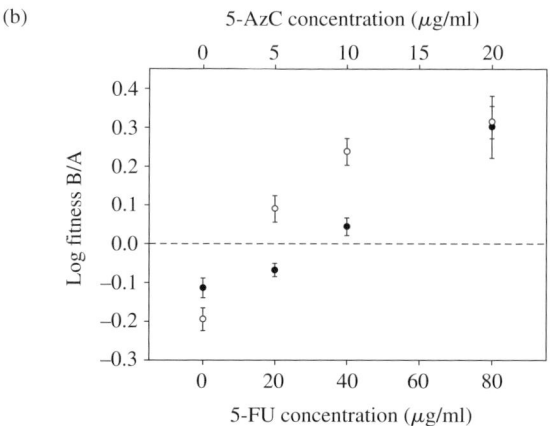

Fig. 4.3 Evidence for the quasispecies effect (survival the of flattest) in experimental populations of VSV. (a) Distribution of 1000 fitness values in experimental populations A (mean log fitness=0.386, variance=2.054) and B (mean log fitness=0.498, variance=0.225). (b) Results of competition experiments at various mutagen doses in which the log fitness of population B is shown relative to that of population A. Two chemical mutagens were used, 5-fluorouracil (5-FU; ○) and 5-azacytidine (5-AzC; ●). Note that the more robust population B is fitter at higher mutagen doses. Taken from Sanjuán et al. (2007) with permission.

of viral polymerases can, to some extent, be altered by a small number of point mutations—which has even been touted as a means to designing better vaccines (Vignuzzi et al. 2008)—and that there was an association between genetic diversity and pathogenesis, whether this behaviour can be attributed to quasispecies dynamics is a rather different matter. Although the authors suggest that there is a 'co-operation' among mutants in the quasispecies, such that the low-fidelity mutant was only able to infect the brain when the high-fidelity mutant was also present, in reality the quasispecies considers the joint effects of mutation and selective competition and says nothing about co-operation (and no mechanistic basis for this co-operation was provided). In addition, it is possible to explain this same behaviour under models that do not invoke quasispecies dynamics: once the pathogenic (i.e. neuro-tropic) strains of poliovirus have breached the physical barriers in the host and debilitated defences, initially non-pathogenic strains can more easily find their way towards the brain, effectively acting as opportunistic infections. Finally, because neuro-tropic strains of poliovirus result in dead-end infections and are not normally transmitted in the population, natural selection will be unable to favour this trait.

4.2.6 Comparative analyses of RNA virus quasispecies

Although studies both *in silico* and *in vitro* suggest that some elements of quasispecies theory are correct, if not the required error rates in nature, the same cannot be said of those comparative studies of quasispecies dynamics undertaken to date. There are a number of pieces of comparative data that argue against the existence of quasispecies in RNA viruses in nature, although none should be considered definitive. One early suggestion was that much of the genetic variation observed in RNA viruses was in fact due to PCR and/or sequencing error (Smith *et al.* 1997). Although it is clear that laboratory error has contributed to the genetic diversity seen in at least some viral populations, widespread genetic variation is such a common observation in RNA viruses that this does not represent a serious challenge to quasispecies theory.

For those RNA viruses where intra-host genetic variation has been studied under natural conditions, such as the results from DENV discussed in Chapter 3, the observed mutant spectrum does not obviously satisfy that predicted under quasispecies dynamics, although this cannot be regarded as a strong test of the model. In particular, rather than comprising a diverse set of inter-linked mutants, there is often a single dominant clone surrounded by off-shoot 'singleton' mutations as expected under conventional survival of the fittest dynamics (see, for example, Fig. 2 of Lin *et al.* 2004). Alternatively, intra-host DENV diversity sometimes appears as a more diverse set of clones, yet ones that do not usually form the inter-connected network (i.e. non-tree-like) structure that might be expected under quasispecies dynamics (Dopazo *et al.* 1993). For example, a diversity of phylogenetic structures are visible in the intra-host populations of DENV-1 studied by Aaskov *et al.* (2006), with networks apparent in only a minority of cases (Fig. 4.4).

Further comparative evidence against the quasispecies as applied to RNA viruses is that most, if not all, cases of positive selection documented in these systems to date involve the fitness advantage of individual mutants over others in the population, and not the propagation of low-fitness mutants surrounded by advantageous mutational neighbours. Take, for example, the case of HIV where the process of natural selection has been particularly well studied, involving the accumulation of mutations that

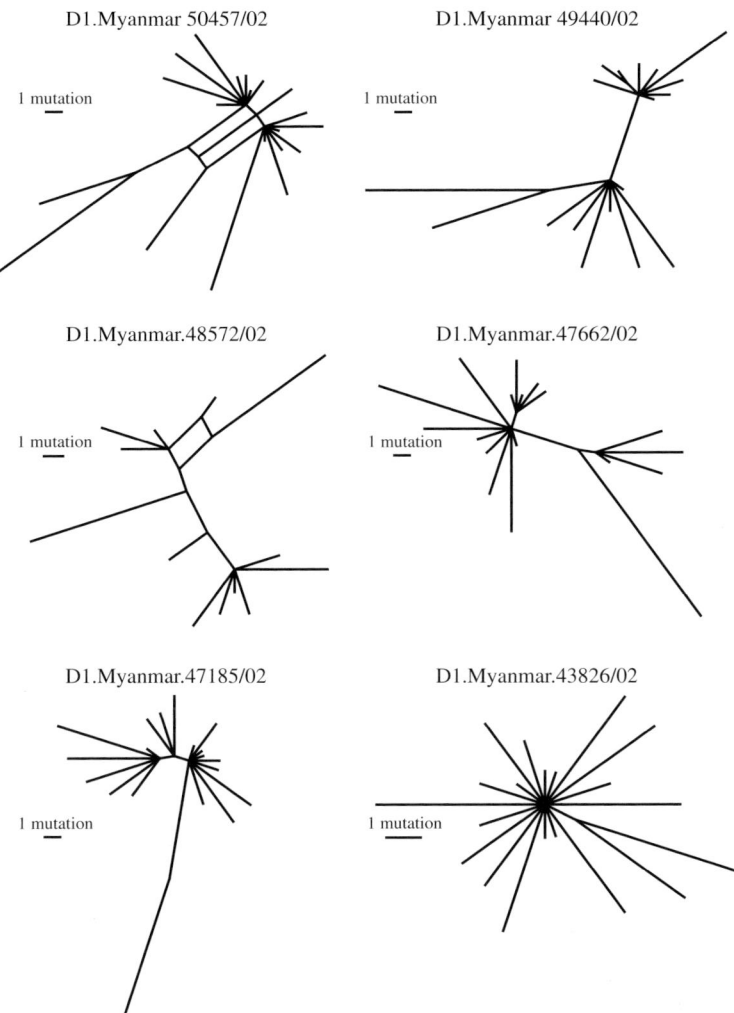

Fig. 4.4 Parsimony splits networks of six intra-host populations of the E gene of DENV-1 (20 sequences from each host) inferred using the SplitsTree4 program (Huson and Bryant 2006). All network edges are drawn to scale. The sequence data—from individuals sampled in Myanmar—was taken from Aaskov et al. (2006). Similar, generally non-complex, networks were observed for the other nine intra-host populations described in Aaskov et al. (2006).

evade either cellular or humoral immunity, or which provide resistance to antiviral drugs. Despite intensive study there is no clear evidence of quasispecies dynamics in this system. If there is any unpredictability in evolutionary dynamics, this seems most likely due to small effective population sizes and the action of genetic drift (see section 7.2 for a detailed discussion of this subject). As a specific example, consider the evolution of resistance to the drug AZT, the first antiviral agent, an RT inhibitor, used to treat HIV in the days of monotherapy. In this case resistance involves a specific group of mutations that tend to evolve in a similar order and without the clear suppression of those of high fitness (Larder and Kemp 1989; Kellam et al. 1994). The same story can be told for a multitude of other RNA viruses. I would even go so far as to say that, at present, there is no definitive example from a natural RNA virus population where selection has been shown to act on the group rather than the individual. Although Wilke (2005) very reasonably argues that quasispecies dynamics are not expected in situations of strong positive selection, as is clearly the case with drug resistance, these also represent the best opportunities to see the process of adaptation in action. More broadly, if quasispecies dynamics can never be 'seen' to occur in viral populations, one must question their importance. This also highlights a generic criticism of quasispecies models in virology: by attempting to explain every observation in RNA virus evolution, the theory will fall into the realm of untestable hypotheses. Examining the evolutionary dynamics of mutants that experience more subtle differences in fitness should therefore be a major subject for future study, and not simply from the perspective of the viral quasispecies.

As a brief digression, it is interesting to note that HIV is often cited as an archetypal quasispecies, largely because individual infected hosts exhibit extremely high levels of genetic diversity (Yuste et al. 1999; Domingo 2002). However, as noted in section 3.1, the intrinsic error rates associated with the use of RT may be five times lower than those observed in RdRp-utilizing viruses, and so well below the rates needed for quasispecies formation; indeed, the per-replication error rate in HIV may even be lower than the *high-fidelity* mutant of poliovirus generated by Vignuzzi et al. (2005). Hence, it is the additional processes of strong natural selection and frequent recombination that produce the huge levels of genetic diversity seen in this virus. Paradoxically, then, if there is one virus that demonstrably does not form a quasispecies, it is HIV.

A related feature that is claimed to be characteristic of quasispecies dynamics is that rates of adaptive evolution are highest on the 'periphery' of the fitness landscape, as this is where potential fitness gains are greatest (Eigen 1992; Biebricher and Eigen 2005). A slightly different formulation of this concept has a long history in population genetics and was used to counter early arguments for the neutral theory of molecular evolution, which predicted that the highest rates of evolutionary change in proteins occur in the least functionally constrained regions (King and Jukes 1969). However, that those viruses subject to continuous immune selection are often characterized by 'ladder-like' phylogenetic trees, such as the epidemiological-scale evolution of the HA gene of human influenza A virus, or the intra-host evolution of the *env* gene of HIV-1, implies that it is usually the centre of the mutant distribution—the fittest type—that is also most likely to give rise to mutations that confer the greatest fitness gains (Grenfell et al. 2004). Again, though, understanding the true contribution of the periphery of the fitness landscape to viral evolution will require the analysis of mutants with more subtle differences in fitness than usually measured in comparative studies.

4.2.7 Recombination and the quasispecies

One of the most interesting aspects of quasispecies theory, and surely one of the most controversial had it been given major attention, is the idea that recombination is in some respects detrimental for RNA virus evolution because it means that any error threshold is encountered at lower mutation rates (Boerlijst *et al.* 1996; G.M. Jenkins and E.C. Holmes, unpublished results). This sits in marked contrast to other evolutionary models in which recombination is considered a beneficial trait that is favoured by natural selection (see section 3.2). The very different evolutionary behaviour in RNA viruses may be because the mutation rate in this case is so very high that recombination cannot effectively purge deleterious mutations, although the mechanistic basis to this theoretical result has not yet been fully explored (Bull *et al.* 2005).

If the role of recombination in quasispecies dynamics is as claimed, then it is theoretically possible (although currently unstudied) that natural selection has acted to *reduce* the rate of recombination in RNA viruses, in contrast to all other biological systems studied to date. In fact, it is interesting that in many cases rates of recombination in RNA viruses are rather low (section 3.2), which could, very tentatively, be argued as indirect support for the idea that they have been minimized by natural selection. Similarly, the highest rates of recombination are observed in retroviruses such as HIV which, intriguingly, also have lower rates of mutation than RNA viruses replicating with RdRp. Following the same train of thought, this could mean that more recombination is permitted in the case of retroviruses because of their lower mutation rates. Alternatively, and evidently more likely, the high rate of recombination in retroviruses may simply reflect the peculiarities of their biology. In short, the role of recombination in the RNA viruses quasispecies currently raises more questions than answers, although its implications are fascinating.

4.2.8 'Memory' in viral quasispecies

Some discussions of the viral quasispecies seem to engender them, probably unwittingly, with properties that seem almost mystical. In most cases this is simply due to a rather ill-advised choice of terminology, as the concepts discussed are usually entirely reasonable. A high-profile case centres around the idea that quasispecies are able to maintain a 'memory' of their past evolutionary history, first demonstrated in populations of FMDV (Ruiz-Jarabo *et al.* 2000), but later applied to other viruses including HIV (Briones *et al.* 2003). Although it is natural that the use of the term 'memory' should set off a chorus of alarm bells for evolutionary biologists, and the authors do draw analogies to aspects of neurological memory, there is nothing heretical, or even controversial, in the science relating to viral memory. The concept is simply that the selective process acting on a viral population as it adapts to a particular environment has a profound affect on allele frequencies, altering the frequency of many mutations in the viral population, if not pushing them to fixation. In short, natural selection changes the whole mutant distribution, which is not the same as saying that it acts on the mutant distribution as a whole! Although the frequencies of these mutations drop in non-permissive environments, forming minority subpopulations, they increase in frequency when the viral population is again allowed to colonize the environment where they are found to be beneficial (Ruiz-Jarabo et al. 2000, 2002, 2003b; Arias et al. 2004). More importantly, aside from an arguably poor lexicography, there is nothing in the research of this *evolutionary memory* that provides definitive evidence for the existence of RNA virus quasispecies. However, these experiments do make two invaluable points about the nature of viral evolution: that evolutionary history is important when discussing the adaptive process, so that there is a strong historical contingency, even for organisms that evolve as rapidly as RNA viruses, and that it is again crucial to look beyond the consensus sequence.

4.3 Error thresholds, extinction thresholds, and error catastrophes

One of the most important, yet potentially confusing, aspects of quasispecies theory is the relationship between error rate and viral extinction, and the various terms used to describe it. Not only is it important to be semantically correct, but a proper understanding of the relationship between mutation rate and population extinction will go a long way to explaining the workings of a major new form of drug treatment. As I will outline below, the development of antiviral therapies based on the counter-intuitive concept of 'lethal mutagenesis' is perhaps the greatest practical achievement of research on the RNA virus quasispecies and one which beautifully illustrates the importance of evolutionary biology in medicine (for which the originators deserve great credit). It is therefore immensely ironic that there is growing body of thought which suggests that the true explanation for the success of lethal mutagenesis does not residue with the quasispecies (Bull et al. 2005, 2007).

Central to this particular section are the correct definitions of three terms, following the lead of Bull et al. (2005) and shown schematically in Fig. 4.5. (i) The *error threshold* can be defined as the point at which populations experience a phase transition from individual-based evolutionary dynamics to a situation where the fittest genotype suffers so many deleterious mutations that it cannot sustain itself in the population and is therefore only regenerated by back mutation from other variants in the population. Hence, this marks the point when selection favours genotypes that have lower individual fitness but increased mutational robustness (Bull et al. 2005). Importantly, the 'best conditions for evolution' (Eigen 1992, p. 84) are considered to be in the region just below the error threshold, as this is where the greatest number of mutants are produced, facilitating adaptation yet without an excessive cost of deleterious mutations. Although a concept that is often cited, it is also important to note that a threshold *per se* will only arise given a specific fitness function (Wiehe 1997), and is dependent on the sign of epistasis (Sanjuán 2008). (ii) Breaching the error threshold then leads to an *error catastrophe*, again reflecting the loss of the fittest genotype through deleterious mutation, and which is often touted as the explanation for the

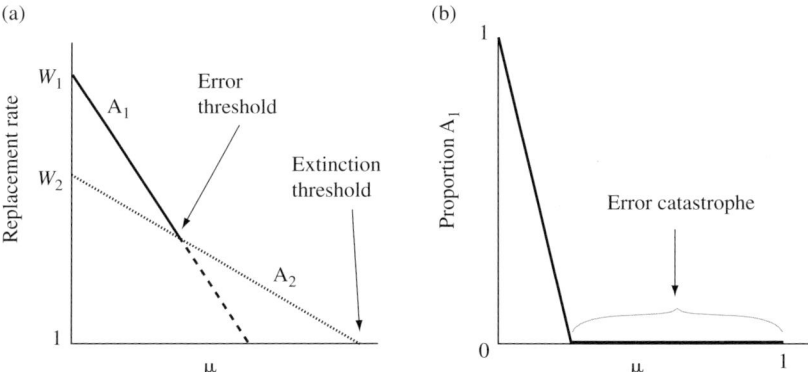

Fig. 4.5 Schematic representation of the error threshold, extinction threshold, and error catastrophe. This simplest possible example considers two genotypes: A_1 with fitness W_1 and A_2 with lower fitness W_2, and which is produced by mutation from genotype A_1. The replacement rate is fitness (W) multiplied by the proportion of mutation-free progeny, where μ is the mutation rate. Under this model (a), the error threshold marks the point (mutation rate) beyond which the fittest genotype A_1 disappears because it has a lower replacement rate than genotype A_2 (hence the dashed line). The extinction threshold marks the point at which the entire population goes extinct because the replacement rate is less than 1.0. (b) Rather than being a dramatic transition, as implied in the term error catastrophe, there is in fact a gradual decline in the frequency of the A_1 genotype (and which may not be linear). Taken from Bull *et al.* (2005) with permission.

action of lethal mutagenesis. Finally (iii) there is the *extinction threshold*, at which point there is a complete loss of the viral population because deleterious mutations accumulate faster than they can be eliminated by natural selection, thereby leading to a deterministic fitness decline and, finally, to population extinction.

The importance of the first two of these terms is that, as fundamental aspects of quasispecies theory, they have been claimed to form the intellectual bedrock of lethal mutagenesis (Domingo *et al.* 2005). The idea (as claimed) behind this form of drug treatment is that viral populations can be driven into error catastrophe through the use of mutagens such as 5-fluorouracil and ribavirin, although often in combination with more standard antiviral inhibitors (Loeb *et al.* 1999; Crotty *et al.* 2000, 2001; Sierra *et al.* 2000; Pariente *et al.* 2001; Ruiz-Jarabo *et al.* 2003a; Anderson *et al.* 2004). This technique has been successful both *in vitro* and *in vivo*, and applied to such infectious agents as FMDV, HCV, HIV, and lymphocytic choriomeningitis virus (LCMV), although different viruses respond in rather different ways. Although determining which evolutionary theory correctly explains lethal mutagenesis has proven controversial (Summers and Litwin 1996; Bull *et al.* 2005, 2007; Zeldovich *et al.* 2007), there is no doubt that its effects can be remarkable. For example, the deployment of 5-fluorouracil in combination with antiviral inhibitors such as guanidine hydrochloride resulted in the systematic extinction of various clones of FMDV, although this effect was not seen with each drug individually (Pariente *et al.* 2001) (Fig. 4.6). As expected, viral populations subject to this particular form of antiviral therapy show an increase in the complexity of mutational diversity (Domingo *et al.* 2005). However, that mutational diversity is not necessarily a predictor of extinction (Grande-Pérez *et al.* 2002), such that extinction can occur without the generation of hypermutated genomes (Grande-Pérez *et al.* 2005), argues against the action of an error catastrophe (see below).

Although the results of lethal mutagenesis are extremely exciting, it is important to exercise some caution. In particular, it possible that the application of mutagens will impose a strong selective pressure for the evolution of resistance (Sanjuán *et al.* 2007), in the form of an increase in overall replication fidelity, the selective exclusion of the mutagen from the active site of the viral polymerase, or increased robustness, although this has been debated (Martin *et al.* 2008). In addition, and of more direct importance for this book, it is another matter to say that the results of lethal mutagenesis can be unequivocally explained by quasispecies theory, although this question is probably irrelevant for the potential use of these drugs. Indeed, a strong case can be made that rather than being due to error catastrophe, as predicted under quasispecies theory, lethal mutagenesis instead involves breaching the extinction threshold. To put it another way, whereas an error catastrophe involves 'an evolutionary shift in genotype space' that is independent of population size (Bull *et al.* 2007), extinction requires a drop in *population numbers* and so explicitly involves viral demography. Hence, crossing the extinction threshold means that the viral population size will

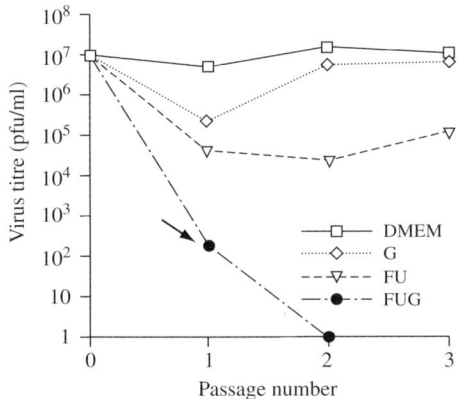

Fig. 4.6 The theory of lethal mutagenesis applied to RNA viruses. The graph shows the results of an experiment in which the titre of a strain of FMDV (C-S8C1) is exposed to different drug regimens: DMEM, Dulbecco's modified Eagle's medium; FU, 5-fluorouracil; G, guanidine hydrochloride; FUG, 5-fluorouracil and guanidine hydrochloride. Note that the strongest effect on viral titre is seen when a mutagen (5-fluorouracil) is used in combination with a replication inhibitor (guanidine hydrochloride). pfu, plaque-forming units. Taken from Pariente *et al.* (2001) with permission.

ultimately decline (although this process can be reversed), and that this will occur irrespective of the initial population size. The key parameters setting this threshold are the mutation rate and the viral yield per infection cycle. That lethal mutagenesis works best in combination with antiviral inhibitors—that must act to reduce viral yield—supports this view.

4.4 Concluding remarks

If its underlying assumptions are met, particularly that of an error rate high enough to ensure mutational coupling, then the dynamics specified in quasispecies theory are a predictable evolutionary outcome. It is even likely that these assumptions match the conditions of the primordial RNA world for which the theory was originally derived. In laboratory systems it is clearly possible to generate artificial conditions that match those required for quasispecies formation, largely through the administration of mutational agents. However, by definition these are artificial systems requiring elevated mutation rates and therefore may not be directly applicable to RNA viruses as they evolve in nature (and viral evolution today is not the same as that of the RNA world).

Reading between the lines of this chapter it is obvious that a large part of the uncertainty over the practical applicability of quasispecies theory arises from an imbalance toward *in vitro* studies of this particular aspect of viral evolution. To truly determine the practical value of quasispecies theory it is essential that more analysis be conducted on RNA virus evolution in nature, with its myriad of complex interactions. Indeed, I would strongly argue that too few acute RNA viruses have been studied in sufficient detail through clonal sequencing at specific times in infection to determine whether they form quasispecies. As argued in section 3.6, the analysis of intra-host diversity in natural systems should be a major element of future studies in viral evolution.

Although there is still considerable uncertainty as to whether quasispecies theory correctly describes the evolutionary behaviour of RNA viruses in nature, the importance and value of the theory as a means of introducing evolutionary ideas into virology cannot be denied. However, rather than accepting the theory blindly whenever genetic variation is encountered in an RNA virus, as is the current vogue, I contend that its most important prediction—that natural selection acts on groups of viral genomes—still needs to be verified for natural populations of RNA viruses. Claus Wilke has stated that, '... we currently have no evidence (theoretical or experimental) that contradicts the existence of quasispecies effects in finite populations of RNA viruses, but we also have no experimental evidence in favor of it' (Wilke 2005). While I hope I have shown that there is more evidence against the viral quasispecies than Wilke might believe, this is a remarkable admission given how frequently RNA viruses are claimed to form quasispecies.

5

Comparative genomics and the macroevolution of RNA viruses

5.1 The evolution of genome architecture in RNA viruses

One of the most interesting and important, yet understudied aspects of RNA virus evolution are the processes responsible for the diverse array of genome architectures employed by these infectious agents. Nestled within this general topic are some of the most intriguing of all questions raised in this book. To give a few specific examples: what explains the range of genome sizes in RNA viruses? What forces led to the evolution of segmented genomes? Why are some RNA viruses positive-sense and others negative-sense? Although of immense importance, these questions have, with few exceptions (notably Reanney 1982), rarely been addressed. The aim of this chapter is to suggest answers, albeit tentative ones, to these and a variety of other questions relating to the comparative genomics of RNA viruses. Although my conclusions are unlikely to be definitive, they should at least provide hypotheses for future testing. In doing so this chapter will also emphasize the evolutionary consequences of possessing highly restricted genome sizes: the small-genome dynamics I mentioned at the outset of this book.

5.1.1 The evolution of genome size

One of the most obvious, and therein important, biological features of RNA viruses are their small genomes. This also represents a natural place to start on our quest to understand the evolution of viral genome architecture. As noted in Chapter 1, genome sizes in RNA viruses vary in size by a little over one order of magnitude, irrespective of what host they infect. The smallest infectious agent that may be considered a true RNA virus (rather than a viroid) is *Ophiostoma novo-ulmi* mitovirus 6-Ld, weighing in at only 2343 nt (all members of the *Narnaviridae*, including the mitoviruses, are very small). In fact, there are very few RNA viruses with genomes smaller than 4000 nt (Fig. 5.1). At the other end of the scale, the largest RNA viruses are the coronaviruses (and their relatives the roniviruses), which have genome sizes of approximately 30 000 nt, with murine hepatitis virus the largest at 31 526 nt. Across RNA viruses as a whole, the mean genome size is approximately 10 000 nt. A far wider range of genome sizes are seen in DNA viruses, from a mere 1758 nt

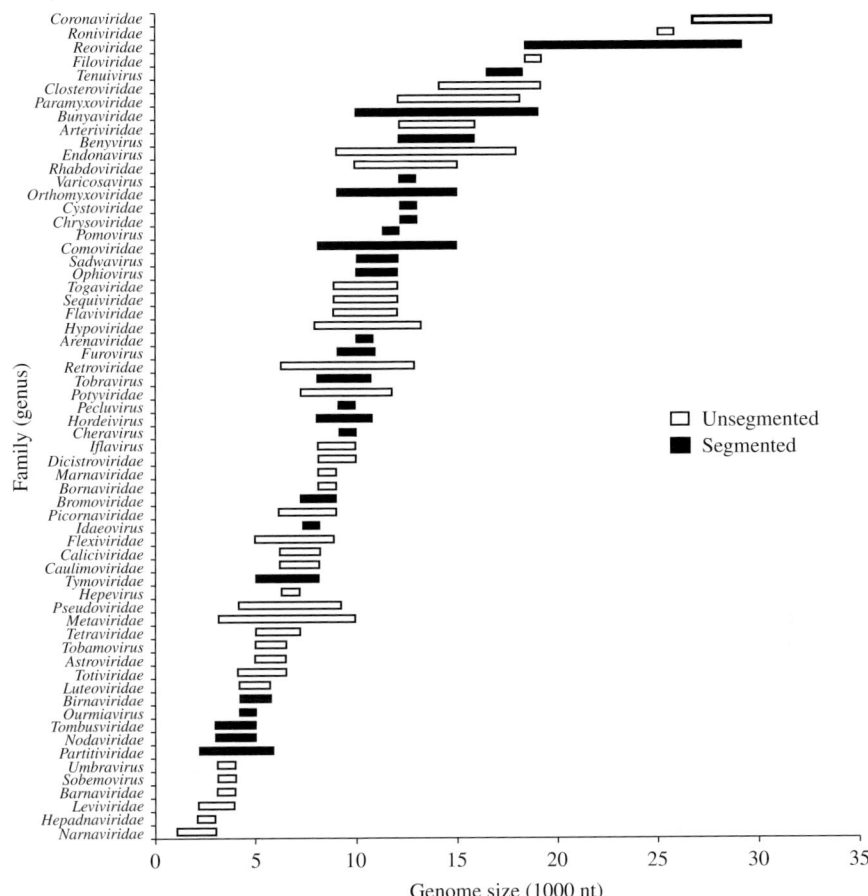

Fig. 5.1 The distribution of genome sizes among different families (and some genera) of RNA viruses. Segmented (closed bars) and unsegmented (open bars) genomes are indicated. Data taken from the 2005 ICTV classification of viruses (Fauquet *et al.* 2005), adjusted to correct a few known errors. The characteristics of these viruses are given in Fig. 1.1.

(porcine circovirus; *Circoviridae*; ssDNA) up to 1 181 404 nt (*Acanthamoeba polyphaga* mimivirus, with the related 'mamavirus' perhaps even larger; La Scola *et al.* 2008). It is also highly significant that the genomes of all ssDNA viruses are small, with none larger than 11 000 nt (with the segmented nanovirus milk vetch dwarf virus the largest at 10 958 nt). This size restriction reinforces a central argument of this book: that ssDNA viruses are very RNA-like in their evolution. As a brief aside, a partial reason for the very large genome sizes of some dsDNA viruses is that they have captured host genes that allow them to modulate host immune responses (reviewed in Shackleton and Holmes 2004). Obviously, the major constraints imposed on genome size mean that this option is closed to RNA viruses (see section 5.2). Also of relevance is that the genomes of segmented RNA viruses are, on average, a little larger than those of unsegmented RNA viruses (mean values of ≈11 000 and 9000 nt, respectively, but excluding the *Coronaviridae* and *Roniviridae*), with the reoviruses, with genomes up to 29 000 nt and 12 segments, the largest. The highly informative exceptions of the *Coronaviridae* and *Roniviridae*, which are both long and unsegmented, are discussed in more detail below.

A variety of hypotheses can be put forward to explain why the genome sizes of RNA (and ssDNA) viruses are so small. One, seemingly reasonable, idea is that genome sizes are constrained by the maximum size of the genetic material that can be packaged within a single virion. Hence, viral genomes are small simply because they are unable to 'fit in' any more genetic material. Interestingly, segment number does not seem to vary extensively in nature for individual segmented viruses, suggesting that there must be major costs to packaging more than the required number of segments, although these costs do not necessarily relate to genome size. For example, in the case of influenza A virus which has been studied in detail, virions that possess more than the normal eight segments have only been observed very rarely and usually *in vitro* (Enami *et al.* 1991), and packaging may not be a random process (Duhaut and McCauley 1996). However, while the mechanics of packaging are only known in a few cases (Qiao *et al.* 1997), there are important reasons to doubt the packaging argument. In particular, dsDNA viruses, which must also be subject to these same packaging constraints, are able to attain far larger genome sizes than RNA viruses.

Although the constraints of packaging seem easy to dismiss, other structural features of RNA virus genomes may play a more important role in regulating genome sizes. In particular, an important topological constraint facing all RNA viruses is the requirement to unwind potentially long regions of dsRNA during replication (Reanney 1982). For those RNA viruses with longer genomes, this process is mediated by a distinct helicase (HEL) domain. The potential importance of this constraint is apparent in the idea that in ssRNA+ viruses there is a strong association between presence of a HEL domain and genome sizes of more than 6000 nt (Gorbalenya and Koonin 1989). It has therefore been suggested that the acquisition of a HEL domain represented a major transition in viral evolution, as it allowed genome sizes to increase beyond a previous threshold level and hence generate more phenotypic diversity (Gorbalenya *et al.* 2006). However, while the ability to unwind long RNA molecules clearly plays some role in setting genome size (and may explain why most dsRNA viruses are segmented; see below), it cannot explain why ssDNA viruses also possess small genomes. Similarly, that long RNA molecules are fragile and so liable to breakage may also constrain the length of ssRNA viruses, although this does explain the equally restricted genome sizes of dsRNA viruses.

It is also possible that the restricted genome sizes of RNA viruses are determined by a requirement to replicate quickly (see section 3.1). Hence, because RNA viruses are competing with both hosts and each other for cellular resources, the smaller the genome, the faster the virus will be able to replicate, conferring on it a selective advantage. However, while rapid replication offers clear benefits to RNA viruses, this theory is unable to explain the enormous range of genome sizes occupied by dsDNA viruses, even though as obligate parasites they might be expected to be under considerable pressure to replicate rapidly. In addition, there is no clear relationship between replication rate and genome size in those RNA viruses studied to date.

The final, and I think most likely, explanation for the small genome sizes of RNA viruses is that the amount of genomic material is limited by the intrinsically error-prone process of replication. As noted elsewhere in this book (especially section 3.1), if there is a roughly constant rate of mutation per nucleotide, then the longer the genome the greater the number of mutations produced, most of which will be deleterious. The power of this explanation is that it derives directly from one of the most fundamental aspects of RNA virus biology: their highly error-prone replication. In contrast, the higher-fidelity dsDNA viruses with far lower mutation rates are able to achieve much larger genome sizes. Importantly, this theory can also be extended to ssDNA viruses, which are both small and have mutation rates—and more notably substitution rates—that are far higher than those seen in dsDNA viruses (Fig. 3.1). The remaining puzzle is that ssDNA viruses have rather lower mutation rates per nucleotide than RNA viruses, but not larger genomes. One possible explanation that merits further investigation is that mutation rates in ssDNA viruses are rather higher than measured in experimental assays, perhaps because of additional deamination mutation (Duffy and Holmes 2008).

5.1.2 The exceptions: coronaviruses and roniviruses

Biology being what it is, there are exceptions to the broad-scale generalizations made above. In the case of the evolution of genome sizes the exceptions, which I believe help prove the rule, are the coronaviruses and roniviruses, which together with the rather smaller arteriviruses make up the order *Nidovirales*. From an evolutionary perspective, the *Coronaviridae* and *Roniviridae* are remarkable because their unsegmented ssRNA+ genomes have sizes—roughly 26 000–32 000 nt—that exceed those observed in other RNA viruses, often substantially. They also encode a polypeptide close to 7000 amino acid residues in length, almost twice the maximum ORF size seen in any other viral family. In fact, the genomes of the coronaviruses and roniviruses are essentially twice the size of their relatives the arteriviruses (13 000–16 000 nt). How is such a size increase possible without a mutational meltdown?

The expanded genome size in the coronaviruses and roniviruses is principally due to the presence of a very large (>20 000 nt) replicase gene, and one which is composed of multiple functional domains (Gorbalenya *et al.* 2006) (Fig. 5.2). A number of these domains are very familiar, such as the RdRp and the HEL. Others, however, are unique. Most interesting from the perspective of this book is the ExoN domain, which encodes a 3′-to-5′ exoribonuclease. Remarkably, the ExoN domain exhibits distant similarities to host cellular proteins of the DEDD superfamily of exonucleases (Snijder *et al.* 2003; Minskaia *et al.* 2006) which, among other things, are involved in proofreading and repair. This hints that coronaviruses and roniviruses are able to reduce the error rate associated with the RdRp through some sort of repair function, possibly involving proofreading activity of the 3′-to-5′

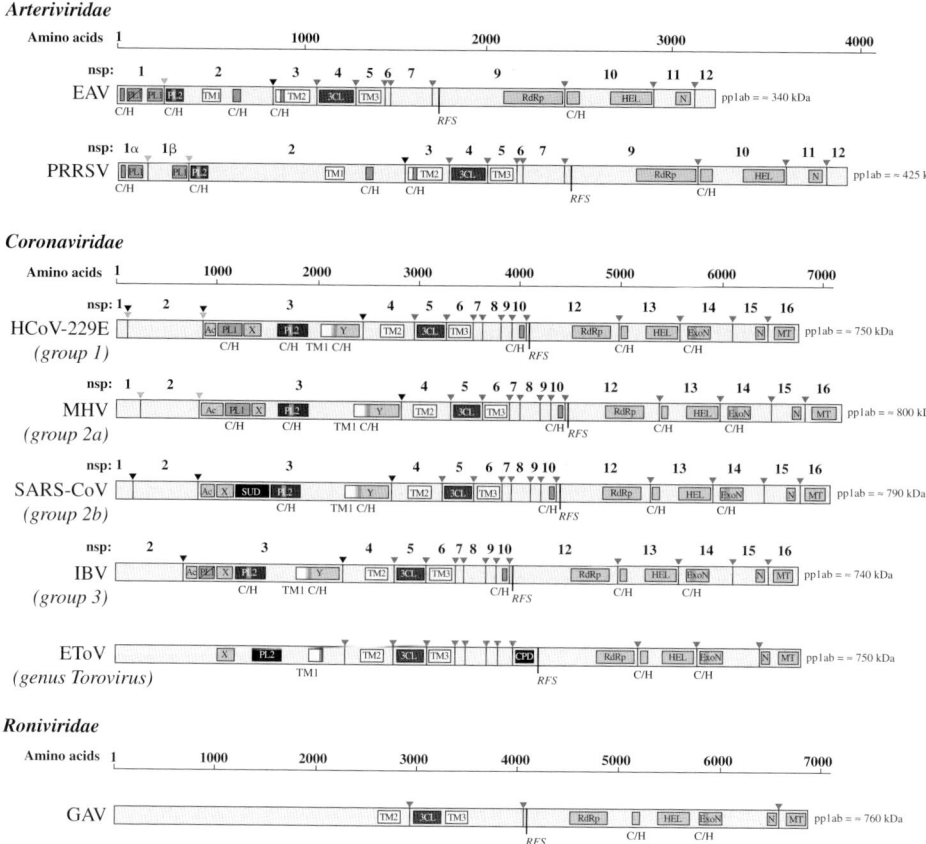

Fig. 5.2 The extremely large replicase genes of coronaviruses and roniviruses (>20 000 nt) in comparison to that of their relatives the arteriviruses (<12 000 nt) (not drawn to scale). All these viruses are classified in the order *Nidovirales*. Note the additional domains in the coronavirus and ronivirus replicases, particularly the ExoN domain. Other domains of note are the RdRp and the helicase (HEL). EAV, equine arteritis virus; EToV, equine torovirus; GAV, gill-associated virus; IBV, avian infectious bronchitis virus; MHV, mouse hepatitis virus; PRRSV, porcine respiratory and reproductive syndrome virus. More details of all the domains present are provided in the original publication. Taken from Gorbalenya *et al.* (2006) with permission.

exoribonuclease (Minskaia *et al.* 2006). This, in turn, will reduce mutational load and allow larger genome sizes. Indeed, such a dramatic increase in genome size is unprecedented in the recent history of RNA virus evolution, although it is important to note that arteriviruses also have genomes that are rather larger than normally seen in unsegmented RNA viruses. However, the consequences any putative repair function has on evolutionary rates are unclear. Whereas some analyses have documented rather lower substitution rates in coronaviruses compared to other RNA viruses (Jenkins et al. 2002), as well as a reduced burden of deleterious mutation (Pybus et al. 2007), evolutionary rates in SARS-CoV are as high as those seen in more rapidly evolving RNA viruses (Hon et al. 2008), although this could be a function of positive selection associated with the emergence of SARS in humans.

As discussed in more detail below, the replication strategy of the *Coronaviridae* and *Roniviridae* is also unique among RNA viruses, and which again seems to be largely a function of possessing a very large genome. In particular, the extremely large replicase gene of these viruses consists of two large ORFs denoted 1a and 1b. A −1 ribosomal frameshift, mediated by a 'slippery sequence' and an RNA pseudo-knot, is required to express the ORF1b polypeptide, which occurs just upstream of the ORF1a stop codon (Gorbalenya et al. 2006). In addition, although the RNA polymerase activity of the coronaviruses is due to non-structural protein 12 (nsp12) that contains the usual conserved RdRp domain (Fig. 5.2), a second RdRp is contained within the nsp8 protein, and which may have arisen from nsp12 by gene duplication (Imbert et al. 2006).

5.1.3 The evolution of genome organization: an overview

Although the genomes of RNA viruses are small and might appear to look rather similar to the unitiated, this apparent simplicity hides a truly remarkable amount of genomic complexity. In this section I will argue that to explain the evolution of these diverse genome organizations requires us to understand the evolution of mechanisms to control gene expression—considered here as the processes of transcription and translation—as I believe that the two are inextricably linked. Specifically, I will suggest that many of the most interesting aspects of genome organization in RNA viruses represent solutions to the fundamental problems of controlling gene expression in an environment of small-genome dynamics (see Jaspars 1974 for an early exposition of this idea).

Central to this task is my classification of the genome organizations and replication strategies in RNA viruses into six general categories. Although this undoubtedly hides a great deal of biological detail, and some viruses cannot be easily categorized in this manner, I believe that these categories do highlight fundamental biological differences among viruses that have a profound impact on their evolution. Our goal is therefore to understand *why* these different organizations exist. These six categories are: (i) ssRNA− unsegmented, (ii) ssRNA− segmented, (iii) ssRNA+ unsegmented, single polyprotein, (iv) ssRNA+ unsegmented, subgenomic RNAs, (v) ssRNA+ unsegmented, ribosomal frameshift, and (vi) ssRNA+ segmented (and usually multicomponent) (Fig. 5.3). Because dsRNA viruses are mechanistically similar to ssRNA− viruses in that transcription occurs before translation, and comprise both segmented (the majority) and unsegmented genomes, I will consider them with ssRNA− viruses. Similarly, the replication strategy of retroviruses is so obviously different (i.e. involving cellular integration of a DNA intermediate) it will not be dealt with here.

The simplest argument to explain such a diverse array of genome organizations, although not one that I subscribe to, is that each represents (or at least most represent)

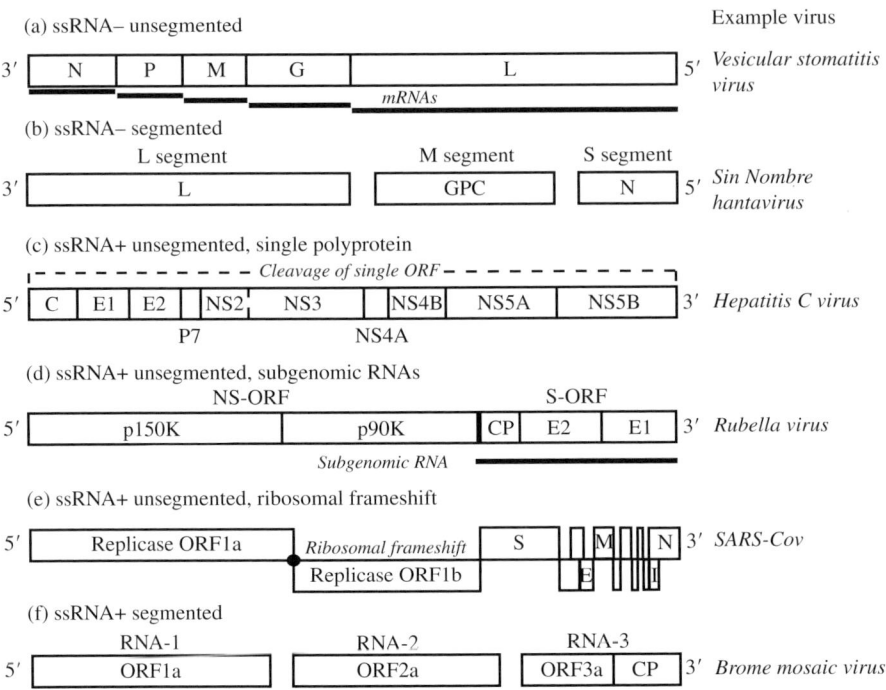

Fig. 5.3 Schematic representation of six major types of genome organization (coding regions only) and replication strategy observed in RNA (RdRp) viruses, with example viruses given in each case. Each of these organizations results in a different way to control gene expression. Gene and segment sizes are drawn approximately to scale within each of the six organizations, but not among them (for example, coronaviruses are much longer than any of the other RNA viruses depicted here and the names of some of their small 3' ORFs have been excluded). Similarly, the different types 5' and 3' terminal sequence/structure have been excluded for the sake of simplicity. It is important to note that these types of organization are not mutually exclusive. For example, polyprotein cleavage occurs in many different types of RNA virus and coronaviruses produce multiple subgenomic RNAs. dsRNA viruses can be considered in the same manner as ssRNA− viruses.

an independent origin as an escaped cellular gene. If true, there may be no clear selective benefit to any particular organization, as each is the end-product of a different evolutionary history. However, as I laid out in Chapter 2, what little evidence there is tentatively supports a pre-cellular theory of RNA virus origins. In addition, it is easy to imagine how, for example, molecules with a negative-sense genome orientation could evolve from ssRNA+ viruses as the former are a natural outcome of replication by the latter (while dsRNA viruses could be derived from dsRNA replication intermediates). Similarly, these different genome organizations could simply represent 'frozen accidents': neutrally evolving traits that have little impact on viral fitness. While this is theoretically possible for those genome organization/replication strategies that are very rare—such as those of the coronaviruses and roniviruses—the take-home message of RNA virus genome organization is that there is little in the way of randomness, which strongly argues against a major role for neutral evolutionary processes. This is an issue discussed in more detail in section 5.2.

5.1.4 The evolution of genome segmentation

Explaining why some RNA viruses have segmented genomes, while others do not, is the most discussed issue in the genome-scale evolution of RNA viruses, although even this has received relatively sparse attention from evolutionary biologists. On one hand it is easy to imagine how segmented viruses could arise when two (or more) viruses co-infect a single cell, particularly if subgenomic RNAs are produced routinely. For example, given their restricted distribution as ssRNA+ viruses of plants, it seems likely that multicomponent viruses were generated when individual segments from different ssRNA+ viruses, and contributing different functions, co-infected a single plant cell and evolved to function together through complementation. The catholic host tastes of many plant RNA viruses (Reanney 1982) as well as their often high MOI suggests that such mixed infections occur on a regular basis, as indeed they must for this form of genome organization to be successful. In this case the evolution of genome segmentation is therefore likely to have occurred concurrently with the development of a replication cycle involving multiple virus particles. However, the necessity for multiple infection must also put an upper limit on the number of particles present in a successful virus.

On the other hand, understanding the evolutionary reasons *why* such a segmented genome might be selectively favoured is an entirely different, and more difficult, question. The most commonly stated idea is that reassortment, an inherent property of viruses with segmented genomes, is a form of sexual reproduction and, as such, is favoured by natural selection in the same why that 'true' sexual reproduction is favoured and maintained in other organisms (Pressing and Reanney 1984; Chao 1988). For example, reassortment has been proposed as a way in which RNA viruses avoid the deleterious consequences of mutation accumulation (Chao 1990, 1994; see also Pressing and Reanney 1984 for a rather different formulation of this idea).

As I discussed the evolution of recombination and reassortment in RNA viruses in detail in section 3.2, I will only recap a few general points here. Specifically, while it is clear that RNA viruses undoubtedly fulfil some of the criteria necessary for natural selection to favour reassortment as a way of avoiding the accumulation of excessive numbers of deleterious mutations, available data suggest that this explanation is unlikely to be correct. Most importantly, the burden of deleterious mutation appears to be high regardless of genome structure or the propensity to recombine/reassort (Pybus *et al.* 2007), and there is no evidence for frequent synergistic epistasis in viral genomes. Similarly, there is no good evidence that reassortment increases the rate at which advantageous genetic configurations are generated.

What then explains the relatively frequent occurrence of genome segmentation in RNA viruses (as well as its rarity in DNA viruses)? One interesting idea is that segmentation in multicomponent viruses is the result of intracellular selection for smaller RNAs, as these have an advantage in either replication or encapsidation; that is, they represent selfish RNAs (Nee 1987). However, as noted by Chao (1988), left to its own this devices this theory predicts that progressively smaller RNAs are favoured, eventually resulting in defective interfering particles, when in fact segment sizes in these viruses are similar to those observed in segmented viruses that utilize a single virus particle. Similarly, this theory cannot readily explain why multicomponent viruses are nearly all restricted to plants.

Another possibility that I have already touched upon is that, by reducing mutational load to some extent, genome segmentation allows RNA viruses to acquire larger genomes than their unsegmented cousins, and larger genomes obviously mean more functional diversity. However, although the average genome size in families of segmented RNA viruses is a little larger than that of unsegmented RNA viruses, the difference is not significant, the overlap is considerable, and the largest RNA viruses are unsegmented (Fig. 5.1). In addition, arguments based on genome size cannot explain why the number of segments does not exceed 12 (although it may be that segmentation allows more efficient viral packaging; Froissart *et al.* 2004). As an interesting aside, the maximum ORF sizes in RNA viruses are rather less variable than both the maximum segment and genome sizes, with an upper limit of approximately 4000 amino acids. Notably, this 4000 amino acid maximum also applies to unsegmented viruses which encode multiple ORFs (or subgenomic RNAs), with the only exception again provided by the coronaviruses and roniviruses.

As mentioned at the start of this section, I propose that the most likely explanation for the evolution of genome segmentation in RNA viruses is that it offers greater control over gene expression. One challenge faced by all RNA viruses is that they need to exert both quantitative and temporal control over the levels of each protein they produce. In the case of ssRNA+ viruses this control over gene expression must usually occur at the level of translation (rather than transcription) as this is the first step in the virus life cycle (Fig. 1.3). An additional problem for ssRNA+ viruses is that the ribosomes of eukaryotes recognize the 5' regions of mRNA molecules, so that internal start codons (i.e. AUG) are not utilized, IRES sequences are often located in 5' UTRs, and mRNAs are usually monocistronic. Therefore, in the case of unsegmented ssRNA+ viruses, it is usually not possible to translate individual proteins downstream of the initial 5' AUG codon. As a consequence, many ssRNA+ viruses are 'forced' to translate a single polyprotein that is then proteolytically cleaved into individual protein products. Although such a genome organization is undoubtedly streamlined, it (in theory) comes at the cost of producing essentially equimolar amounts of each protein, even though producing more copies of specific structural proteins may be beneficial. Any difference in protein abundance must then be achieved through differential cleavage of the original polyprotein.

The most obvious way to overcome such constraints on gene expression is to divide up the viral genome into what can be thought of as separate 'transcriptional units', in which transcription (and translation) can occur at different rates. It is just such a division that I believe explains many of the large-scale patterns of genome organization in RNA viruses. Because the problem of controlling gene expression is most severe for the unsegmented ssRNA+ viruses, it should come as no surprise that they employ at least three strategies to produce distinct transcriptional units: (i) the division of the viral genome into multiple segments, (ii) the use of subgenomic RNAs to transcribe downstream ORFs, which is a common feature of alpha-like and carmo-like viruses, and (iii) the use of a −1 ribosomal frameshift to produce multiple ORFs in the case of the coronaviruses and roniviruses (Fig. 5.3). It is also important to remember that these strategies are not mutually exclusive. For example, coronaviruses employ the ribosomal frameshift and encode multiple subgenomic RNAs.

The division of the viral genome into multiple segments naturally results in independent transcriptional units, in turn enhancing control over gene expression. It is this feature that I believe explains why genome segmentation is so commonly observed in RNA viruses as a whole. Indeed, although all higher-order phylogenetic trees of RNA viruses are riddled with uncertainty (see Chapter 2), it is likely that segmentation has evolved multiple times in ssRNA+ viruses as viruses with this form of genome organization do not form a single monophyletic group (Goldbach and de Haan 1994) (Fig. 2.6). This in turn suggests that there is a continual selection pressure for segmentation in these viruses. Finally, as bacteria are able to produce polycistronic RNAs, there would be less requirement for additional transcriptional units in bacteriophage. This may explain why all ssRNA+ bacteriophage are members of a single family, the *Leviviridae*. Similarly, it is striking that only a single family of bacteriophage possess segmented genomes—the dsRNA *Cystoviridae*—and it cannot be excluded that these were originally derived from eukaryotic viruses.

5.1.5 The evolution of genome orientation and dsRNA viruses

Other avenues for controlling gene expression are open to those viruses with ssRNA− genomes. In particular, because the first step in the life cycle of these viruses is transcription rather than translation, ssRNA− viruses are also able to control gene expression at the level of transcription. For example, in VSV transcription results in five different mRNAs, each of which can be thought of as a natural transcriptional unit (Fig. 5.3). Given such an inherent ability to control transcription, it might also be expected that ssRNA− viruses are subject to less selection pressure to evolve segmentation than ssRNA+ viruses. In support of this idea is tentative phylogenetic evidence that segmentation has only evolved once in the ssRNA− viruses, in contrast to its frequent generation in ssRNA+ viruses (Vieth *et al.* 2004). In addition, that the same template is used for transcription and translation in ssRNA+ viruses requires that both processes be perfectly timed. The evolution of negative-sense viral genomes constitutes a viable solution to this problem, as distinct forms of RNA template are used for transcription and translation in ssRNA− viruses. As a consequence, both processes can proceed concurrently without interfering with each other. It therefore seems likely that the ability to better control RNA transcription is the most likely explanation for why some RNA viruses evolved negative-sense genome orientations in the first instance.

However, ssRNA− viruses could still subject to an important constraint on gene expression that may have favoured the evolution of segmentation in this group of viruses: that the rate of primary transcription is heavily dependent on genomic position, so that the first (i.e. 3′) mRNA is produced more frequently than the last (5′) mRNA (Fig. 5.3). In the case of the *Mononegavirales* this means that more of the N protein (nucleocapsid) is produced than the L protein (RNA polymerase), presumably because the replicatory function of the RdRp means that fewer copies are required. Indeed, this 'transcriptional gradient' is the likely reason why the genes of the *Mononegavirales* are ordered as they are, reflecting the different amounts of protein product required. In support of this idea are experimental studies of VSV which show that changing gene order reduces fitness (Novella *et al.* 2004a).

The ability to undertake transcription before translation—and so better control gene expression—may also explain the existence of dsRNA viruses, although the difficulties in unwinding long stretches of dsRNA are likely have exerted an additional selection pressure for multiple, and hence shorter, segments in this group. Finally, enhanced gene expression may also offer a partial explanation for the highly unusual stop codon lineage of DENV-1, and which is most likely maintained by frequent complementation (Aaskov *et al.* 2006). In this case the proteins upstream of a stop codon in the E gene—the capsid (C) and membrane (M)—are intact and appear functional, whereas the downstream (nonstructural) proteins contain numerous deleterious mutations. This may therefore represent a case of incipient segmentation: the C and M proteins are on their way to become a different segment (or subgenomic RNA) of DENV, presumably because more of these protein products are required than those of nonstructural proteins. Indeed, the idea that complementation may be a critical step in the evolution of genome segmentation has also been derived from *in vitro* studies of RNA virus evolution (García-Arriaza *et al.* 2004).

5.1.6 The evolution of overlapping reading frames

One interesting facet of genome organization commonly used by RNA viruses is that of overlapping ORFs, sometimes called overprinting. At its most basic, this is surely an evolutionary strategy to increase the amount of protein diversity encoded by a single nucleotide sequence in a world of small genomes, again allowing more control over gene expression (although others have argued that it constitutes a more general strategy for the generation of evolutionary novelty; Keese and Gibbs 1992).

The use of overlapping ORFs is not quite a defining feature of RNA viruses. Belshaw *et al.* (2007) recorded 819 cases of gene overlap among 701 reference RNA virus genomes, with 56% of viruses showing some degree of overlap. Of these, nearly all (≈99%) involved a +1 (forward) or −1 (backward) frameshift. As a simple example, the OP (overlapping/movement protein) gene of tymoviruses (ssRNA+), such as turnip yellow mosaic virus, is entirely encoded within the same sequence utilized by the RP (replicase) gene, and covering approximately one-third of the latter's sequence. There are also a variety of mechanisms that can lead to gene overlap, including ribosomal frameshifting, the use of non-AUG start codons, and RNA splicing (reviewed in Belshaw *et al.* 2007).

Aside from being a simple way to increase phenotypic diversity in limited genomic space, there are a number of other interesting evolutionary aspects to the use of overlapping ORFs. First, that synonymous mutations in one frame are likely to be nonsynonymous in another complicates some aspects of evolutionary analysis (Hein and Støvlbæk 1995), and can lead to the false-positive inference of positive selection using d_N/d_S (Holmes *et al.* 2006). Second, the use of overlapping ORFs can be thought of as an extreme form of pleiotropy, as every nucleotide site located within the overlapping region is expected to have a major impact on fitness. This, in turn, will be costly for the evolutionary flexibility of individual nucleotide sites. It is therefore no surprise that in viruses such as HBV, where overlapping ORFs are abundant (in this case representing approximately 50% of the viral genome; Fig. 5.4), lower rates of evolutionary change are observed in overlapping compared to non-overlapping regions (Zhou and Holmes 2007). Third, that HBV, as well as ssDNA viruses, show extensive gene overlap yet have rather lower per-nucleotide mutation rates than RdRp-replicating viruses suggests that gene overlap is not simply a function of possessing high deleterious rates (because it reduces the amount

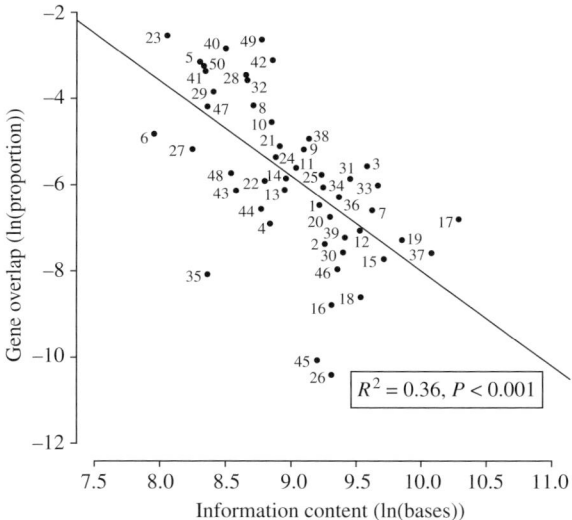

Fig. 5.4 The relationship between the natural logarithm of gene overlap (as a proportion of information content) and 'information content' in RNA viruses. Information content is defined as genome length plus overlap length. Each numbered point refers to a different virus, the details of which are given in the original publication. The extremes of the extents of gene overlap are represented by the retro-transcribing hepadnaviruses (such as HBV, point 23) and the *Hypoviridae* (dsRNA viruses of fungi, point 26). Taken form Belshaw *et al.* (2007) with permission.

of genomic material need to encode all necessary functions; Belshaw *et al.* 2007). Therefore, it is perhaps more likely that gene overlap is favoured simply as a way to increase functional diversity within the limits of already highly constrained genome sizes, but with the trade-off that rates of deleterious mutation are elevated. This may also explain the fascinating observation that RNA viruses with longer genomes tend to show less gene overlap than shorter RNA viruses (Belshaw *et al.* 2007) (Fig. 5.4), presumably because there is less urgency to create protein diversity in the former group.

5.2 The processes of genome evolution

As well as describing the different genomic architectures exhibited by RNA viruses, it is important to discuss, in general terms, the evolutionary processes by which they were generated. In its simplest form, this can also be set within the debate over the respective roles of natural selection versus genetic drift in molecular evolution. In particular, Mike Lynch (Indiana University) has argued in a series of elegant and provocative papers that large-scale features of genome organization in eukaryotes and prokaryotes, such as the numbers of duplicated genes, mobile elements, and introns, are the result of random (or, in his terminology, 'nonadaptive') processes, reflecting the inability of natural selection to shape patterns of genetic diversity when $N_e s$ is low (Lynch and Conery 2000, 2003; Lynch 2007). Irrespective of whether these ideas are correct, they show that the 'great obsession of population genetics' applies equally to large-scale elements of genomic architecture as individual point mutations.

Although it may be dangerous to draw strong conclusions, the data presented so far provide little evidence that genetic drift has played a major role in shaping the genomic architecture of RNA viruses. Most fundamentally, RNA viruses, even those with the largest genomes, contain little in the way of non-functional inter-genic DNA, pseudogenes, or introns. Rather, the evolutionary strategy employed by RNA viruses is to utilize their inherently constrained genomes with as much efficiency as possible. For example, gene orders often seem to reflect function, such as the general tendency to group structural genes into one region of the viral genome and non-structural genes into another (Fig. 5.3), and which I think reflects natural selection for the control of gene expression. Indeed, given that many ssRNA+ viruses encode their genes in a single polyprotein it is difficult to imagine how genetic drift could greatly influence genome evolution in these circumstances.

Moving on from the debate over selection versus drift, there are four modes of evolutionary change that may explain the range of genome architectures seen in RNA viruses: (i) that they have been produced as the long-term result of simple divergent evolution, stemming from the accumulation of point mutations from a common ancestral genome that already possessed the 'core' genes required by RNA viruses (such as the RdRp and the capsid); (ii) that they have been produced by overprinting, resulting in the frequent use of overlapping ORFs; (iii) that they were produced by a series of gene (and even genome) duplications, much in the way that eukaryotes have generated evolutionary novelty; or (iv) that they have experienced successive LGT events, involving genetic material from other RNA viruses or host genomes, and similar to the manner in which bacteria often generate genomic diversity (Ochman et al. 2000). Given that 'simple divergent evolution' is the default model of viral evolution, and overprinting is discussed in section 5.1, I will devote my attention here to the respective roles played by gene duplication and LGT.

5.2.1 Gene duplication in RNA virus evolution

Although the divergent nature of viral sequences makes the analysis of the processes of genomic evolution extremely difficult, particularly when considering viruses assigned to different families, gene duplication appears to be a relatively uncommon occurrence in RNA viruses. In fact, one of the most tangible differences between the genomes of RNA viruses and those of other organisms is a lack of multigene families, the most noticeable outcome of gene duplication. Although gene duplication must be responsible for some of the genome-size variation observed in RNA viruses, particularly during the early diversification of viral genes (as the original viral genomes were surely smaller than they are now), there are sound reasons why we should not expect frequent gene duplication in these organisms. In particular, given the ceiling on genome sizes imposed by high mutation rates, processes that increase evolutionary novelty through the expansion of genome size should be disfavoured compared to strategies that create novelty from existing genetic material. In general terms, this prediction seems to be true, as demonstrated by the relatively high frequency by which gene duplication is observed in large dsDNA viruses (Shackelton and Holmes 2004; Hughes and Friedman 2005) compared to its rarity in RNA viruses and ssDNA viruses (see below). In addition, those cases of gene duplication documented in RNA viruses all seem to involve the action of some form of either homologous or nonhomologous recombination. That these processes may occur only sporadically, if at all, in some RNA viruses further argues that gene duplication cannot be frequent.

Those gene duplication events documented to date in RNA viruses seem to fall into a small number of different types. The first category might be considered as short duplications that occur within the untranslated regions that flank viral genomes and which have been documented on a number of occasions (for example, Panavas et al. 2003; Gritsun and Gould 2006). A second general class of gene duplications are those that involve short intra-genic regions (Zlateva et al. 2007), and which often result in defective viruses (Nagai et al. 2003; Cao et al. 2008). A third, and final, class are those cases where gene duplication events have resulted in two complete ORFs within a viral genome (original and copy) and that are sometimes tandemly repeated. In some cases the sequences of the ORFs in question are highly divergent, because the gene duplication events occurred so long ago (Liljas et al. 2002; Imbert et al. 2006). For example, that the VP1, VP2, and VP3 proteins of picorna-like viruses share the same capsid structure suggests strongly that they are related (Rossmann et al. 1985), and may have arisen through gene duplication. Indeed, it is this form of gene duplication that may have characterized the early evolution of RNA viruses. Although this process is extremely common in eukaryotes, it has been documented relatively infrequently in RNA viruses (Forss and Schaller 1982; Tristem *et al.* 1990; Boyko *et al.* 1992; Walker *et al.* 1992; Wang and Walker 1993; Karasev *et al.* 1995; LaPierre *et al.* 1999; Baroth *et al.* 2000; van Hulten *et al.* 2000; Peng *et al.* 2001; Valli *et al.* 2007).

5.2.2 LGT among viruses and hosts

As with gene duplication, the process of LGT has only been reported sporadically in RNA viruses, so much so that it cannot be regarded as a major evolutionary mechanism. Similarly, the occurrence of LGT is also dependent on recombination (or reassortment) which, as discussed in section 3.2, is a process that is largely dictated by genome structure. Perhaps the clearest case of LGT in RNA virus evolution concerns the haemagglutinin esterase (HE) gene of coronaviruses that was acquired from influenza C virus, and perhaps on multiple occasions (Luytjes *et al.* 1988). However, given that coronaviruses are extremely unusual in their capacity to increase genome size (see section 5.1), this is unlikely to serve as a general example. On a deeper timescale, the similarities among the protein domains of otherwise divergent viruses might also be argued as evidence for LGT (see section 5.2.3, on modular evolution), although the difficulties in resolving phylogenetic relationships at this level make it difficult to test the validity of this idea.

There have also been relatively few reports of LGT between RNA viruses and the genomes of cellular organisms. In these cases, viruses may act as either the donor or recipient. One of the most remarkable observations in viral evolution of recent years was that copies of virus genomes closely related to the flaviviruses Cell Fusing Agent virus and Kamiti River virus (ssRNA+) were found to be inserted into the (dsDNA) genomes of their *Aedes* mosquito vectors (Crochu *et al.* 2004). Not only does this represent a striking example of LGT, but it also requires a RT reaction, either involving a cell-derived version of RT, or through a co-infecting retrovirus. Similarly, a high level of sequence similarity has been observed between the *env* genes of Cer retroelements of *Caenorhabditis elegans* and phleboviruses (*Bunyaviridae*), indicative of LGT from phleboviruses (Malik *et al.* 2000), while the integration of potato virus Y into several varieties of grapevine has also been proposed (Tanne and Sela 2005). As a final interesting example, when reticuloendotheliosis virus (REV) and Marek's disease virus (MDV) co-infect an individual avian host, the former retrovirus can integrate into the genome of the latter dsDNA virus (Isfort *et al.* 1992).

In the other direction, RNA viruses have to shown to transiently incorporate host genome sequences. The temporary integration of cellular sequences, such as ubiquitin into the genomes of bovine viral diarrhoea virus (*Flaviviridae*), is particularly well documented (Meyers *et al.* 1989). The recombination between RNA viruses and host genome sequences has also been recorded in cell culture on a number of occasions (for example, Khatchikian *et al.* 1989; Charini *et al.* 1994), although not in natural populations of these viruses. More interesting is the observation of clear sequence similarities between the 65 kDa protein of closteroviruses (ssRNA+) and the cellular heat-shock protein Hsp70 (Agranovsky *et al.* 1991). Similarly, the nucleoside triphosphate-binding motif (NTBM)-containing proteins of flaviviruses and potyviruses appear to share amino acid motifs with the NTBM proteins of prokaryotes and eukaryotes (Laín *et al.* 1989). As a final example, some isolates of Scottish potato leafroll virus (*Potyviridae*) possess sequence regions sharing strong similarity with a small exon of tobacco chloroplast DNA (Mayo and Jolly 1991). However, the reality is that there are few reports of the stable integration of cellular sequences into the genomes of RNA viruses, with the ExoN domain of coronaviruses one of the best documented (see section 5.1). This again suggests that there are major fitness costs associated with possessing overly long genomes.

5.2.3 Modular evolution

One theory of viral genome evolution that directly relates to the process of LGT, and which is of great historical importance, is that of 'modular evolution' (Botstein 1980) (Fig. 5.5). Although originally developed in the context of DNA bacteriophages, this theory can easily be extended to RNA viruses as a whole (an idea also suggested by

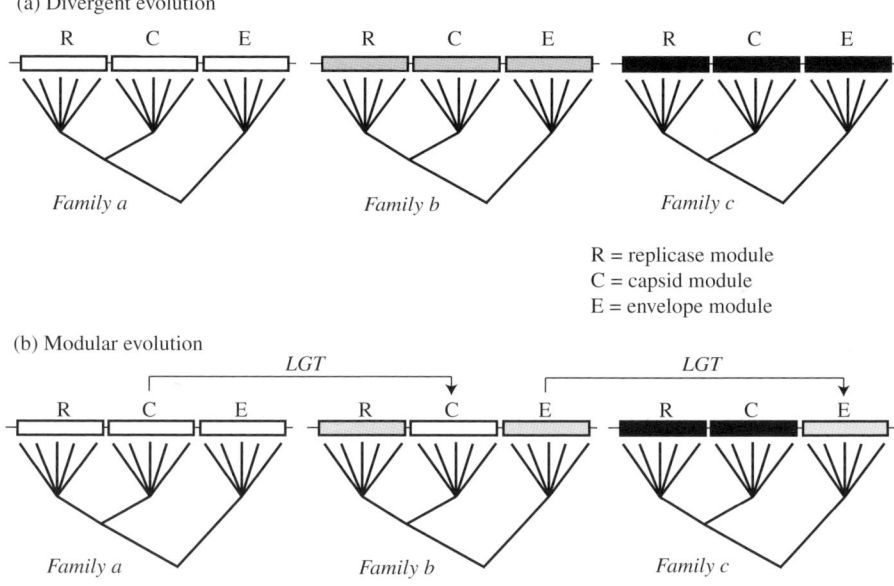

Fig. 5.5 Schematic representation of the theory of modular evolution for RNA viruses. Three hypothetical viral families—a, b, and c—each possess genomes comprising replicase (R), capsid (C), and envelope (E) modules. Under 'standard' divergent evolution (panel a), each module in each family evolves independently. In contrast, under modular evolution (panel b), there is LGT such that family b acquires its capsid module from family a, while family c acquires the envelope module from family b.

the original author). The basic concept underlying modular evolution is that viral genomes can be thought of as comprising a series of functional modules (capsid, envelope, polymerase, etc.) that can be exchanged through recombination/reassortment, thereby creating new types of virus. Suggested evidence for this process is that some plant viruses seem to be composed of proteins (for example, the polymerase and movement proteins) that have very different evolutionary origins, including those that may be closely related to the related proteins of animal RNA viruses (Gibbs 1987). To take a specific example, there is a proposed sequence similarity between the polymerase read-through domains of multicomponent plant viruses of the family *Bromoviridae*, unsegmented plant viruses of the genus *Tobamovirus*, and the unsegmented alphaviruses of animals, such as sindbis virus (Haseloff *et al*. 1984). Although these similarities are compelling, subsequent analyses indicated that these viruses can all be classified within the 'alpha-like' supergroup (Goldbach and de Haan 1994) (Fig. 2.6), in which case this evolutionary pattern may reflect common ancestry rather than cross-family LGT. Similarly, it has been proposed that the genomes of potyviruses are compromised of genes obtained from cauliviruses, flaviviruses, picornaviruses, and tobamoviruses (Goldbach and de Haan 1994), although this remains to be formally tested. More recently, sequence and structural similarities between the IRES domains of unrelated flaviviruses, pestiviruses, and picornaviruses have been documented, again suggestive of cross-virus recombination events involving a key functional sequence (Hellen and de Breyne 2007).

Although an intriguing idea, there is currently little to support a major role for modular evolution in RNA viruses. Other than the examples described above, some of which are debatable, there are relatively few unequivocal demonstrations of either recombination/reassortment among divergent RNA viruses, or of functional modules that possess a phylogenetic pattern that does not match that of the virus genome from which they were derived. Further, many of the recombination events described in the genomes of RNA viruses do not involve intact genes (see Ohshima *et al*. 2007 for an instructive example), let alone functional modules. On the other hand, evidence for recombination reflecting some aspects of genome modularity has been observed in small DNA viruses (Martin *et al*. 2005; Lefeuvre *et al*. 2007; Zhou and Holmes 2007), indicating that this phenomenon should be investigated further. As discussed at length in Chapter 2, a major hindrance to resolving ancient evolutionary events in RNA viruses, implicit in the theory of modular evolution, is that distant relationships are extremely hard to determine among highly divergent amino acid sequences.

5.3 Patterns and processes of macroevolution in RNA viruses

The vast majority of studies in RNA virus evolution undertaken to date, including those covered in this book, may be regarded as exploring the realm of microevolution, focusing on either the processes by which genetic variation is generated in RNA viruses, or on short-term phylogenetic patterns, manifest as molecular epidemiology. Far less attention has been devoted to exploring inter-specific evolution in RNA viruses, at least in part because of the inherent difficulties associated with analysing highly divergent gene sequences. However, if we are to fully understand RNA virus evolution it is essential that we consider the patterns and processes of change at all scales. That such studies may be informative is hinted at by cases such as the genus *Flavivirus* (ssRNA+), an important group of animal viruses and where there is a strong association between inter-species phylogeny and major aspects of viral phenotype, including associated disease syndrome, mammalian host, and vector species (Gaunt *et al*. 2001) (Fig. 5.6). Similarly, it is interesting to ask why some families of RNA virus, such as the *Potyviridae*, are particularly speciose, while others like the *Roniviridae* seemingly contain just a few taxa. For want of a better term, I will refer to this as the 'macroevolution' of RNA viruses.

5.3.1 Speciation in RNA viruses

Although various definitions of virus species exist (Gibbs 1987; Gibbs and Gibbs 2006a), placing divisions between these infectious agents is often an arbitrary exercise. While it may appear odd to consider the question of 'speciation' in organisms where it is unclear exactly what a species is, in reality my aim is simply to explore the processes that shape phylogenetic diversity in RNA viruses, particularly in trees that attempt to connect different viruses within a single family, or different viral families. To put it another way, why do the family (or higher) level phylogenetic trees of RNA viruses look like they do?

There are a variety of evolutionary processes that explain patterns of lineage differentiation in RNA viruses. In this context it is essential to think of viruses as organisms that infect different cell types, as well as different hosts. This realization allows an analysis of whether the speciation process in RNA viruses is dominated by jumping to novel hosts—as is manifest in the process of emergence (see Chapter 6)—or adapting to new cell types *within* an individual host species. To put it another way, despite its in vogue status, cross-species transmission and emergence is only one of the macroevolutionary processes exhibited by RNA viruses. Stretching the point a little further, it is also possible to fit, perhaps a little forcibly, the speciation patterns observed in RNA viruses into the classic division between allopatric and sympatric speciation commonly used in evolutionary biology. Hence, 'allopatric' speciation occurs when viruses jump to new host species, while 'sympatric' speciation reflects the adaptation to different cell types, or other aspects of the viral life cycle, within individual hosts (Fig. 5.7). Given this division, the key question then becomes whether allopatric or sympatric speciation is the dominant mode of speciation in RNA viruses.

Under a simple allopatric model viral lineages could separate, eventually forming distinct species, following a jump in host species that renders them ecologically distinct. In accordance with standard speciation theory this can also be regarded as an essentially neutral evolutionary process as there is no selective requirement to fix

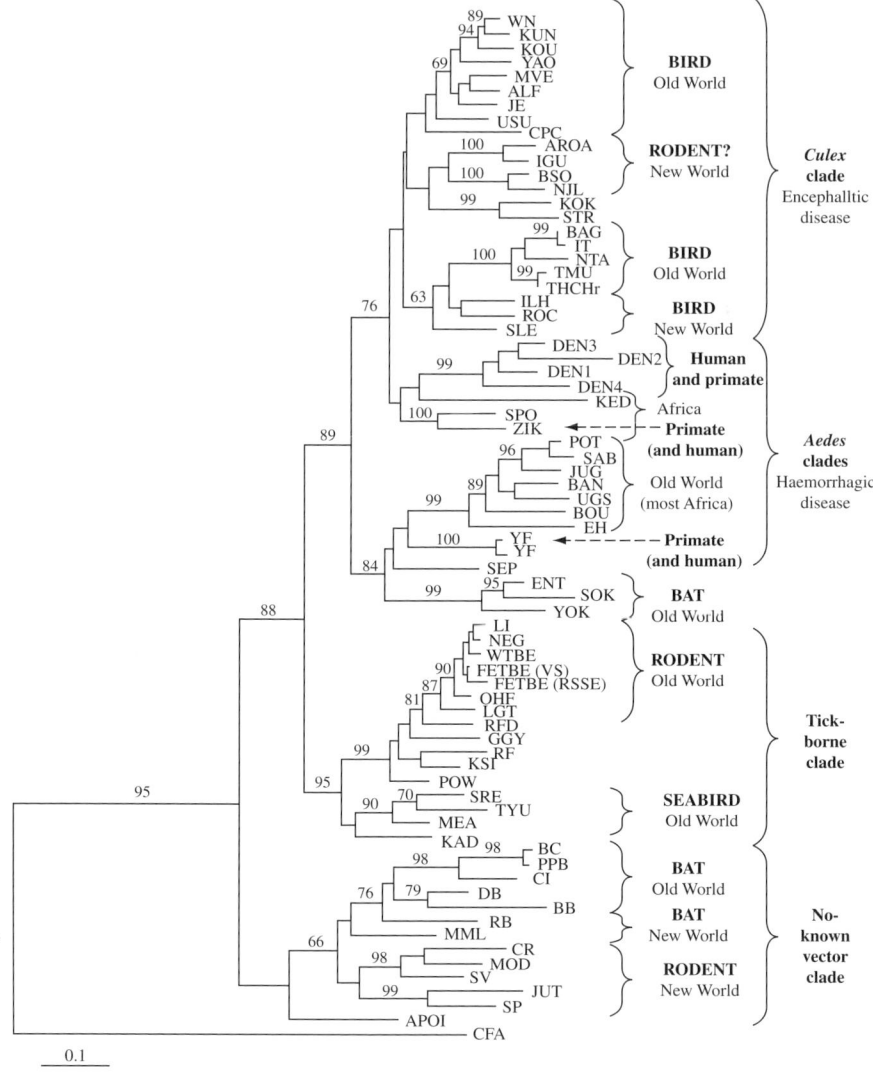

Fig. 5.6 Inter-species phylogenetic tree of the genus *Flavivirus* based on the NS5 gene (which includes the RdRp). A variety of phenotypic characters are added to each cluster: host species, vector (where applicable), geographic location, and disease association. The major division by vector species is clear. Virus abbreviations are defined in the original publication (fig. 1, p. 1871). Taken from Gaunt *et al.* (2001) with permission.

mutations in either lineage that render them 'reproductively isolated'. Hence, although natural selection may have fixed mutations that allow viruses to adapt to their own host species (an issue discussed in detail in section 6.3), these mutations were not selected to 'make' these lineages genetically different. As such, viral speciation—the

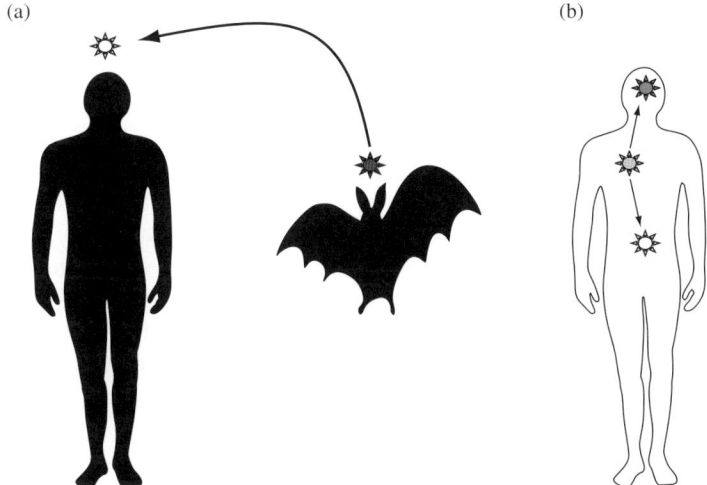

Fig. 5.7 Models of 'speciation' in RNA viruses. (a) 'Allopatric' speciation in which viruses jump species boundaries and establish themselves in new hosts (i.e. cross-species transmission). (b) 'Sympatric' speciation in which viruses adapt to different cell types within an individual host, perhaps by utilizing different cellular receptors. That allopatric may occur more frequently than sympatric speciation is signified by the thicker arrow.

generation of independently replicating viral lineages—is simply a *consequence* of genetic divergence.

Alternatively, 'sympatric' models of viral speciation require the action of natural selection to reinforce their separation. In this case, the diverging viral lineages would remain pathogens of the same host but adapt to different ecological niches, or lifestyles, within that host. Such separation would be favoured if it reduced the extent of inter-pathogen competition (i.e. there is character displacement). For example, viruses undergoing sympatric speciation may infect different cells, tissues, or organs, infect different host age groups, peak in different seasons, or evolve unique routes of transmission. As the intra-host selective regime will be intensive, particularly given cross-protective immune responses, the sympatric speciation of viruses within individual hosts is likely to involve the acquisition of a limited number of critical mutations that immediate cause 'speciation'. Under this model speciation is then the *cause* of lineage differentiation.

As constraints and trade-offs appear to be commonplace in viral evolution, it will often be easier for a virus to jump species boundaries and replicate in a familiar cell type than to change its lifestyle within the same host. Hence, cross-species transmission and emergence should not be considered unusual traits. This is satisfyingly counter-intuitive. For closely related hosts where cellular and immunological landscapes are likely to be similar, such as different species of primate, allopatric speciation may therefore simply require sufficiently frequent contact to allow viruses to jump among them, with few adaptive mutations (Brault *et al.* 2002; see section 6.3 for an extensive discussion of this subject). If this process is able to occur frequently, it may explain why some viral families, such as the *Picornaviridae*, are more speciose than others. On the other hand, changing cellular or tissue tropism, as may occur in sympatric speciation, is likely to require fundamental changes in viral biology. For example, the structure of the capsid and/or envelope protein may need to change to complement a different array of cell receptors, while a change in transmission route may require additional changes in the virion's response to differences in optimal temperature and humidity. Although there are currently few data by which to test these theories, a previous analysis of the paramyxoviruses revealed that tissue tropism changed less frequently than host species (Taber and Pease 1990), exactly as expected if sympatric is more difficult than allopatric speciation. Similarly, the case of the flaviviruses discussed at the outset of this section highlights the relative evolutionary stability of some phenotypic traits.

5.3.2 A birth-death model of viral evolution

The macroevolutionary process in RNA viruses is reflected in the branching structure of phylogenetic trees that attempt to link viruses assigned to different families. These phylogenies possess a striking topological pattern: (relatively) short clusters of intra-family branches that are linked by very long inter-family branches of perhaps indeterminate length (see Fig. 5.8 for a schematic representation and Zanotto *et al.* 1996a for a real data presentation). To describe this pattern in another way, there is a marked lack of viral lineages that fall 'mid-way' between the tips and the root of the tree. The aim of this section is to explain, if only generally, what evolutionary processes might have given rise to this phylogenetic pattern.

It is clear that extensive multiple substitution, which will lead to a huge underestimation of phylogenetic distances at the inter-family level, is in part responsible for such a distinctive tree structure. However, this saturation effect will largely act to artificially *reduce* the length of the already long inter-family branches. In addition, I will shortly show that the loss of 'intermediate' lineages also occurs in far more closely related sequences, so that multiple substitution is only part of the story. Similarly, the huge under-sampling that plagues studies of viral evolution should affect all parts of the tree, and not just those lineages that fall between distinct families.

I therefore propose that the macroevolutionary phylogenetic pattern observed in RNA viruses—short family-level clusters joined by very long inter-family branches—depicts a true evolutionary phenomenon, albeit one that is massaged by frequent multiple substitution. I also propose that this phylogenetic pattern reflects the continual birth and death of viral lineages, and how we sample both extant and extinct members of these lineages. Specifically, under a simple birth-death model viral lineages are born under the processes of allopatric or sympatric speciation described above, and die when they fail to infect a sufficient number of hosts such that the basic reproductive number, R, is less than unity. The macroevolution of RNA viruses is therefore one marked by a continual lineage turnover. Critically,

5.3 Patterns and processes of macroevolution • 125

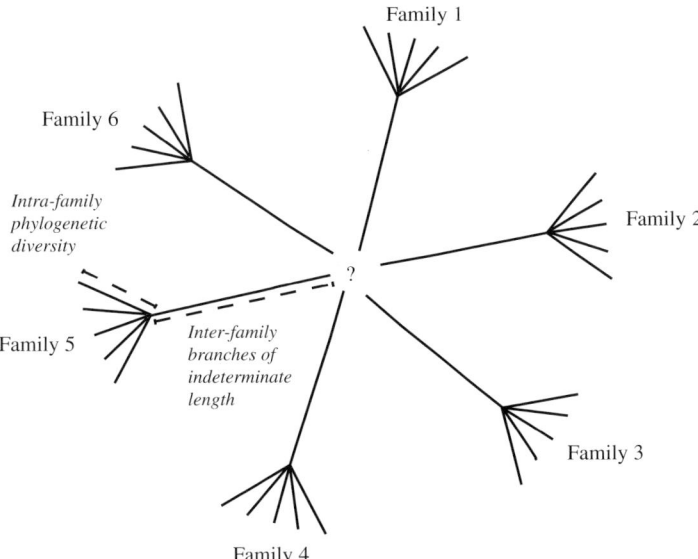

Fig. 5.8 Schematic representation of the distinctive 'macroevolutionary' phylogenetic trees of RNA viruses, characterized by relatively short intra-family branches and very long inter-family branches, whose true length is extremely hard to assess because of frequent multiple substitution. Extreme levels of inter-family divergence also make it impossible to infer the origins of viruses based on sequence data alone, hence the ? symbol at the root. The branch lengths shown here are for stylistic purposes only and should not be regarded as drawn to any scale.

under a constant rate of lineage birth and death, and sampling only extant lineages, we would expect trees to be relatively 'tip-heavy'—resembling those of RNA viruses inferred at the inter-family level—as lineages that arose in distant past are more likely to have gone extinct than those that arose more recently (Nee et al. 1994b) (Fig. 5.9). In contrast, if we were able to include extinct lineages in such a phylogeny then a more even distribution of branches across the tree would be observed. In short, rather than representing an anomaly, the distinctive inter-family phylogenetic trees of RNA viruses can be thought of as the expected outcome of a simple birth-death evolutionary process, particularly if there is little recombination among lineages. Such models also predict that we are massively under-sampling viral biodiversity (Pybus et al. 2002).

Evidence for the applicability of such a model comes from those RNA viruses where sequences have been sampled over sufficiently long time periods to witness lineage birth and death in action. Take DENV, where the birth-death process is well documented (see section 7.3). In this case viruses have been isolated over a roughly 60-year period, spanning the 1940s to the present day, with DENV-2 (sampled between 1944 and 2004) a typical example. A phylogenetic tree of these DENV-2 sequences has a number of 'intermediate' lineages from root to tip, usually those

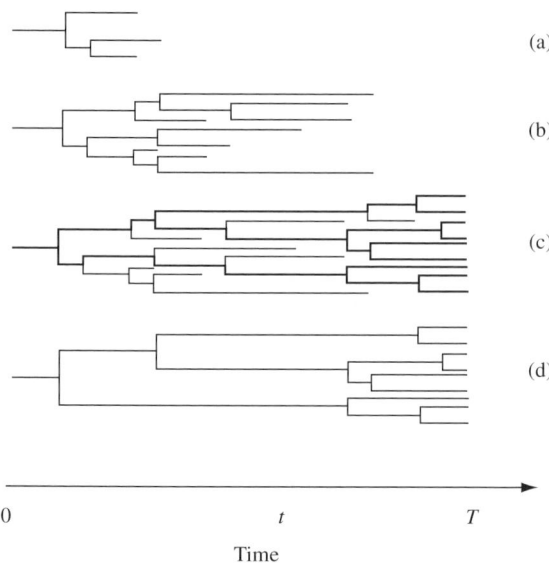

Fig. 5.9 A model of RNA virus macroevolution based on lineage birth and death. (a) A process of lineage birth, but where all lineages die before time t. (b) Some lineages survive to time t, but not to time T (the present). (c) A continual birth-death of lineages, and with some surviving to the present (shown in bold). (d) A redrawing of (c) depicting only those (bold) lineages that have survived to the present (i.e. excluding extinct lineages). Notice the general similarity between panels (c) and (d) here and the real data example of DENV-1 shown in Fig. 5.10. Taken from Nee *et al.* (1995) with permission.

that were sampled some decades ago, so that they can be considered extinct (grey lineages, Fig. 5.10a). However, although such temporally expansive trees are drawn routinely in studies of viral evolution, they paint a misleading picture of evolutionary dynamics as many of the lineages included are extinct (hence matching tree c in Fig. 5.9). As 60 years of DENV evolution is equivalent to roughly 60 million years of mammalian evolution, this is rather like reconstructing a tree of mammals to estimate current biodiversity but including a selection of fossil forms. Indeed, if the same DENV-2 tree is drawn only including those (extant) sequences sampled from the year 1984 to the present day, intermediate lineages are preferentially lost because they have suffered extinction (Fig. 5.10b; and see the resemblance to tree d in Fig. 5.9). I propose that this same effect is played out across RNA viruses as a whole, but that it is rarely observed because the norm is for extinct viral lineages to be included in phylogenies as if they were extant.

A hugely important side effect of this birth-death model is that estimates of viral divergence time based on contemporary lineages may have little bearing on the true

Fig. 5.10 The birth-and-death evolution of DENV-2. (a) Maximum likelihood phylogenetic tree of the E gene of 341 isolates of DENV-1 sampled between 1944 and 2004. Note that viruses sampled prior to 1984—shaded grey—tend to fall deep in the phylogenetic tree and most likely represent extinct lineages. (b) Equivalent phylogeny of the E gene of 294 isolates of DENV-2 sampled between 1984 and 2004. Note the loss of 'intermediate' (extinct) lineages. All branch lengths are drawn to a scale of nucleotide substitutions per site, and the tree is mid-point rooted for purposes of clarity only.

age of the virus in question, but instead correspond to the date of the last lineage turnover event. To make a similar point in a different way, molecular clocks only provide information on the age of the virus in the *sample* analysed (i.e. when the lineages sampled last shared a common ancestor), and not necessarily on the absolute age of that virus. For example, that molecular clocks suggest that the available lineages of measles virus share an ancestor within the last 200 years does not mean that this virus is only 200 years old, but rather that we have not sampled older, extinct lineages (Pomeroy *et al.* 2008). In fact, it is likely that measles began spreading endemically at the time of the first cities (see section 6.4). Similarly, the correct interpretation of molecular clock studies that place age of the genus *Flavivirus* in the last 10 000 years or so (Zanotto *et al.* 1996b) is simply that these *particular* species of flavivirus arose within this time period. The ultimate ancestry of the flaviviruses is likely to be far older—and perhaps measured on timescales of millions of years—but a process of continual lineage birth and death has resulted in a much shallower phylogenetic distribution of extant lineages. In section 7.2 I will suggest that this same birth-death model can be used to explain the anomalously shallow common ancestry of the primate lentiviruses.

5.3.3 The birth and death of endogenous retroviruses

Although I believe that the simple lineage birth-death model outlined above offers a meaningful explanation for the macroevolutionary patterns observed in exogenous RNA viruses, perhaps the best evidence for its occurrence are the endogenous retroviruses (ERVs): the (usually) defective copies of retrovirus genomes that have integrated into the germline of eukaryotes and since been inherited in a vertical manner. The importance of ERVs is that they effectively constitute a 'fossil record' of past viral infections that cannot be inferred from an analysis of contemporary phylogenetic trees alone. Specifically, once ERVs become integrated into host genomes they cease to evolve like retroviruses and instead assume the evolutionary dynamics of their hosts: they no longer replicate using the highly error-prone RT, but instead employ the far higher-fidelity host DNA polymerases. Hence, ERVs can be thought of experiencing something of a 'phase transition' in their evolutionary dynamics. It is this property that allows inference of their age: if the mutational differences between ERVs are known to occur post-integration, such as those observed between the LTRs of a single ERV, or between duplicated ERVs, then divergence times can be estimated using host substitution rates. As an important case in point, while phylogenetic analysis suggests that the current diversity of primate lentiviruses was generated in the past few thousand years at most (see section 7.2), the discovery and dating of endogenous lentiviruses using host substitution rates places the ancestry of this family as a whole back several million years (Katzourakis *et al.* 2007; Gifford *et al.* 2008). This, in turn, suggests that we have only sampled the modern representatives of a continual birth-and-death process. Although I propose that similar processes apply to RNA viruses as a whole, ERVs provide the smoking gun. In one case—that of koala retrovirus (KoRV)—it has even been possible to observe the process of endogenization in action. KoRV most likely appeared recently in koalas following cross-species transmission in a zoo setting from gibbons that carried a closely related retrovirus (gibbon ape leukaemia virus; GALV) (Hanger *et al.* 2000). Yet KoRV is also an endogenous retrovirus, transmitting vertically in the koala population, although it is still to reach some isolated koala populations in southern Australia (Tarlinton *et al.* 2006) (Fig. 5.11).

ERVs are also remarkable in their abundance. For example, approximately 5–8% of the human genome is composed of human endogenous retroviruses (HERVs), comprising at least 31 distinct families present in around 100 000 copies (Katzourakis and Tristem 2005). Although some HERV families contain intact genome sequences suggesting that they are still active, and most notably members of the HERV-K superfamily (in particular HERV-K(HML2)), these are relatively rare. It has also been proposed that ERVs are able to block the replication of related exogenous retroviruses (Arnaud *et al.* 2007), although this is an issue that clearly needs to be explored further.

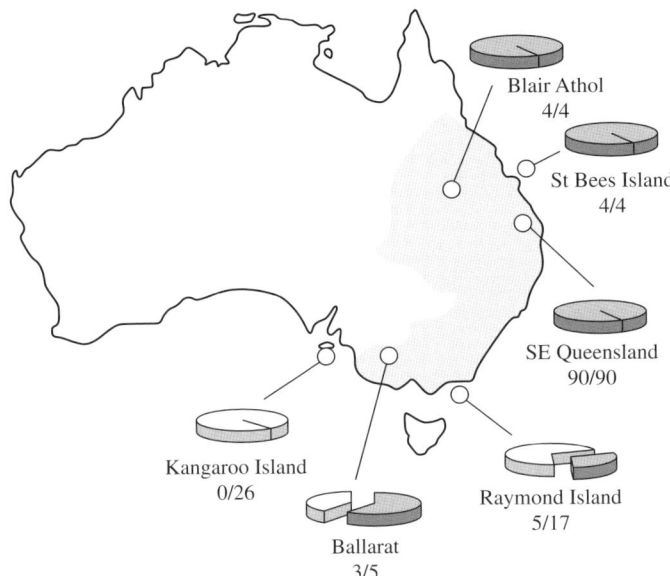

Fig. 5.11 Distribution of koala retrovirus (KoRV) in koalas in Australia. Prevalence was determined by PCR. Numbers indicate numbers infected/numbers sampled. Note their absence from the geographically isolated Kangaroo Island. Taken from Tarlinton *et al.* (2006) with permission.

The 31 families of HERVs have been further classified into three major groups based on their similarity to the seven genera of exogenous retroviruses (a classification scheme that applies to ERVs generally): class I HERVs are similar to gammaretroviruses and epsilonretroviruses, class II to lentiviruses, alpha- beta-, and deltaretroviruses, and class III to spumaretroviruses (although distantly) (Gifford *et al.* 2005; Katzourakis and Tristem 2005). Dating the appearance of these families again provides an important insight into evolutionary dynamics. Of particular note is that the exogenous ancestors of some HERVs may have existed long ago in primate evolution (Katzourakis and Tristem 2005). For example, Katzourakis and Tristem (2005) suggest that the prototype HERV-A element existed between 57 and 92 million years ago, the ancestral HERV-K at between 51 and 83 million years ago, and the prototype HERV-L at between 62 and 100 million years ago. Indeed, most ERVs are not closely related to exogenous retroviruses.

The abundance and distribution of ERVs in a wide range of taxa indicates how frequently animal populations have been subject to retroviral infections. However, infection and integration of an exogenous (i.e. infectious) retrovirus is not the only mechanism that generates ERVs, as they can also arise following retrotransposition from an existing endogenous copy. Importantly, birth-death models of their intra-genic proliferation have been developed (Katzourakis *et al.* 2005), and are mechanistically similar to that which I outlined earlier in this chapter. In some cases the process of lineage birth also appears to be episodic. For example, there may have been a significant and independent expansion of the retroviral content of chimpanzee and gorilla genomes some 3–4 million years ago (Yohn *et al.* 2005), although the precise reasons why are unclear.

6

The molecular epidemiology, phylogeography, and emergence of RNA viruses

In some respects, this is a chapter about viral 'life history'. It is certainly the most ecological of any chapter is this book, although doubtless many ecologists would object to my classifying it so. My aim is to document the epidemiological patterns, both temporal and spatial, exhibited by RNA viruses in nature, and to explore aspects of their emergence. In keeping with the overall theme of this book, the phylogenetic analysis of viral gene sequences will be the main analytical tool, often set within an approach that has been termed phylodynamics (Grenfell *et al.* 2004). Given that human viruses have been the most actively studied in this respect, I will necessarily focus on their evolution. However, the generalities drawn from this chapter should be applicable to a broad range of RNA viruses.

It is arguable that the most common form of evolutionary study performed on RNA viruses is molecular epidemiology, in which phylogenetic methodology is used to infer the origins, emergence, spread, and dynamics of viral epidemics. Such studies can be broad-scale, examining the global distribution of viral genetic variation, or highly localized, in which the precise network of transmission is reconstructed. Those studies which explicitly consider the process of emergence have attracted most attention. The success of this endeavour is due in part to the rapid rate of evolutionary change in RNA viruses, such that the epidemiological processes that shape their genetic diversity act on approximately the same timescale as mutations are fixed in viral populations (Holmes 2004). Indeed, the rapidity of RNA virus evolution is such that phylogenetic relationships can often be resolved among isolates that have been sampled only a few days apart (Yeh *et al.* 2004; Cottam *et al.* 2006, 2008), so that the analysis is sometimes hardly 'retrospective' at all. More tentatively, by revealing the 'rules' of viral evolution it might also be possible to make some general statements about what sorts of infections, from what reservoir species, and in what locations, will emerge in human populations in the future (Kilpatrick *et al.* 2006; Jones *et al.* 2008).

6.1 Phylodynamics: linking viral evolution at the phylogenetic and epidemiological scales

A key theme of this chapter is that there is a direct link between the population dynamics of RNA virus diseases, manifest as changes in prevalence and/or incidence through time, and their molecular evolution. Although it is easy to make this link in abstract terms—RNA viruses evolve as they infect hosts—it is a very different matter to make a formal, more quantitative, connection between these two scales of analysis (Fig. 6.1). However, this synthesis is the central goal of the emerging discipline of phylodynamics (Grenfell *et al.* 2004).

The basis of the phylodynamic approach is that epidemiological processes, namely rates of population growth and decline, the extent of population subdivision, and patterns of viral migration, are written into gene sequences and can be recovered using a suite of phylogenetic techniques, with the coalescent approach paramount among them (Grenfell *et al.* 2004). As well as considering explicitly epidemiological processes, the phylodynamic approach can also be employed to study aspects of intra-host

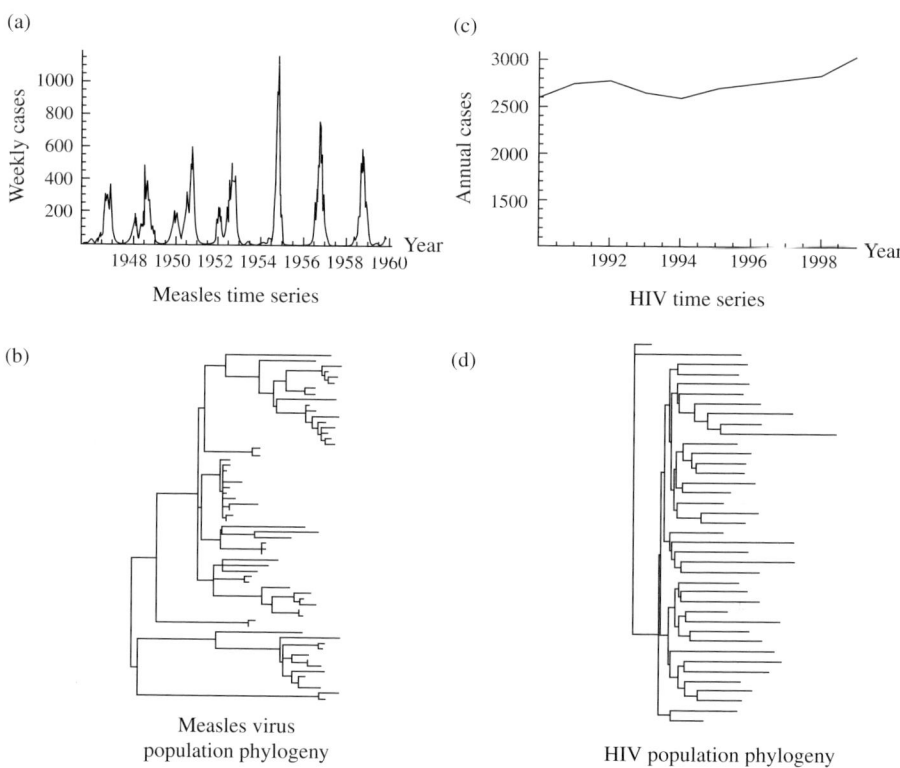

Fig. 6.1 The phylodynamics approach. The link between the epidemiological and phylogenetic scales are shown for measles virus (a and b), which exhibits rapid (recurrent) epidemiological dynamics, and HIV-1 (c and d), which exhibits slow epidemiological dynamics. The measles time series represents weekly case reports from Leeds, UK, whereas the HIV-1 time series depicts annual diagnosed cases in the UK. The measles virus phylogeny is that of the N gene (63 sequences, 1575 nt), whereas 39 subtype B sequences (2979 nt) were used in the HIV-1 tree. Adapted from Grenfell *et al.* (2004) with permission.

viral evolution (at least in the case of chronic RNA virus infections), simply by thinking in terms of infected cells, rather than infected hosts.

The coalescent is crucial to phylodynamics because it represents a direct link between the phylogeny (or genealogy) of gene sequences sampled from a viral population, and the demographic history of that population (see Rambaut et al. 2008 for an important illustration concerning influenza A virus). Given sufficiently large data sets sampled with adequate temporal resolution, the coalescent will even allow the estimation of key epidemiological parameters, the most important of which is R. Although the coalescent may never replace simple case counts or serological surveys as a way of estimating the incidence and prevalence of an infectious disease, it crucially allows epidemiological dynamics to be inferred on a lineage- (or strain-) specific basis. Indeed, it is the ability to focus on genetically defined viral types that represents the true power of the coalescent: techniques such as serological assays only allow identification of major viral types—such as the four serotypes of DENV—rather than the multitude of lineages contained within each of these serotypes, and which may have very different epidemiological dynamics.

Important early work in this area was undertaken by Paul Harvey and Sean Nee at Oxford (Nee et al. 1992; Harvey et al. 1994a, 1994b; Nee et al. 1994a, 1994b, 1995). The major insight from these studies was that aspects of population history, notably rates of lineage birth and death (the latter of which also informs on the rate of extinction), can be revealed through the analysis of the branching structure of phylogenetic trees. The main analytical tool developed during this period was the 'lineage-through-time' plot, which depicted changing temporal patterns of observed (and expected) lineage birth and death (Fig. 6.2). Although this technique could in principle be applied to any phylogenetic pattern, it became clear that very different interpretations of the results were required if the phylogeny in question was likely to represent a nearly complete sample of available taxa, such as when studying macroecological patterns in vertebrates, or a very small sample, as is undoubtedly the case when dealing with viral populations (Nee et al. 1995). It is this reliance on small samples that links the analysis of lineages-through-time with the coalescent (and which has a longer history in population genetics; Kingman 1982), a marriage that eventually saw the development of a variety of models of viral demography and the estimation of R directly from gene sequence data (Pybus et al. 2000, 2001).

6.1.1 Coalescent approaches to viral epidemiology

In recent years there have been many important developments in the use of coalescent methods to infer aspects of viral demography from sequence data, including relaxation of the earlier restrictive assumption of a 'strict' molecular clock. Indeed, the development of 'relaxed' molecular clocks, in which lineages can vary in their substitution rates, is arguably one of the most important advances made in modern molecular evolutionary genetics (Drummond et al. 2006). Of equal note was the movement away from basing inferences on a single representation of phylogenetic history, to those based on a very large sample of plausible phylogenies and usually set within

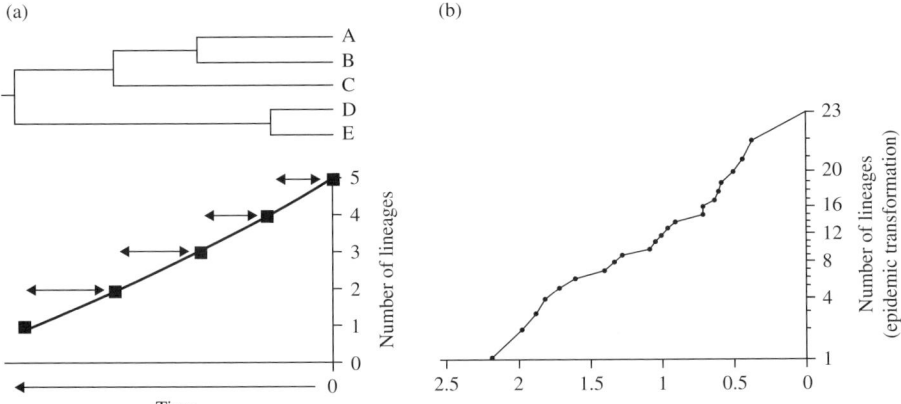

Fig. 6.2 The lineages-through-time approach to inferring the population dynamics of RNA viruses. (a) The relationship between an ultrametric (i.e. constant-rate) phylogenetic tree (top panel) and the number of lineages through time (bold line, bottom panel). Arrows show the times between coalescent events. (b) 'Epidemic transformation' of the number of lineages through time for 24 *env* gene sequences of HIV-1 (see the original publication for more details of this transformation). An approximately straight line under this transformation, as here, is indicative of exponential population growth. Adapted from Nee *et al.* (1995) with permission.

a Bayesian Markov chain Monte Carlo (MCMC) framework, thereby minimizing the impact of phylogenetic error (Drummond *et al.* 2005). Finally, the development of the Bayesian 'skyline plot' allows changing patterns of genetic diversity through time to be analysed without specifying a demographic model, and which can be interpreted as a measure of N_e under strictly neutral evolution (Drummond *et al.* 2005).

Despite these advances, there are still important limitations in the coalescent approach as applied to RNA viruses, and hence to the phylodynamic approach in general. First, on the theoretical side, it is clear that the analytical models currently available to describe the demographic history of RNA viruses are limited in their scope, despite the development of the Bayesian skyline plot. Although simple demographic models like constant population size, or exponential or logistic growth, might apply to viruses with 'simple' population dynamics, such as HIV and HCV, they are woefully inadequate for viruses like influenza or measles which exhibit more complex epidemiological dynamics. Evidently, for the field of phylodynamics to advance, it is crucial that we develop more realistic models of demographic history.

A second major theoretical limitation of the coalescent is that it is compromised, to some extent, by the occurrence of a number of other population processes, namely recombination, natural selection, and population subdivision. As incorporating these processes into a coalescent framework is an active and complex research area, I will not discuss this any further here. However, given that RNA viruses spread on a spatial plane, and this process is often recorded in the structure of phylogenetic trees (see Biek *et al.* 2007 for an interesting example as well as section 6.4), it is clear that building spatial dynamics into the coalescent will be hugely beneficial.

The final, and practically the most important, limitation of the phylodynamic approach is the lack of availability of gene sequence data of the right temporal resolution to enable detailed demographic inference. For those viruses with slow population dynamics, such as HIV or HCV, and where there is often an extended waiting time between transmissions, the epidemiological signals of interest can usually be recovered with samples collected on only a yearly basis. However, for those viral epidemics that have a more distinct periodicity, such as annual cycles, it is clear that a far more fine-grained temporal sampling is required to extract all epidemiological information from sequence data. At the time of writing, the only RNA virus where sufficient data are available to infer precise population dynamics is influenza A virus, where genome sequence data have been collected on a daily basis from diverse populations (Ghedin *et al.* 2005). Coalescent analyses of these data have provided important insights into the population dynamics, and interactions, among the H1N1 and H3N2 subtypes of influenza A virus (Rambaut *et al.* 2008) (see Fig. 3.5 in this volume). It is hoped that similar large-scale genome projects, with linked epidemiological information, will now be undertaken for other acute RNA viruses.

6.2 Cross-species transmission, co-divergence, and emergence

Understandably, most definitions of emerging viruses focus on the issue of disease. This means that there is often no distinction between those viruses that spread efficiently among us, and those that only result in sporadic infection. Indeed, it seems that many, if not the majority, of the emerging diseases of humans represent dead-end infections with little or no onward transmission. This may then represent the natural, 'background' dynamics of cross-species transmission: human populations are continually exposed to new pathogens, but few ever become established. For example, the vast majority of avian-to-human transmissions of influenza A virus result in dead-end infections, and only very occasionally in major pandemics (Taubenberger *et al.* 2005). The problem that faces those working in this area is identifying and quantifying the factors that determine whether a new disease will survive and grow into a fully-fledged epidemic, or will fade out with no subsequent transmission. To understand emergence it is therefore crucial to understand how and why only some viruses are able to establish long-term transmission networks (Wolfe *et al.* 2007).

6.2.1 The RNA/DNA divide again

Perhaps the most powerful general 'rule' relating to the evolution of viral emergence is that there is a major division between those viruses that frequently jump species boundaries, and that those that co-diverge with their host species over longer stretches of evolutionary time. This generally mirrors the distinction between

Table 6.1 The major evolutionary and epidemiological properties of RNA and DNA viruses

Property	RNA virus-like	DNA viruses
Genome type	RNA; ssDNA	dsDNA
Genome size	Small (<32 000)	Small or large
Substitution rate	High (10^{-2}–10^{-5} subs/site/year)	Low (10^{-7}–10^{-9} subs/site/year)
Infection type	Generally acute	Generally persistent
Emergence	Cross-species transmission	Long-term co-divergence
Transmission mode	Aerosol; body fluid; faecal–oral; vector-borne	Sexual; vertical
Critical community size	Large	Small
Virulence	Can be high	Often low

Table adapted from Holmes (2008) with permission.

RNA and DNA viruses (Holmes 2008) (Table 6.1). Specifically, RNA (and ssDNA) viruses tend to (but certainly not always: think SFV) establish only short-term acute infections in their hosts and evolve by a mechanism of cross-species transmission, whereas dsDNA viruses tend to result in persistent infections and experience long-term virus-host co-divergence, in which the evolutionary history of the virus tracks that of its host species over many millions of years. As RNA viruses are the most common cause of emerging diseases in humans, so host jumping is the principle mechanism of viral emergence (Woolhouse 2002). Understanding the mechanisms that determine cross-species transmission is therefore fundamental to understanding the nature of viral emergence. In fact, the ability of viruses to jump between hosts of different species can be thought of as an extension of the 'normal' mode of transmission in most RNA viruses, in which viruses move horizontally among members of the same host species.

The evolutionary division between host-jumping RNA viruses and co-diverging DNA viruses correlates with a number of other key characteristics of these pathogens, particularly their mode of transmission, their virulence, and their rates of evolutionary change (Table 6.1). To summarize, acute RNA viruses tend be transmitted horizontally by aerosols, body fluids, faecal material, or vectors, can result in infections of extremely high virulence, and can evolve with great rapidity. The end result of occupying this part of *epidemiological* parameter space is that acute RNA viruses often require large and well-connected host populations to survive: a reduction in the number of susceptible hosts to below a virus-specific critical community size (CCS) will result in $R<1$ and thus extinction. In contrast, chronic dsDNA viruses tend to be transmitted either sexually or vertically (that is, between mother and offspring), are often of lower or delayed virulence, as the virus is required remain in the host for extended time periods to ensure transmission, and evolve more slowly. The evolution of persistence in viral infections may therefore often require a mode of transmission—sexual or vertical—that facilities long-term co-divergence (see below). Early evidence for this fundamental ecological division among viruses were the classic studies of Francis Black who noted that indigenous Amerindian populations do not carry acute RNA viral infections such as measles and mumps endemically because their populations are insufficiently large to allow $R>1$ (Black 1975). In contrast, these same Amerindians frequently carried persistent infections, such as various herpesviruses, presumably because these require a smaller CCS to sustain themselves at $R>1$. The root cause of the division between RNA and dsDNA viruses most likely rests with their very different rates of evolutionary change (section 3.1; see below).

6.2.2 Inferring co-divergence

The principal evidence for co-divergence between viruses and their hosts is a significant match in their tree topologies, and which is currently fulfiled by only a small number of RNA viruses (Jackson and Charleston 2004; Switzer et al. 2005; section 3.1). In addition, it is also the case that a variety of other ecological processes can produce viral phylogenies that superficially match those of their hosts, wrongly leading to the conclusion that there has been long-term co-divergence, particularly when sample sizes are small. First, an important rule of thumb from those studies of viral emergence undertaken to date is that the closer the host species involved, the more likely that viruses are able to successfully jump between them (DeFilippis and Villarreal 2000; Holmes and Rambaut 2004; see below). Played out in full, this phylogenetic effect—which can also be thought of as 'preferential host switching' (Charleston and Robertson 2002)—may result in a pattern of pure cross-species transmission that generally mirrors that of co-divergence. It is just such a process that may explain the current biodiversity of the primate lentiviruses (see section 7.2). Briefly, although the discovery of endogenous lentiviruses indicates that these viruses have been in circulation for millions of years (Katzourakis et al. 2007), including in lemurs (Gifford et al. 2008), there is little good evidence for the co-divergence of the simian lentiviruses and their host species (Wertheim and Worobey 2007), and the human lentiviruses (HIV-1 and HIV-2) clearly represent recent cross-species transmissions. A second reason why virus/host co-divergence is not necessarily indicative of an ancient relationship centres around the mode of speciation. If closely related host species tend to live sympatrically (that is, with overlapping geographical ranges), which is to be expected if they also experienced sympatric speciation, then cross-species transmission is again predicted to generate phylogenies that often match those of their host species. Indeed, phylogenetic trees of feline immunodeficiency virus (FIV) from free-ranging cat species show a better match to geographic region than to host phylogeny (Troyer et al. 2008). As such, it is dangerous to draw strong conclusions on the antiquity and mode of viral evolution simply from an examination of current species distributions. However, quantifying the relative rates of cross-species transmission versus co-divergence as modes of RNA virus evolution is clearly an important area for future study. Similarly, the current paucity of our sampling of viruses in nature, particularly from species other than humans, means that statements on the occurrence of either process should be made with caution.

6.2.3 The evolution of persistence in RNA viruses

As noted above, the majority of RNA viruses cause acute infections. Despite this important generality, a subset of RNA viruses do manage to generate persistent infections in their hosts and some normally acute infections occasionally become persistent, often resulting from a major reduction in the extent of cell lysis. For example, although measles virus usually causes a well-known acute infection, it sometimes results in subacute sclerosing panencephalitis (SSPE), a rare encephalitis that afflicts 1 in 100 000 of those infected and which is associated with persistence. Similarly, persistent infections with enteroviruses have been mooted as the cause of some chronic diseases (Tam and Messner 1999). A brief discussion of the evolution of persistence is therefore merited. Importantly, comparative analyses can assist in determining the causes of persistence. For example, in the case of feline calicivirus (FCV), a detailed molecular phylogenetic analysis revealed that 'persistence' in this virus is in fact largely explained by continual reinfection (Coyne *et al.* 2007). It is therefore possible that a number of other cases in which viruses are claimed to cause persistent infections are in fact better explained by the reinfection of the same hosts in the absence of strong protective immunity, particularly in hosts that live at high population densities.

It is meaningful to discuss persistence both in terms of the intra-host mechanisms that allow its occurrence, and the broader evolutionary processes by which it might be favoured. Within hosts, persistence can be thought of as a particular form of virus–host interaction. For RNA viruses this often involves either immunotolerance, in which viruses do not stimulate a strong immune response against them, or immune escape, such that persistence is associated with active evasion (reviewed in Oldstone 2006). Immunotolerance may explain persistence in viruses such as GBV-C, in which long-term infection is not associated with an accumulation of antigenic variation, including an absence of the 'hypervariable' regions of antigenically variable sites observed in the related flavivirus HCV (Zampino *et al.* 1999; Sheridan *et al.* 2004). Indeed, HCV represents a classic case in which the evolution of persistence clearly results from immune evasion (Grakoui *et al.* 2003; Cox *et al.* 2005). An additional form of persistence occurs in those normally acute RNA viruses that are not cleared by hosts that are severely immunocompromised, most commonly AIDS patients (for example, Klimov *et al.* 1995, and see section 6.5 in this volume). That persistence in this case is associated with a weakened immune system again showcases the importance of the interaction between virus and host immune response. However, this is not always so. For example, in the case of HTLV-I, persistence may simply be a consequence of more rapid viral replication than immune clearance (Asquith and Bangham 2008). Similarly, defective interfering particles have been implicated in developing persistence both *in vitro* and *in vivo* (Bangham and Kirkwood 1990). Persistence may also be associated with widespread tissue diffusion, as appears to be the case in some hantaviruses (Botten *et al.* 2003), including migration to non-dividing cells, as has been shown in coxsackievirus (Tam and Messner 1999). In the latter case persistence is also associated with a noticeable lack of intra-host viral evolution.

At the epidemiological scale persistent infection presents a number of properties that might be favoured by natural selection. First, when host populations are small and sparse, persistence provides viruses with a better chance of infecting new hosts, thereby increasing R. It is therefore no surprise that the majority of the infections that characterize small Amerindian populations are persistent ones (Black 1975). Similarly, persistence is central to sexual and vertical (mother-to-child) transmission, as these obviously require hosts to carry viruses for extended time periods. In other cases, however, the evolution of persistence may be 'short-sighted' at the epidemiological scale such that it is out of reach of natural selection. For example, each strain of SSPE is generated, independently, from a normal (non-SSPE) isolate of measles virus, and is unable to complete a full infectious cycle (Hirano 1992). As SSPE isolates are not transmitted among hosts they fall in diverse positions on phylogenies of measles virus (Woelk *et al.* 2002). This is likely true of other cases where persistence is associated with the production of defective viruses.

6.2.4 Host phylogeny and viral emergence

One of the most frequently cited ideas in the realm of viral emergence is that there are phylogenetic 'constraints' to the process of cross-species transmission, such that the more closely related the host species in question, the greater the chance that viruses will be able to jump between them and successfully establish productive infections (DeFilippis and Villarreal 2000). For example, under this model humans are most likely to acquire new viral infections from other species of primate. However, this simple evolutionary rule ignores the fact that it is usually difficult to disassociate the probability of successful transmission from the probability of exposure: although humans are obviously more closely related to other primates than to rodents, the global human population is more often exposed to the latter, enhancing their role as reservoir populations. In addition, rodents (and bats) also live at high population densities which, by allowing more viruses to attain their CCS, entails that they are able carry both a greater number and more virulent pathogens.

There are some data that support the 'phylogenetic distance' theory, although much of it is indirect. Anecdotally, there is no evidence that viruses from organisms as divergent as plants, fish, reptiles, or amphibians have ever been able to infect humans (Holmes and Rambaut 2004), even though in the case of plant viruses exposure might occur on a regular basis through the consumption of infected food; as a case in point, perhaps 10% of cabbages and 50% of cauliflowers are infected with cauliflower mosaic virus (Hull *et al.* 2000). Rather, the majority of human viruses are of mammalian origin, with an occasional few, and famously influenza virus, coming from birds (although WNV is also an avian infection, human-to-human transmission is not thought to occur). In some cases experimental studies have even shown that viruses from non-human primates, such as sylvatic dengue, have the capacity to replicate in human cells, thereby facilitating their emergence (Vasilakis *et al.* 2007b). In addition, while it is the case that insect viruses often infect human populations (that is, the arboviruses), these always jump from another mammalian species, where they act as transmission vectors, rather than directly from insects. Arboviruses as a whole are in fact less able to evolve sustained transmission cycles in new hosts than those viruses transmitted by other means (Woelk and Holmes 2002; section 3.3).

More direct evidence for a phylogenetic component to emergence was a large-scale comparative analysis which found that evolutionary relationship was the best predictor of whether primate species shared pathogens (Davies and Pederson 2008). Intriguingly, however, geographical overlap was a better predictor than phylogeny in the case of viruses, again highlighting the importance of ecological contact in disease emergence (although this study did not make the vital distinction between RNA and DNA viruses). There are also some important examples demonstrating how frequently RNA viruses jump between primate species. As well as HIV-1 and HIV-2, whose ultimate origins lie with chimpanzees (*Pan troglodytes troglodytes*) and sooty mangabey monkeys (*Cercocebus torquatus atys*), respectively (section 7.2), a variety of other major human RNA viruses seem to have their origins in non-human primates. These include DENV, GBV-C, HTLV-I and -II, and YFV. These cases also illustrate how human encroachment into the virgin forest homes of primate species has had a major bearing on disease emergence (Wolfe *et al.* 2005). It is also likely that HBV falls into this category, although 'reverse' transmission from humans to apes cannot be excluded as the genetic diversity present in ape HBVs is set within that sampled from humans on phylogenetic trees (Starkman *et al.* 2003; Holmes 2008). Indeed, there are a number of examples of humans transmitting their viruses to other primates, and often with serious consequences, as appears to be the case for measles (Ferber 2000), other paramyxoviruses (Köndgen *et al.* 2008), and TTV (Okamoto *et al.* 2000). A combination of increasing population size and changing agricultural practices makes humans the perfect vector for viral diseases.

There may also be a simple mechanistic basis for the relationship between phylogenetic distance and the likelihood of viral emergence. Specifically, as the ability to recognize, infect, and be released from host cells is a key component of cross-species transmission (Baranowski *et al.* 2001; Woolhouse 2002; Parrish *et al.* 2008), then viruses should be better able to jump between phylogenetically related host species as these will usually harbour more closely related cell receptors. A reliance on conserved cell receptors may explain the very high numbers of host species (>200) that some plant viruses are able to infect (Reanney 1982). However, the rapid rate of RNA virus evolution also predicts that highly dependent relationships between viruses and cell receptors will be established very quickly—viruses rapidly become specialists unless they are continually exposed to alternating hosts—so that the probability of successful cross-species transmission will decrease with increasing phylogenetic distance (Fig. 6.3). Such a rapid evolution of host specificity has been observed in experimental populations of RNA phages (Duffy *et al.* 2007). In the same vein, the

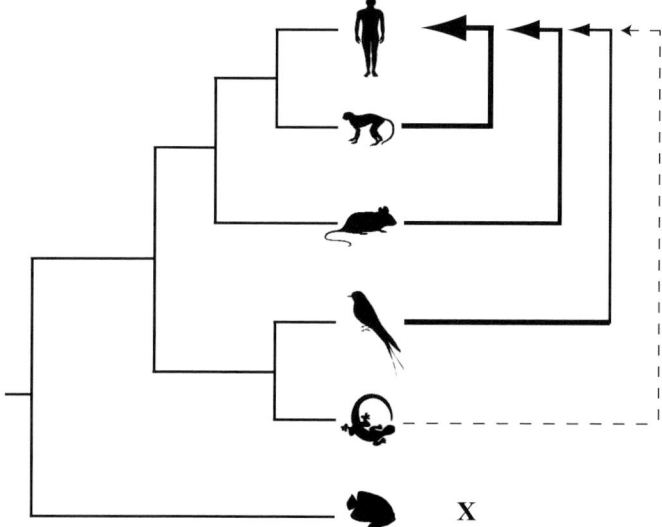

Fig. 6.3 The phylogenetic distance model of cross-species transmission in which the closer the phylogenetic relationship among the host species, the more likely they are to share their RNA viruses. The probability of successful cross-species transmission is reflected in the thickness and solidity of the lines with arrows. At great phylogenetic distances no successful cross-species transmission is expected. Although this model explains some of the viral data, any phylogenetic effect is greatly mediated by the differing probabilities of exposure between different host species.

more slowly evolving dsDNA viruses should initially be able to jump wider phylogenetic boundaries—because host specificity will take longer to evolve—but, when they do fully adapt to their host, dsDNA viruses will find it more difficult to make subsequent species jumps. This may in part explain the tendency for dsDNA viruses to co-diverge rather than move horizontally among species (Holmes 2004), and more generally why dsDNA viruses are less often a cause of emerging disease than RNA viruses (Woolhouse et al. 2001; Woolhouse 2002).

Of course, the general principles developed here should never be treated as anything more than rules of thumb, particularly as on occasion RNA viruses are able to switch among more divergent host species (Woolhouse 2002). To reiterate, ecological opportunity is also critical to the process of emergence. For example, bats are the most likely reservoir species for both EBOV (Leroy et al. 2005) and SARS-CoV (Li et al. 2005). It is also possible that closely related host species may share some of the alleles that determine immune responses to specific pathogens, and which may impact on the likelihood of cross-species transmission. This phenomenon is well documented in the case of major histocompatibility complex (MHC) loci, where some allelic lineages are present in multiple species (Figueroa et al. 1988).

6.3 The evolutionary genetics of viral emergence

Another of the most important questions in the evolution of viral emergence is whether, following cross-species transmission, emergent viruses must actively adapt to develop sustained transmission in their new species, or that the process of emergence is essentially neutral, simply reflecting differences in the frequency of exposure, with little post-transmission adaptation (Fig. 6.4).

6.3.1 Adaptation and emergence

According to one model, adaptation to a new host species during the early period of an epidemic is essential to viral emergence, because by elevating $R>1$ sustained transmission networks can be established (Antia *et al.* 2003) (Fig. 6.4). Because selection for enhanced transmission cannot occur without at least some transmission already in place, this adaptive process is thought to occur during the 'stuttering chains of transmission' that might characterize the initial appearance of a virus in a new host species (Antia *et al.* 2003; Woolhouse *et al.* 2005). The small clusters of H5N1 influenza A virus transmission in humans (Kandun *et al.* 2006) may constitute exactly these stuttering chains.

It is very reasonable to assume that cross-species transmission routinely involves adaptive evolution as a number of key barriers must be breached for any emerging virus, including (i) evasion of both innate and adaptive immunity, (ii) infection of a host cell, (iii) replication utilizing host factors, (iv) exit from the host cell, and (v) successful transmission to other individuals in the population (reviewed in Webby *et al.* 2004). Clearly, adaptive evolution could occur at any, and perhaps all, of these stages. Empirical evidence for just such an adaptive process is provided by the carnivore parvoviruses (ssDNA viruses), one of the best-documented cases of emergence. The origins of canine parvovirus (CPV), which resulted in a major epidemic in dogs during the 1970s, lie with the feline parvoviruses (FPV) that were described previously in cats (Parrish 1990). That this species transfer involved direct adaptation is manifest in a greatly elevated rate of nucleotide substitution, including $d_N > d_S$, on the branch linking FPV and CPV (Shackelton *et al.* 2005). More importantly, these fixed mutations have direct and measurable effects on binding to the canine transferring receptor (Hueffer *et al.* 2003). Interestingly, the earliest CPV isolates lost their ability to infect cats, although this was regained in those sampled at later time points (Truyen *et al.* 1996), arguing that these mutations have important pleiotropic effects. However, what proportion the mutations associated with adaptation in CPV arose in dogs, as opposed to being seeded from cats, is unclear.

Direct adaptation to a new host species also seems to have been central to the emergence of Venezuelan equine encephalitis virus (VEE), where a single positively selected amino acid change—position 213 of the E2 envelope glycoprotein—appears to be responsible for the adaptation of rodent (reservoir) viruses to horses (Brault *et al.* 2002; Weaver and Barrett 2004; Anishchenko *et al.* 2006). Similarly,

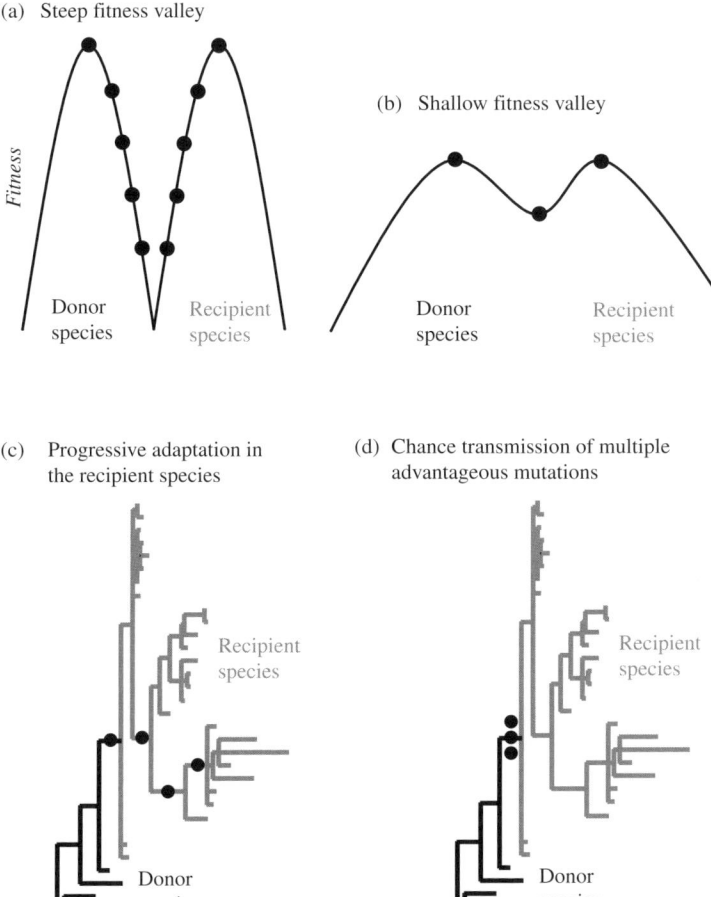

Fig. 6.4 Evolutionary models for the cross-species transmission of RNA viruses. (a) The donor (black) and recipient (grey) species represent two distinct fitness peaks separated by a steep fitness valley. Multiple adaptive mutations are therefore required for the virus to successfully establish onward transmission in the recipient host. (b) The donor and recipient species are separated by a shallower fitness valley, for example because the host species are relatively closely related. This facilitates successful cross-species transmission as a smaller number of advantageous mutations are required. (c) If multiple advantageous mutations are required for a virus to adapt to a new host it is possible that these evolve progressively in the recipient species, although this requires some onward transmission. (d) Under an alternative model, many of the mutations required for adaptation to a new host species pre-exist in the donor species and transmitted, by chance, in a single event. This will accelerate viral emergence. Intermediate hosts may also facilitate cross-species transmission. In each case mutations are represented by closed circles. Also see Fig. 6.6. Adapted from Kuiken *et al.* (2006) with permission.

adaptation to the principle vector of DENV in an urban human setting, the *Aedes aegypti* mosquito, may have been central to enabling its sustained transmission in humans (Moncayo *et al.* 2004), although sylvatic (i.e. monkey) DENVs themselves have little trouble replicating in human cells (Vasilakis *et al.* 2007b). A role for adaptation in viral emergence can also be seen in some plant viruses. For example, in the case of *Pelargonium* flower break virus (PFBV) five amino acid changes, which appear convergently, are required to adapt this virus to *Chenopodium quinoa* hosts (Rico *et al.* 2006). Finally, it has recently been demonstrated that a single amino acid change to arginine at position 31 in the *gag* gene has occurred independently on the three main lineages of HIV-1 (M, N, O) following its jump from chimpanzees (SIVcpz) (Wain *et al.* 2007). Intriguingly, this mutation allows HIV-1 to replicate better in human than chimpanzee cells, although the precise mechanistic basis for this is uncertain. This strongly suggests that the transition from SIV to HIV involved adaptive evolution (see also Heeney *et al.* 2006), a hypothesis compatible with a change in d_N/d_S on the branches leading to human isolates (Sharp *et al.* 2000), and that the suppression of T-cell activation by the lentivirus Nef protein—a crucial component of the anti-HIV response—was lost on the branch leading to HIV-1 (Schindler *et al.* 2006).

6.3.2 'Off-the-shelf' emergence

A competing (although not mutually exclusive) model for viral emergence is that rather than the emergent virus adapting to the new host species following exposure, successful emergence will only occur if the recipient host is exposed to a virus that already possesses at least a subset of the necessary mutations, such as those for receptor binding. In other words, successfully emergent strains are those that are in some sense *pre-adapted* to establish productive infections in the new host species, so that the probability of emergence then becomes a function of the frequency of exposure. Mark Woolhouse has cleverly termed this 'off-the-shelf' adaptation compared to the 'tailor-made' model of emergence described above. Indeed, that the majority of emerging infections (at least in humans) result in dead-end infections implies that even short transmission chains are difficult to establish for most viruses. Moreover, for the majority of emergent viruses it has been difficult to show conclusively that cross-species transmission is associated with adaptation in the recipient host, although this in large part may reflect the complexities in undertaking analyses of this sort. To take one high-profile example, although sequence analyses suggest that SARS-CoV was subject to positive selection during its brief stay in the human population (Yeh *et al.* 2004; Kan *et al.* 2005), it is unclear whether this reflects adaptation to facilitate transmission in the new host species, via utilization of the angiotensin-converting enzyme 2 (ACE2) and CD209L receptors (Li *et al.* 2003; Jeffers *et al.* 2004), or selection for immune escape. Interestingly, although adaptation to ACE2 may have been fundamental to the successful cross-species transmission of SARS-CoV from bats to humans via Himalayan palm civets (*Paguma larvata*) (Li *et al.* 2006), it is also used as a receptor for other human coronaviruses (Hofman *et al.* 2005).

The complexities of deciphering the evolutionary genetics of viral emergence are clearly seen in influenza A virus. A necessary requirement for the successful infection of any host species is binding to the sialic acid cell receptors found on cell-surface oligosaccharides (although mutations in other genes, notably the polymerase protein PB2, also play a key role; Matrosovich *et al.* 1997; Taubenberger *et al.* 2005; Hatta *et al.* 2007). In humans, influenza viruses typically infect cells in the nose, throat, and lungs, preferentially binding to a sialic acid in an α-2,6 linkage to galactose. In contrast, avian influenza viruses typically infect cells in the gastrointestinal tract, preferentially binding to an α-2,3-linked sialic acid (Carroll *et al.* 1981; Rogers and Paulson 1983) (Fig. 6.5). This specificity is determined by different amino acid changes in different HA subtypes; for example, HA1 positions 226 and 228 in subtypes H2 and H3, and position 190 in subtype H1 (Horimoto and Kawaoka 2005; Stevens *et al.* 2006), whereas analyses of H5N1 viruses have identified a role for changes at HA1 positions 182 and 192 (Yamada *et al.* 2006). Such binding differences represent an important barrier to the infection of humans and other mammals by avian influenza viruses (Horimoto and Kawaoka 2005). However, in the case of H5N1, binding to sialic acid in the (avian) α-2,3 linkage does occur in cells of the lower respiratory tract of humans (Shinya *et al.* 2006; van Riel *et al.* 2006), although replication in cells of the upper respiratory tract is considered necessary for sustained transmission. The key question, therefore, is whether the critical receptor-binding mutations appear *de novo* in humans, in the short transmission networks of people who initially suffer avian influenza, or pre-exist in the avian population. Although

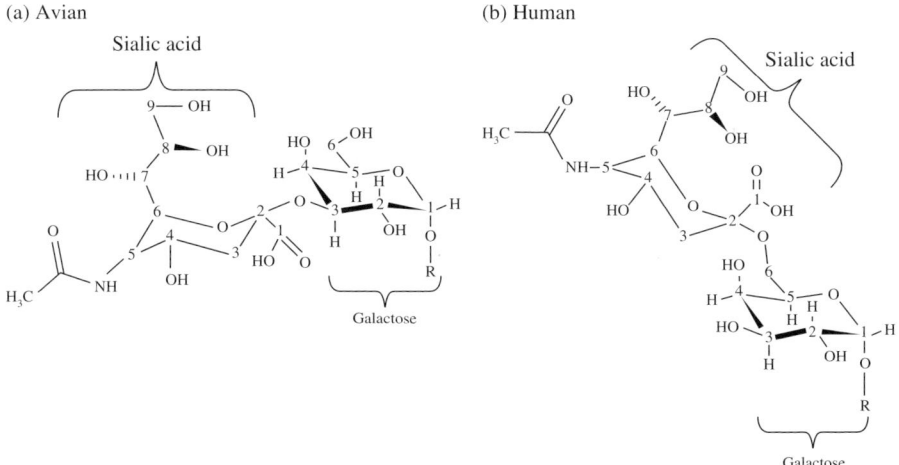

Fig. 6.5 Patterns of α-2,3- and α-2,6-sialic acid linkage in influenza A viruses from birds and humans. In the avian pattern the sialic acid is attached by its carbon 2 (C-2) to the C-3 of the galactose molecule, whereas in humans the sialic acid attaches to the C-6 of galactose. Adapted from Palese and Shaw (2007) with permission.

human-like receptor-binding mutations are undoubtedly deleterious in birds, they will be regularly produced by mutation and there are obviously many infected birds (including perhaps 10% of mallard ducks; Olsen et al. 2006). More intensive surveys of intra-host genetic variation in avian species are clearly a central requirement to answering this question.

As should be evident from the discussion so far, not only is there a great deal of uncertainty in the evolutionary processes that underpin emergence, but a number of advances are need to elucidate the role of adaptive evolution in viral emergence. My particular hobby-horse is that despite the growth of genome sequence data from RNA viruses, there are remarkably few examples of where such data are available from *both* the donor and recipient species. As a case in point, although dengue is one of the most important emerging diseases of humans and the genomic database of DENV isolates is growing rapidly, only a small number of viruses have been isolated from the most likely donor species, Old World monkeys (Wang et al. 2000; Vasilakis et al. 2007a, 2008). Improvements are also needed in the analytical methods available to detect positive selection at sporadic amino acid sites along a single lineage, particularly as this may be the adaptive process most often associated with viral emergence. Indeed, given the frequency with which adaptive evolution is observed when RNA viruses are passaged in different cell types (section 3.1), it seems naïve at best to think that successful cross-species transmission does not involve at least some selectively driven optimization.

6.3.3 The fitness landscapes of emergence

One potentially useful conceptual framework for the evolutionary genetics of cross-species transmission may come from earlier studies of the evolution of Batesian mimicry in species like butterflies, in which the mimic exhibits similar 'signals' to the model but does not possess the unpalatability that repels predators. Although the evolution of mimicry and emergence may at face value appear to have little in common, they share some theoretical common ground as both deal with movement between distinct fitness peaks separated by potentially steep-sided fitness valleys. While an intermediate mimic is of no use, as it can still be recognized by a prey organism, so is an RNA virus that is intermediate between being able to replicate in two different host species.

Although worthy of a book in itself, a general conclusion from explorations of the evolution of mimicry is that mutations of major effect, which allow a large leap across the adaptive valley although not to a new fitness peak, represent an important evolutionary pathway. Once this major fitness jump has taken place, additional mutations with smaller phenotypic effects are able to push the incipient mimic up to the new adaptive peak (Fig. 6.6). More generally, the notion that adaptation involves a few mutations of large effect and more frequent mutations of small effect corresponds to commonly used models for the genetic basis of quantitative traits (Barton and Turelli 1989). In the context of the cross-species transmission of RNA viruses this model is equivalent to suggesting that host specificity is due to a limited number of critical mutations, with the remainder adjusting fitness, likely through epistatic interactions.

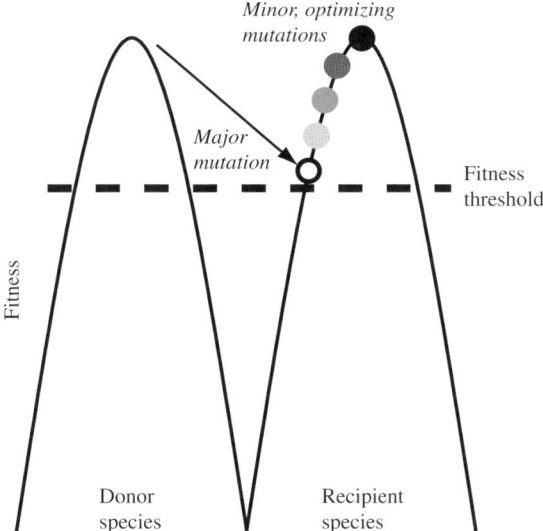

Fig. 6.6 The two-step process for the evolution of host specificity in viruses and extending the model of host adaptation developed in Fig. 6.4. A mutation of major effect allows the virus to make the leap across the steep fitness valley, breaching a notional fitness threshold below which mutations are insufficiently beneficial to allow host adaptation. Multiple mutations of minor effect then act to optimize fitness in the new host, pushing the virus to the top of an adaptive peak. This model is analogous to models developed for the evolution of Batesian mimicry. Mutations are represented by circles of different shades.

Hence, of the 13 genome-wide mutations that have been proposed as required for the adaptation of avian influenza viruses to humans (Finkelstein et al. 2007), most of the variance in fitness may be contributed by only a small number.

Another way in which RNA viruses could potentially cross major valleys in fitness is through the use of intermediate hosts, which act to make the fitness valley shallower (Fig. 6.4). For example, it has been proposed that pigs represent an intermediate 'mixing vessel' for avian influenza viruses to acquire the mutations that enable the productive infection of humans (Scholtissek 1987). The Himalayan palm civet may have played such a role in the emergence of SARS-CoV (Li et al. 2006). However, in many cases intermediate hosts species do not exist, or have not been detected, so that they do not appear to constitute a general mechanism allowing viruses to jump to new hosts.

6.3.4 Recombination, reassortment, and viral emergence

Because recombination (and reassortment) is a process that potentially increases fitness by creating advantageous genetic configurations, it might also be supposed that it can assist the process of emergence (Burke 1998) (although recombination is as likely to destroy advantageous genetic configurations as create them). In particular, recombination may allow viruses to acquire the suite of necessary host-adapting mutations more rapidly than through mutation alone. As a simple case in point consider the primate lentiviruses. Not only are rates of recombination very high in these viruses, with multiple template-switching events occurring during each replication cycle, but recombinant viruses seem to be associated with many cases of cross-species transmission, with the hybrid viruses found in chimpanzees, which then made their way into humans, being a powerful example (Bailes et al. 2003; section 7.2). Similarly, the cross-species transmission of influenza A virus from birds to humans is often associated with reassortment among the haemagglutinin (HA) and neuraminidase (NA) subtypes (Webby and Webster 2001; section 7.1). Indeed, reassortment with a co-circulating human influenza A virus is perhaps the most likely way in which avian influenza viruses can acquire the suite of mutations that facilitate transmission in humans.

Recombination has also been suggested to have played a central role in the emergence of SARS-CoV, particularly as coronaviruses as a family experience relatively frequent recombination (Lai 1996). Specifically, it was proposed that SARS-CoV is a recombinant between diverse avian and human coronaviruses (Stavrinides and Guttman 2004), which may have allowed the virus to acquire the critical amino acid changes required to cause infection in humans (Stanhope et al. 2004). However, more detailed analyses of the relevant sequence data argues against such deep recombination events (Gibbs et al. 2004; Holmes and Rambaut 2004). A recent phylogenetic analysis provided more convincing evidence for a recombination event involving a bat SARS-CoV and as yet unidentified lineage of SARS-CoV (Hon et al. 2008). But as this recombination event clearly occurred before SARS-CoV started spreading in humans, and perhaps before it emerged in the palm civet, it would be wrong to conclude that recombination was essential for SARS emergence. As a final example, recombination is postulated to have played role in the generation of the strain of rabbit haemorrhagic disease virus (RHDV) responsible for the deaths of millions of rabbits globally, although in this case recombination has perhaps changed virulence, rather than host range (Forrester et al. 2008).

Another reason to doubt that recombination is somehow a prerequisite for successful emergence in general is that, as discussed in Chapter 3, the rate of recombination, per nucleotide, in RNA viruses is usually very much lower than that of mutation. Mutation is therefore a more efficient way to create evolutionary novelty than recombination. Indeed, unless the required mutations are already in the population, recombination will be of no consequence to adaptive evolution. In sum, although the occasional recombination event may have kick-started the process of emergence in some viruses, it does not appear central to cross-species transmission.

6.4 The phylogeography of human viruses

The rapid rate of RNA virus evolution makes it natural to investigate their temporal dynamics. However, RNA viruses also exist on a spatial plane, reflecting the movement and growth of their host species. As a consequence, no study of RNA virus evolution is complete without some consideration of their phylogeography. Those studies of human RNA viruses undertaken to date have revealed a variety of different phylogeographic patterns (reviewed in Holmes 2008). These patterns are presented here largely as a framework to understand the factors structuring viral diversity at the epidemiological scale, rather than as an exhaustive survey of the phylogeography of human RNA viruses. Indeed, the barriers between these patterns are often fluid, and most research that touches on viral phylogeography is usually directed towards broader aspects of molecular epidemiology. I will focus my discussion on five phylogeographic patterns which seem to describe a number of RNA viruses, and which are presented schematically in Fig. 6.7: (i) no spatial structure, such that there is complex, and even random, mixing among isolates sampled from different geographical locations and indicative of frequent viral traffic among localities, (ii) 'wave-like' transmission, in which viruses move outwards from a specific starting point, therein generating a relatively simple relationship between genetic and geographical distance, including isolation by distance, (iii) 'source–sink' (or 'core–satellite') models in which viruses flow from one or a limited number of so-called source populations to other sink populations where they may only survive in the short term (for example, with a strong seasonal basis), and which also may generate regular transmission waves, (iv) 'gravity-like' dynamics (Xia et al. 2004), in which patterns of viral transmission are driven by major population centres, which act as gravity (mass) attractors and perhaps following human working patterns (rather like airline routes passing through major or minor hubs), and (v) strong spatial subdivision, in which phylogenetically distinct viral isolates circulate in different geographical localities, with little evidence of movement among them. Importantly, as long as viral sampling is dense enough, and associated geographic and/or demographic data are forthcoming, all these patterns can potentially be distinguished from gene sequence data using appropriate phylogenetic techniques (some specific examples are presented in Chapter 7).

6.4.1 Viruses differ in phylogeographic pattern

Although limited, those analyses undertaken to date show that a variety of human viruses follow each of these patterns, reflecting the relative rates of viral traffic and also the mode of transmission. It is also possible, if not commonplace, for single viruses to exhibit multiple phylogeographic patterns depending on the spatial scale under consideration; for example, comparing single cities to entire continents. A good example is provided by human influenza A virus. In this case phylogeographic structure reflects the fluid movement of the virus among human populations, clearly a consequence of its infectiousness (see section 7.1). On a global scale, influenza A virus seems to fit a source–sink model, with East and South-east Asia acting as a source (Russell et al. 2008), and populations in the northern and southern hemispheres, where the virus is highly seasonal, representing sinks (Rambaut et al. 2008) (Fig. 6.8). This also means that there is little, if

150 • *6 Molecular epidemiology and phylogeography*

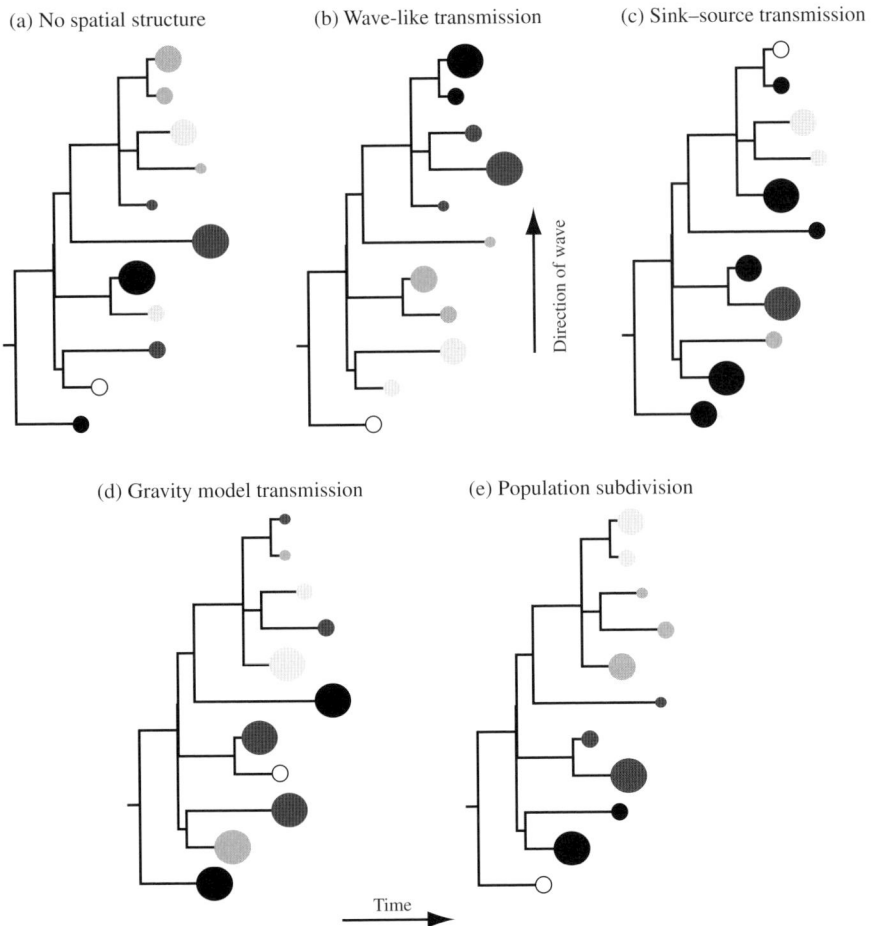

Fig. 6.7 The phylogeographic patterns of human viruses. The figure shows a schematic representation of some of the different phylogeographic patterns exhibited by human RNA viruses. Each population is represented by a different shaded circle, with the size of the circle reflecting the size of the population. The darker the shade, the closer the populations are in space. (a) No spatial structure, with a random mixing of populations; (b) wave-like transmission, so that viruses move in a wave from the root to the tips of the tree; (c) source–sink transmission, in which the black populations act as the source for those viruses seen in all other populations; (d) gravity dynamics, in which viruses first move among the largest populations before diffusing to smaller populations; (e) strong spatial subdivision, with no clear evidence of migration among populations. Based on Holmes (2008).

any, persistence of the virus within individual localities in temperate regions over the summer epidemic trough (Nelson *et al.* 2007, 2008), although there is more continual transmission in tropical regions (Viboud *et al.* 2006a; Finkelman *et al.* 2007). However, both wave-like and gravity-like dynamics are observed at other

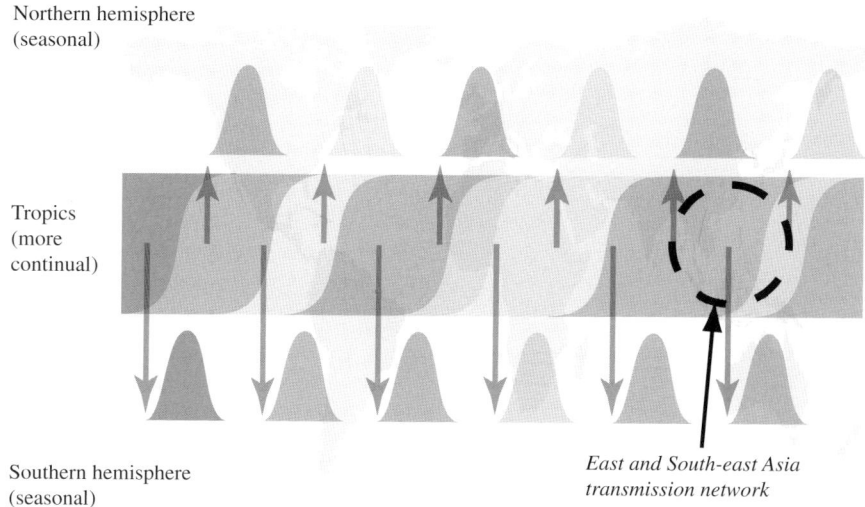

Fig. 6.8 A sink–source model for the phylogeography of human influenza A virus. The putative global source of antigenic diversity in this virus—East and South-east Asia (identified by Russell *et al.* 2008)—is also indicated. This region of Asia is the most likely global source for influenza because it is densely populated and highly connected, allowing natural selection to act more effectively. Each season viruses are exported to the sink populations in the northern and southern hemispheres. Shading represents different antigenic variants of the virus. Adapted from Rambaut *et al.* (2008) with permission.

spatial scales. For example, in the USA influenza seems to satisfy the conditions of a gravity model, largely following movement patterns set by adult workflow (Viboud *et al.* 2006b), whereas in Brazil influenza seems to move southwards in a traveling wave (Alonso *et al.* 2007). This being said, it is important to recall that the evidence for a number of these spatial patterns comes exclusively from indirect epidemiological data—combined pneumonia and influenza mortality in the case of influenza—as gene sequence data with the appropriate scale of resolution have yet to be obtained. Indeed, there is currently no phylogeographic study of a human virus that is clearly compatible with a gravity model, although this undoubtedly reflects a lack of data. Melding epidemiological and genetic data to decipher the spatial dynamics of RNA viruses within a single geographical region remains an important task for the future.

In contrast, the phylogeographic structure of HCV is often characterized by relatively strong spatial subdivision, such that genetic diversity is partitioned into a series of discrete clades, referred to as types and their component subtypes, many of which also have a distinct geographical and risk group association (Simmonds 2004; Holmes 2008) (Fig. 6.9). The difference with influenza most likely stems from the reliance on lower-rate blood-borne transmission. Populations of injecting drug users in industrialized nations carry subtypes 1a and 3a, which have spread rapidly in these regions during the last 60 years, such that spatial structure has been broken down at

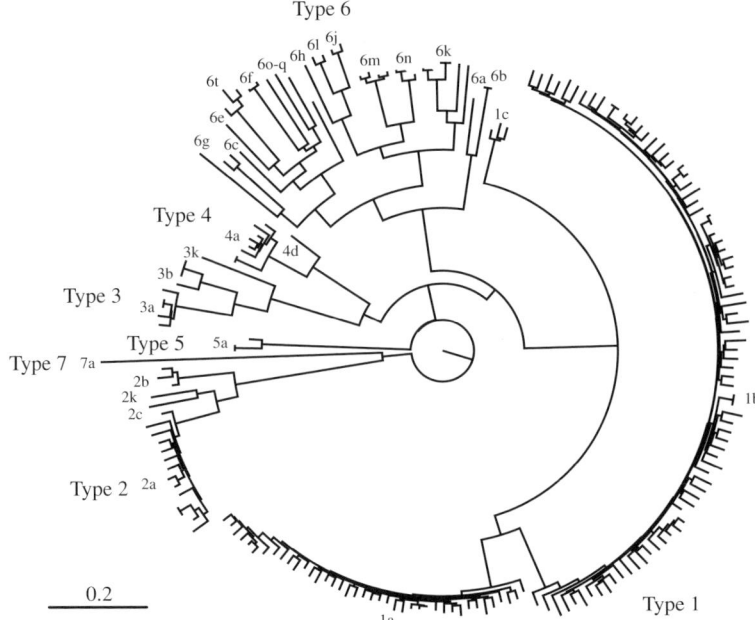

Fig. 6.9 The global genetic diversity of HCV. Subtypes 1a and 1b (and to a lesser extent 3a) dominate infections in industrialized nations, and have been sequenced most frequently, hence their preponderance here. Also note the greater genetic diversity of the type 6 viruses, suggestive of an ancient evolutionary history (most likely in Asia), although the timescale of HCV evolution is uncertain. The tree was estimated using a sample of 200 sequences of the *NS5B* gene (1777 nt) taken from the HCV database (see http://hcv.lanl.gov/content/hcv-index). The tree was estimated using a maximum likelihood method (PAUP*) with branch lengths drawn to a scale of nucleotide substitutions per site.

this level (Simmonds 2004; Pybus *et al.* 2005), while subtype 1b is also very common in industrialized nations. Far greater genetic diversity is present in African and Asian populations, often associated with types 2, 4, and 6, and which may have been circulating for thousands of years, although the timescale of HCV evolution, as well as its ultimate origin, are unclear. Given the probable age of types 2, 4, and 6 it should come as no surprise that they are particularly diverse, containing 18, 18, and 21 subtypes, respectively (see http://hcv.lanl.gov/content/hcv-index). While such a pattern of spatial subdivision is expected if a virus has been associated with humans for extended time periods, it can also be established in very recent time frames given a combination of rapid evolution and strong founder effects. As discussed in section 7.2, it is just such a potent combination that seems to have produced the current global diversity of HIV (Rambaut *et al.* 2001).

Finally, wave-like transmission is best described for measles virus using detailed incidence data (Grenfell *et al.* 2001). Unfortunately, there is currently insufficient gene sequence data available from this virus to determine whether these dynamics are recapitulated at the phylogenetic level. A similar reservation applies to EBOV, where suggestions of a transmission wave in equatorial Africa since 1976 have been proposed using a very small number of gene sequences (Walsh *et al.* 2005). This will need to verified should more sequences from this notoriously elusive virus ever become available, and is complicated by the long-standing debate over the principal reservoir species of this virus. Fortunately, far better phylogenetic data are available for some other systems. For example, transmission waves have been described in rabies virus (RABV), most clearly when the virus enters susceptible populations for the first time, such as red foxes in Europe (Bourhy *et al.* 1999), raccoons in the northeast USA (Biek *et al.* 2007), and foxes in Canada (Real *et al.* 2005), all of which result in isolation by distance. However, at the global scale RABV exhibits strong population subdivision, reflecting the fact that the primary reservoir of RABV worldwide—the dog—does not move large distances (Bourhy *et al.* 2008; section 7.4).

6.5 Major transitions in human ecology and viral evolution

Humans have undoubtedly suffered a heavy burden of RNA viral infection throughout their evolutionary history. Indeed, it is likely that the morbidity and mortality due to RNA viruses was one of the greatest challenges facing modern human populations as they migrated from Africa and achieved global colonization: without the protection of prior adaptive immunity, migrating human populations would have been immunologically naïve and susceptible in every new environment they encountered. Conversely, the major population bottleneck humans seem to have experienced as they migrated out of Africa is likely to have resulted in the stochastic loss of a number of viruses, with a reduced population size meaning that only viruses that required a small CCS may have been lucky enough to survive. It is just such a stochastic purging that may explain the apparent paradox that SFV has co-diverged with non-human primates for many millions of years, yet is absent from humans. Similarly, habitat fragmentation, as has clearly happened in populations of great apes, may also lead to the random loss of viral infections in animal species.

As human ecology has changed through time, major new opportunities for viral infection have also been established. Although the ecological history of humans is complex, herein I will ponder four such major 'transitions' in the epidemiology of human viral infections, in the spirit of the major transitions in evolution outlined by John Maynard Smith and Eörs Szathmáry in their thought-provoking book of the same name (Maynard Smith and Szathmáry 1995), and which increase complexity at each stage. These transitions are (i) the evolution of farming, (ii) the onset of urbanization, (iii) the rise of global travel, and (iv) the modern human world, characterized by major changes in land use exemplified by widespread deforestation, highly connected networks, and widespread immunodeficiency, concurrent with the rise of HIV infection (Holmes 2008).

6.5.1 The transitions

The abandonment of the hunter–gatherer lifestyle for the sedentary ways of farming some 10 000 years ago was clearly a major turning point in human history. This transition would have changed the burden of RNA virus infections on humans in two ways: by increasing population size and density (i.e. the CCS), thereby allowing acute RNA viruses to spread more easily through human populations, and by bringing humans into closer contact with animal species, in turn increasing the likelihood of cross-species transmission (Dobson and Carper 1996; Diamond 2002). Indeed, palaeo-pathological records indicate that the health of the first farmers was often inferior to that of their hunter–gatherer ancestors (Strassman and Dunbar 1999). In a similar fashion, the major effect of urbanization on the epidemiology of human viral infections would have been to greatly increase the number of susceptible hosts, as well as the contact networks among them (and also making these networks more complex), thereby increasing probabilities of transmission. The rise of urbanization during the last 5000 years may therefore mark the point in time when acute RNA viruses were first able to sustain themselves endemically in humans, without continual replenishment from an animal reservoir population. Measles represents a classic example. The CCS for measles has been estimated at between 250 000 and 500 000 (Bartlett 1957), a population size that would not have been established until the rise of cities. Unfortunately, there are currently no examples where the molecular clock analysis of a human RNA virus has unequivocally demonstrated an ancestry that dates back to the rise of cities or of agriculture, although the latter has recently been suggested for families of plant RNA viruses (Fargette *et al.* 2008b; Gibbs *et al.* 2008b).

The obvious importance of global travel as a major ecological transition for human disease is that it allowed the rapid dissemination of viruses to diverse geographical areas. Further, as widespread travel developed relatively recently, the history of human global travel and population subdivision is often written into viral genomes and so relatively easy to reveal through phylogenetic analysis (Holmes 2004). One particularly important historical facet of global travel was the slave trade, which has long been considered as a major factor shaping the spatial distributions of infectious diseases in humans. The importance of this sorry episode in human history is clearly demonstrated in the spread of YFV, long the scourge of humans living in tropical regions (see section 7.2). Phylogenetic analyses provide compelling evidence that YFV moved from Africa to the Americas at the time of the slave trade, in terms of both its pattern of spread and its timing (Bryant *et al.* 2007). The more recent influence of global travel and migration on the spread of human viruses can be seen with HIV-1. In this case, molecular clock analyses suggest that subtype B of HIV-1 (the subtype first described in the 1980s) spread from Africa to the western world during the 1960s, following the return home of Haitian workers formally based in the Belgium Congo (Gilbert *et al.* 2007).

Finally, it is clear that the ecology of contemporary human populations has greatly assisted the origin and spread of new viral infections. After all, it is this that has stimulated the current interest in 'emerging viruses'. There are a range of ecological factors that have facilitated the emergence of new viral infections, or allowed older pathogens to expand their geographical range, including rapid global travel, famine, war, the growth of mega-cities, and changes in agricultural practices and land use such as deforestation. Indeed, it is likely that there is an ecological cause, either direct or indirect, for all emerging diseases of humans (Morse 1995). A powerful example is provided by the emergence of HIV in central West Africa, which is associated with the rise of the logging industry and its impact on the hunting of local bushmeat (Wolfe et al. 2005). This important change in land use meant that humans encroached more on the habitats of those non-human primates that harboured SIVs, resulting in multiple cases of cross-species transmission (section 7.2).

6.5.2 Immunodeficiency and disease emergence

A final aspect of modern human ecology that may, in theory, assist in the emergence of human viral infections is widespread immunodeficiency, closely associated with the AIDS pandemic. Because of a lack of routine anti-retroviral therapy in sub-Saharan Africa where HIV prevalence is at its highest, the majority of those who are HIV infected in this region will go on to develop profound immunodeficiency manifest as full-blown AIDS. As well as having a major impact on human mortality, demography, and work patterns, which may itself affect the epidemiology of other pathogens (for example, by changing the age structure of the population), high levels of immunodeficiency might also impact on the process of viral emergence. If successful emergence is dependent on breaching immune barriers as well as simply infecting new cell types, then widespread immunodeficiency may allow other infectious agents that would have otherwise been cleared by healthy immune systems to gain a foothold in the human population (Weiss 2001). Although there is currently no evidence for this process, it is worth recalling that the global spread of HIV has stimulated a resurgence in opportunistic pathogens like *Mycobacterium tuberculosis*, which thrive in an environment of weakened immune systems. Similarly, there is also a growing body of data to suggest that normally acute RNA viral infections, such as influenza virus and poliovirus, can establish persistent infections in the face of immunodeficiency (Evans and Kline 1995; Klimov et al. 1995; Kew et al. 1998; Martín et al. 2000; Boivin et al. 2002). The elongated infectious period associated with immunodeficiency may therefore allow acute RNA viruses jumping from other host species to more readily acquire the array of mutations required for human adaptation.

7

Case studies in RNA virus evolution and emergence

The aim of this penultimate chapter is to put the previous, somewhat general, discussions of RNA virus evolution into the context of real, at-the-coalface, biology, by describing how a small number of viruses evolve in nature. Each case study will consider an RNA virus with a very different biology, epidemiology, and relationship with its host species. I have also selfishly chosen those viruses to which my own research has taken me. Because of space limitations, and the familiarity of some of these viruses, I cannot cover every aspect of their evolution and emergence, and I will bypass most of their basic biology. Rather, I have focused on those areas that best serve to illustrate the major points covered in this book. When reading this chapter it is also important to keep in mind that our success in unravelling the evolutionary history of RNA viruses has resulted from a happy marriage of increased amounts of gene and genome sequence data with new methods of phylogenetic analysis (for example, Drummond *et al.* 2006).

7.1 The evolutionary biology of influenza virus

7.1.1 The diversity of influenza virus

Influenza viruses are a textbook study in viral evolution, highlighting many of the main issues raised in this book. The viruses themselves possess a segmented ssRNA− genome and cause regular seasonal epidemics in humans, other mammalian species, and birds. They are classified within the family *Orthomyxoviridae* and fall into three 'types'—A, B, and C—with types A and B constituting the most closely related pair (Suzuki and Nei 2002b). As those viruses assigned to type A have the most profound implications for human disease, and exhibit the most genetic diversity, I will only consider their evolution here. Remarkably, the annual mortality caused by influenza is still estimated at between 250 000 and 500 000 globally (World Health Organization 2003), although most deaths are due to secondary bacterial pneumonia. Occasional global pandemics can result in 20–40% of the population being infected in a single year, with normal case fatality rates of approximately 0.1% (Taubenberger *et al.* 2001). As has been well documented, the influenza A virus pandemic of 1918–1919 (the misleadingly named 'Spanish flu') may have caused over 50 million deaths globally, making it the single most devastating outbreak of infectious disease in human history (Johnson and Mueller 2002).

The biodiversity of influenza A virus is usually described in terms of the particular combination of HA and NA surface glycoproteins present in a specific virus. There are currently 16 known subtypes of HA and nine subtypes of NA, all of which are found in wild birds of the orders *Anseriformes* and *Charadriformes*, and which represent the natural reservoir species of influenza A virus (Slemons *et al.* 1974; Webster *et al.* 1992; Fouchier *et al.* 2005; Obenauer *et al.* 2006; Olsen *et al.* 2006; Munster *et al.* 2007; Dugan *et al.* 2008). In total, at least 105 species of wild birds have been identified as harbouring influenza A viruses (Munster *et al.* 2007; Stallknecht and Shane 1988), and 103 of the 144 possible HA and NA subtype combinations have been identified. Influenza virus in wild birds replicates in the intestinal tract and is excreted in faeces in high concentrations for up to 30 days, with contaminated water efficiently transmitting the virus among hosts (Webster *et al.* 1978). Avian influenza viruses (AIVs) are also usually maintained as asymptomatic infections in their hosts, in which case they are referred to as low pathogenic avian influenza (LPAI) viruses, although some behavioural consequences of LPAI infection have been reported (van Gils *et al.* 2007). The importance of AIVs in the context of emerging disease is that they occasionally transmit to other species; either poultry (chickens and turkeys) or mammals (humans, horses, and pigs). These may result in either isolated outbreaks with little or no onward transmission, as has happened a number of times over the last 20 years (for example, involving subtypes H7N7 and H9N2), or less frequently in major human pandemics. Although there has understandably been a great deal of attention devoted to the possibly that H5N1 influenza virus will evolve stable transmission in humans, resulting in a global pandemic, the process of emergence is perhaps better illustrated by the recent cross-species transmission of H3N8 from horses to dogs (Crawford *et al.* 2005) and of H2N3 from birds to pigs (Ma *et al.* 2007).

It is this process of cross-species transmission from an avian reservoir that ultimately led to the global influenza pandemic of 1918, caused by a subtype H1N1 virus (Taubenberger *et al.* 1997). However, the origins of the 1918 virus remain contentious; in particular, whether it jumped to humans directly from an avian reservoir, or first circulated in another mammalian host such as pigs before infecting humans (Taubenberger *et al.* 2005; Antonovics *et al.* 2006; Gibbs and Gibbs 2006b). Although there is no direct phylogenetic evidence to suggest that the 1918 virus jumped into humans straight from birds (and on the current limited data this virus is more closely related to pig than bird viruses), an analysis of mutational patterns provides compelling evidence that the genes of the 1918 virus have an avian-like nucleotide composition (Rabadan *et al.* 2006). Resolution of the ultimate origin of the 1918 pandemic is clearly going to require the isolation and sequencing of more 'archival' viruses, particularly those from the first, less virulent, Spring wave of 1918.

Since 1918, five of the eight genome segments of human influenza A virus have maintained an unbroken evolutionary history without reassortment with viruses from other animal species: those encoding the matrix proteins (M1/2), the nucleocapsid protein (NP), the non-structural proteins (NS1/2), and two of the three polymerase proteins (PB2 and PA) (Taubenberger *et al.* 2005). In contrast, new HA and NA segments, as well as the PB1 polymerase segment, have been periodically acquired through reassortment with AIV. As is well documented, these reassortment events also coincided with global pandemics in humans: HA subtype H2 and NA subtype N2 emerged in 1957 ('Asian flu'), HA subtype H3 appeared in 1968 ('Hong Kong flu'), and a new PB1 segment was acquired in both the 1957 and 1968 pandemics. Although H1N1 viruses re-emerged in humans in 1977 (Scholtissek *et al.* 1978) (most likely due to a laboratory escape) and continue to circulate, the seasonal epidemics of influenza A virus in humans that have occurred since 1968 have been dominated by viruses of the H3N2 subtype (Fig. 7.1).

7.1.2 The evolution of avian influenza virus

Although species jumps of AIV to humans understandably dominate our thinking on influenza emergence, in reality most cross-species transmission events involve wild birds passing their viruses to poultry species. Importantly, these events are sometimes associated with a change in virulence, in which poultry-adapted subtypes, notably H5 and H7, evolve into high pathogenic avian influenza (HPAI) viruses following the insertion of a run of basic amino acids at the HA1–HA2 cleavage site (Baigent and McCauley 2003). Subtypes H7N7 and H9N2 have also been directly transmitted to

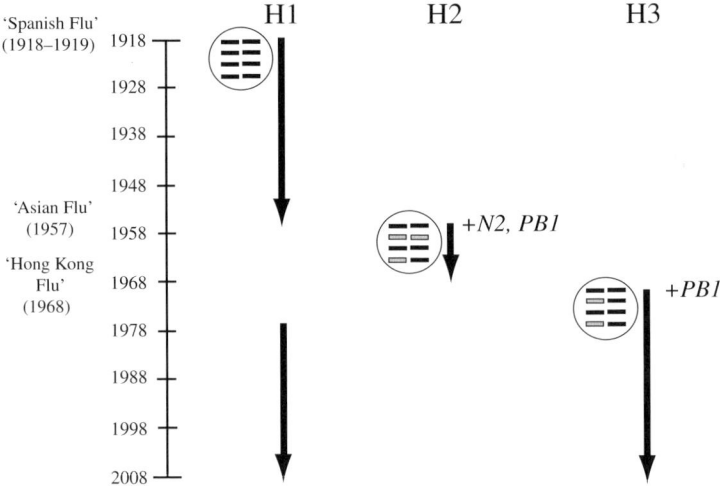

Fig. 7.1 Epidemiological history of influenza A virus in humans. The figure shows the timeline of pandemics and the emergence of different HA and NA subtypes through reassortment; H1N1 (1918, and again in 1977), H2N2 (1957), and H3N2 (1968). The circles represent the segment composition of each genome. Note that new PB1 segments were acquired in both the 1957 and 1968 pandemics, and that all gene segments circulating in humans today have their ultimate origins in the 1918 virus.

humans, causing severe and occasionally lethal disease (Peiris *et al.* 1999; Fouchier *et al.* 2004), although with no onward transmission. The current outbreak of H5N1 is especially notable because this virus, unusually, also causes lethal infections in wild birds as well as humans (Chen *et al.* 2005).

One of the most important concepts in evolutionary studies of influenza is that AIV in its natural reservoir species of wild water birds is in some form of evolutionary 'stasis' (Webster *et al.* 1992). To anyone reading this book, the idea that an RNA virus can exhibit stasis at the genotypic level, particularly in one which such a rapid replication cycle as influenza virus, should be an anathema. However, for those trained in more classical virology, the concept of 'evolutionary stasis' was equivalent to 'antigenic stasis', such that the serotypes of AIV do not make the regular changes in antigenicity that characterize human influenza A virus. This then represents another example of the perennial confusion over the difference between antigenic and genetic evolution mentioned in section 3.1. The recent expansion in genome sequence data indicates that both interpretations are correct. Hence, although overall substitution rates in AIV are not much lower than those observed in mammalian influenza viruses, and certainly within the range seen in other RNA viruses (Chen and Holmes 2006), it is also true that rates of nonsynonymous change are reduced (i.e. lower d_N/d_S values), reflecting a lack of antigenic evolution in the HA and NA and particularly strong selective constraints in the genes that comprise the polymerase complex (Obenauer *et al.* 2006; Dugan *et al.* 2008). In addition, the genetic divergence between the 16 subtypes of HA and the nine subtypes of NA (and also the two major alleles—A and B—of NS1) is extensive, and characterized by a marked lack of intermediate lineages, such that each subtype appears to represent a distinct 'fitness peak' separated by deep fitness valleys (Dugan *et al.* 2008) (Fig. 7.2).

The expanded species and genomic surveys undertaken in recent years have also revealed that AIV genomes in wild birds experience extremely high rates of reassortment. In this respect, AIVs from wild birds can be thought of comprising a pool of functionally equivalent, and so often interchangeable, gene segments that form transient 'genome constellations' (Dugan *et al.* 2008) (Fig. 7.2b). Hence, in contrast to the situation seen in mammals, there does not appear to be a strong selective pressure for specific segment combinations (H1N1, H3N2, etc.; see below). This distinctive fitness landscape is probably generated by complex patterns of cross-immunity—such that natural selection disfavours viruses that are antigenically similar, resulting in the discrete HA, NA, and NS types—in combination with some geographical and ecological partitioning, particularly between birds that occupy non-overlapping flyways.

A very different situation is seen in mammals, including humans. In this case, distinct eight-segment genome configurations of influenza viruses spread through populations, having ultimately jumped from the AIV gene pool. Because a single segment combination dominates global infections these might almost be thought of as 'clonal' outbreaks, although this does not exclude frequent intra-subtype reassortment. In addition, because humans represent a large and spatially mixed population, natural selection is able to act with far greater efficiency on individual subtypes, allowing

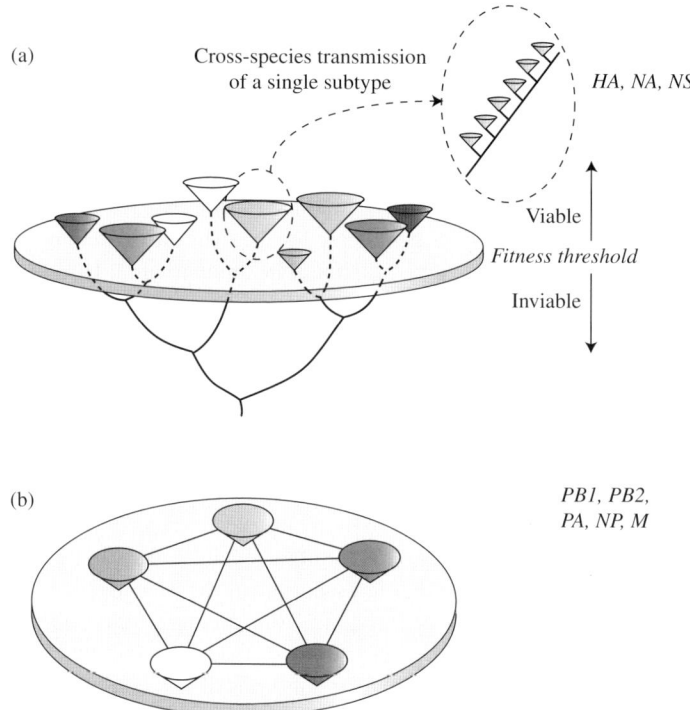

Fig. 7.2 The fitness landscapes of AIV. (a) Fitness landscapes of the HA, NA, and NS segments, and represented here by NA. Each cone represents an individual subtype, which are connected by a bifurcating phylogenetic tree. The lack of 'intermediate' subtypes—those falling below the disc—reflects the presence of steep fitness valleys, most likely due to strong cross-protective immunity. Occasionally, specific viral subtypes cross species boundaries and emerge in humans, where they experience a continue selection pressure and accumulate amino acid substitutions through antigenic drift. (b) The very different fitness landscapes observed in the PB2, PB1, PA, NP, and M segments from birds. In this case, there is little functional difference among the genetic variants of each segment, producing a flat fitness landscape and allowing frequent reassortment (represented by the horizontal lines). Adapted from Dugan *et al.* (2008) with permission.

the HA segment in particular to evolve rapidly at antigenic sites (Fitch *et al.* 1991; Ferguson *et al.* 2003; see below).

The recent emergence of HPAI H5N1 virus represents an important exception to this avian/mammalian divide, as these viruses possess evolutionary rates and d_N/d_S ratios closer to those seen in human than AIVs (Chen and Holmes 2006). This is most likely a function of the fact that H5N1 is spreading in large poultry populations, which allows natural selection to efficiently drive rapid antigenic change (Dugan *et al.* 2008).

7.1.3 Antigenic drift and shift

Despite the growing interest in AIVs, most research has understandably focused on those viruses that infect humans. Perhaps the most famous inference here is that there is a major division between the seasonal or epidemic evolution of the virus, manifest as the 'antigenic drift' of HA (and to lesser extent NA), and the evolution of pandemic forms, in which new viral segments enter the human population through reassortment with AIV. This latter process is usually termed 'antigenic shift'. While there is no doubt that the division between antigenic drift and shift is broadly correct, comparative analysis of genome sequence data has revealed that our understanding of these processes needs some refinement (reviewed in Nelson and Holmes 2007).

Antigenic drift occurs because the human immune response to viral infection is leaky rather than completely cross-protective, so that natural selection favours amino acid variants of the HA and NA proteins that evade immunity, infect more hosts, and hence proliferate (Fitch *et al.* 1997). Although both the HA and NA proteins contain antigenic sites where immune-driven natural selection may occur, the HA1 domain of the HA protein contains the highest concentration of epitopes and correspondingly experiences the strongest positive selection (Ina and Gojobori 1994; Fitch *et al.* 1991, 1997; Bush *et al.* 1999; Suzuki 2006b). This explains why the H3N2 component of the influenza vaccine needs to be updated so regularly. At the phylogenetic scale, this continual selective turnover of amino acid variants produces the distinctive 'ladder-like' (or 'cactus-like') phylogenetic tree of the HA1 domain, in which a single main trunk lineage depicts the pathway of advantageous mutations fixed by natural selection from past to present (Fitch *et al.* 1991, 1997; Ferguson *et al.* 2003) (Fig. 7.3). In contrast, the short side branches stemming from this main trunk represent those viruses that die out either because they were insufficiently antigenically distinct to evade immunity, or because they carried deleterious mutations which seriously hampered their fitness. Although this is an old concept, there is still considerable debate over what virological and/or epidemiological processes so strongly favour the survival of a single HA1 trunk lineage in human H3N2 viruses, while multiple lineages co-circulate more frequently and antigenic evolution proceeds more slowly in populations of human H1N1 (Ferguson *et al.* 2003; Rambaut *et al.* 2008), equine H3N8 (Daly *et al.* 1996), and influenza virus types B (Yamashita *et al.* 1988; Kanegae *et al.* 1990) and C (Buonagurio *et al.* 1985).

Although antigenic changes in the HA are clearly important determinants of viral fitness, this essentially 'progressive' model of influenza A evolution was largely based on studies of the HA1 domain in isolation. As such, the characteristic ladder-like phylogeny depicts the evolutionary history of a single protein, not the entire virus. Indeed, detailed phylogenetic analyses of large-scale genome sequence data from isolates of H3N2 has revealed that the evolutionary pattern observed in the HA1 domain does not always apply to the rest of the genome (Ghedin *et al.* 2005; Holmes *et al.* 2005; Nelson *et al.* 2006). In particular, multiple viral lineages often co-exist on a limited spatial and temporal scale, reflecting the continual importation of genetic diversity

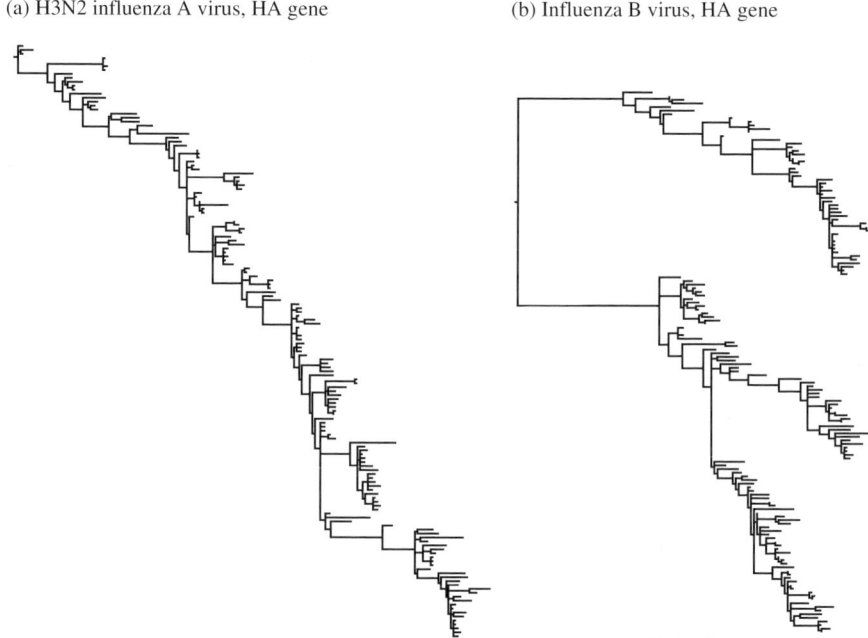

Fig. 7.3 Phylogenetic patterns in human influenza viruses. (a) Phylogenetic tree of the HA gene of H3N2 human influenza viruses (150 sequences) sampled between 1985 and 2005. Note the distinctive ladder-like branching structure indicative of continual antigenic drift. As a comparison, the phylogeny of 150 sequences of the HA gene of human influenza B virus (b) is shown over the same timescale. Note the slightly shorter root-to-tip time depth and the co-circulation of multiple lineages. All trees were estimated using the maximum likelihood method available in PAUP* (Swofford 2003).

into specific localities (Holmes *et al.* 2005; Nelson *et al.* 2006, 2008; Russell *et al.* 2008) (Fig. 7.4). In addition, that different segments can have very difficult phylogenetic histories reflects a high rate of reassortment (Holmes *et al.* 2005), a process that is considered in more detail below.

7.1.4 Antigenic cartography and the punctuated evolution of HA

It is also becoming increasingly clear that the antigenic drift of HA1 occurs in a more punctuated manner than previously realized, with periodic episodes of adaptive evolution that often have a major impact on antigenicity (Smith *et al.* 2004; Koelle *et al.* 2006; Nelson *et al.* 2006; Wolf *et al.* 2006). The episodic nature of the antigenic evolution of HA1 is especially well documented in antigenic maps—so-called 'antigenic cartography'—an innovative approach that has implications beyond its initial application to influenza (Smith *et al.* 2004). Antigenic cartography involves the

Fig. 7.4 Spatial structure of human H1N1 influenza A virus sampled from the 2006–2007 epidemic season in the USA. Eight phylogenetically distinct viral clades are present (A–H), all of which are likely to represent independent entries into the USA. Shaded rectangles contain individual isolates from the region of the USA associated with that shade (see map): region 1 (north-east), region 2 (mid-Atlantic), region 3 (south), region 4 (mid-west), region 5 (south central), and region 6 (west). Taken from Nelson et al. (2008) with permission.

construction of a matrix of haemagglutinin inhibition (HI) assay distances among viral isolates which are then plotted to produce a cartographic surface, analogous to a normal geographical map. Importantly, this map also enables a tentative link to be made between genotype and phenotype (Fig. 7.5). Antigenic maps of HA from H3N2 viruses sampled since 1968 reveal that major jumps between antigenically distinct clusters of sequences occur on a roughly 3-year periodicity (Smith *et al.* 2004). Although these antigenic 'cluster jumps' are also usually apparent as long branches on HA1 phylogenies, small genetic changes sometimes have a major effect on antigenicity. Further, as the cluster jumps tend to correspond to occurrences of vaccine failure (de Jong *et al.* 2000), they clearly represent a better predictor of antigenic novelty than genotypic data in isolation.

Unfortunately, the rules—should any exist—that govern the path that influenza A virus takes across the cartographic surface are still uncertain, and represent a major research goal (Boni *et al.* 2006). Further, and as discussed in more detail below, the

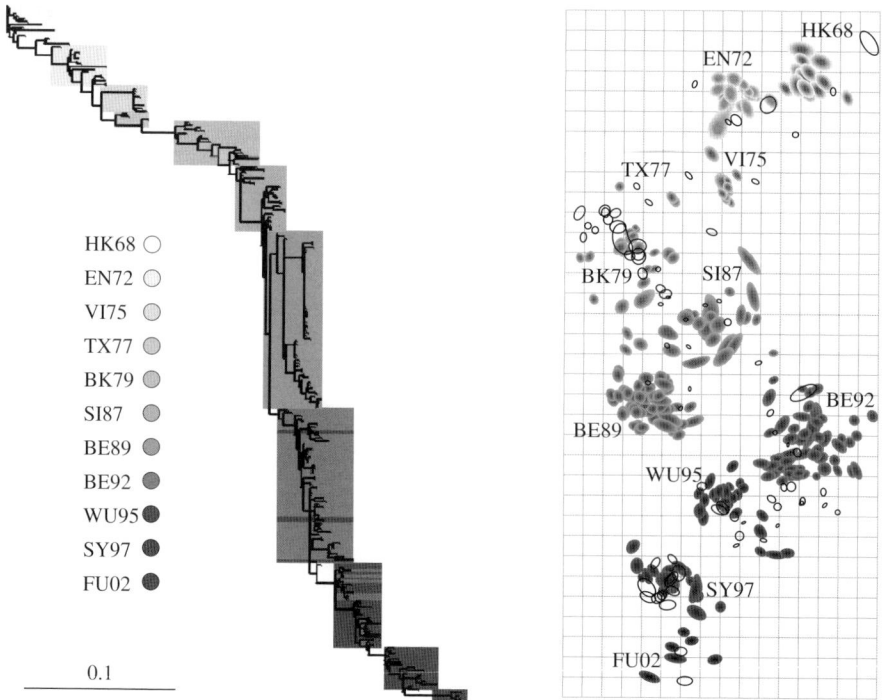

Fig. 7.5 Comparison of the phylogenetic and antigenic evolution of the HA1 domain of human H3N2 influenza A virus from 1968 to 2003. Antigenic clusters, such as FU02, are coded by shade in the same way in both panels. In the antigenic map (right), antigenic (HI) distance is represented by both the horizontal and vertical axes, with each grid square representing one unit of antigenic distance. Adapted from Smith *et al.* (2004) with permission.

antigenic evolution of HA cannot be taken out of context of that which occurs across the virus genome as whole. One very elegant model is that the evolutionary dynamics of HA are shaped by the epistatic interactions among the amino acids that make up this protein, such that they form a 'neutral network' (Koelle et al. 2006). The evolution of HA within the network is characteristically 'epochal', with occasional major bursts of amino acid change which in turn are associated with antigenic cluster jumps. However, there is currently no evidence that influenza, or any other RNA virus, evolves in a manner that conforms to a neutral network. Indeed, a key argument of this book is that the limited genomic space available to RNA viruses, and the complex constraints to which they are subject, make the existence of truly neutral networks unlikely.

There is also growing evidence that antigenic drift does not occur within the time frame of a single epidemic season in a single locality. Specifically, phylogenetic studies of sequences obtained from the USA (Nelson et al. 2006, 2008), France (Lavenu et al. 2006), and Denmark (Bragstad et al. 2008) have shown that few amino acid changes are fixed in HA1 within populations at the seasonal scale, with the importation of viruses a far more important source of genetic diversity. Shortly, I will explain the apparent paradox for why antigenic drift in HA1 is commonly observed a global scale, yet rarely within individual populations.

As has been noted many times previously, severe influenza pandemics may occur following an antigenic 'shift', in which a reassortment event generates a novel combination of HA and NA antigens to which essentially the whole population is immunologically naïve. As noted above, the segmented genome of influenza virus facilitates reassortment between isolates that co-infect the same host cell. There is growing evidence for the importance of reassortment in seasonal influenza, particularly as revealed through the occurrence of phylogenetic incongruence in genome sequence data (Nelson et al. 2006). Further, these estimates of reassortment frequency are likely to represent significant underestimates, because some reassortment events are undetectable by phylogenetic analysis, or result in unfit progeny. It is therefore essential that methods are developed that are better able to estimate the intrinsic rate of reassortment, particularly relative to that of mutation (Nelson and Holmes 2007).

7.1.5 Genome-wide evolutionary processes

Reassortment also appears to be important in the generation of evolutionary novelty in influenza A virus. As a case in point, the antigenic cluster jump between the SY97 and FU02 strains coincided with a reassortment event and resulted in a new antigenic type (FU02) that became globally dominant and led to a major vaccine failure (Holmes et al. 2005). In short, reassortment allowed influenza A virus to place a fit HA segment in a compatible genomic background, thereby increasing fitness. Not to overly labour the point, but this example neatly illustrates why the evolutionary dynamics of the HA must be considered within the context of that occurring at the genomic scale.

The importance of documenting viral evolution at a genomic scale is also apparent when the changing patterns of genetic diversity in H1N1 and H3N2 viruses are analysed through time. A detailed study of many hundreds of complete genome sequences sampled over an approximately 10-year period from the USA and New Zealand, and illustrative of the northern and southern hemispheres, respectively, revealed periodic reductions in genetic diversity, and sometimes involving all eight genome segments, set against a background of frequent reassortment (Rambaut *et al.* 2008) (Fig. 7.6). Consequently, reassortment appears to be the norm, rather than the exception, in the seasonal evolution of influenza A virus. The HA segment seemed to be especially prone to reassortment, events that were sometimes associated with antigenic cluster jumps, again suggesting that there is a major epistatic component to fitness. Finally, despite the seasonal crashes in genetic diversity—which were more dramatic in H3N2 than H1N1 for reasons that are still unclear—the reappearance of past viral lineages strongly suggested that genetic diversity is maintained in a global reservoir population. A combination of antigenic cartography and extensive phylogenetic analysis then revealed that the probable location of this global reservoir population was East and South-east Asia (Russell *et al.* 2008). Specifically, a large-scale transmission network exists in this region, where big and

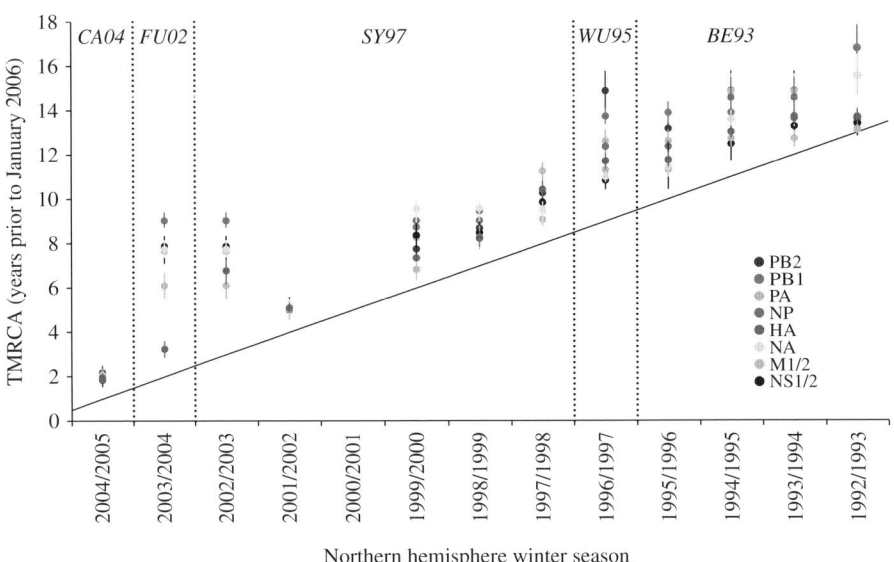

Fig. 7.6 Population genetic history of human influenza A virus. The figure shows the time to the most recent common ancestor (TMRCA) of each genomic segment for isolates of H3N2 circulating each season in New York State, USA. Values shown represent the mean and 95% highest posterior density (HPD) intervals for TMRCAs estimated across the trees sampled using a Bayesian MCMC method. The diagonal line goes through 1 January of each season, approximating the seasonal mid-point in the northern hemisphere. The timescale of major antigenic changes in the USA is also depicted (see Fig. 7.5). Adapted from Rambaut *et al.* (2008) with permission.

well-mixed human populations combined with more annual influenza transmission allow natural selection to proceed with great efficiency, generating the antigenic variants that are then exported worldwide to ignite the influenza epidemics we see each winter. Influenza A virus therefore fits a source–sink epidemiological model at a global scale (see section 6.4), and it is this that explains why antigenic drift is important globally, yet not so at the level of individual populations in the northern and southern hemispheres (Fig. 6.8). Of more tangible benefit, recognition of the importance of genomic scale evolutionary processes, as well as the identification of the global source population, should assist in the surveillance of new antigenic variants and hence in vaccine design.

7.2 The emergence and evolution of HIV

No virus has generated as much interest from evolutionary biologists as HIV, the causative agent of AIDS. Not only has the evolutionary work on HIV provided critical insights into the origin and spread of this virus, but the vast amount of sequence data obtained, combined with rapid evolution, has meant that HIV has become an important testing ground for new methods of sequence analysis (section 3.3). It is even arguable that HIV ignited the current interest in emerging diseases that has shaped much of the research conducted on viral evolution, including that covered in this book. Lastly, the case of HIV illustrates beautifully many of the fundamental concepts in viral evolution, such as the respective roles of mutation, recombination, natural selection, and demographic history in shaping genetic diversity, and the factors that mediate successful cross-species transmission. As a consequence, HIV serves as a rich case study in viral emergence and evolution. However, because of the huge and familiar literature on this subject, I will necessarily restrict my comments to a few key issues that fit the general theme of this book, particularly its emergence and evolutionary dynamics at the intra- and inter-host levels.

7.2.1 A brief history of HIV/AIDS

Although the virus and disease had clearly been in existence for some time, the modern history of HIV/AIDS starts in 1981 when reports of rare opportunistic infections, such as pneumonia caused by the fungus *Pneumocystis carinii*, were recorded in the gay communities of several large US cites (Gottlieb *et al.* 1981). As well as suffering from unusual microbial infections, the individuals in question in fact possessed a severe immunodeficiency, such that the bodies' immune system is unable to fight off many of the pathogens to which we are commonly exposed. Acquired immune deficiency syndrome (AIDS) was born.

Early epidemiological work established that as well as gay men, AIDS was also present in populations of injecting drug users and haemophiliacs. Such clustering provided a strong indication that the disease was caused by the transmission of an infectious agent—such as a virus—and most likely one that was blood-borne. This prediction was confirmed dramatically in 1983 with the discovery of the causative agent—a retrovirus eventually christened the human immunodeficiency virus type 1 (HIV-1) (Barré-Sinoussi et al. 1983). Other significant milestones in the early period of AIDS research were: (i) the realization that rather than being restricted to specific risk groups in the industrialized world, the virus was in fact most commonly found in individuals from sub-Saharan Africa and predominantly transmitted by heterosexual intercourse; (ii) the discovery of a second 'type' of HIV with a rather different genome organization (HIV-2), although one restricted to persons of West African origin; (iii) the discovery of closely related viruses in a variety of African non-human primates, including sooty mangabeys and chimpanzees (Huet et al. 1990); (iv) the identification of the primary cellular receptor of HIV—CD4—although other chemokine co-receptors were indentified in the 1990s; and (v) the development of the first anti-retroviral drug to fight HIV infection—AZT—which, after initially promising results, in reality has little positive impact as a monotherapy.

Since its discovery, the burden of mortality, morbidity, as well as the linked demographic and economic costs, caused by HIV/AIDS has been depressingly immense. The UNAIDS organization estimates that 33.2 million people were living with HIV/AIDS at the end of 2007, some 22.5 million ($\approx$68%) of whom reside in sub-Saharan Africa (UNAIDS 2007). The adult (age range 15–49) prevalence of HIV in this region reaches 40%, an increase from less than 1% in the early 1980s. Other geographic regions suffering a major burden of HIV are South-east Asia (4 million HIV carriers), Latin America (1.6 million), and Eastern Europe and the former Soviet Union (1.6 million), with the latter largely associated with injecting drug use. Overall, some 2.5 million people were newly infected with HIV in 2007, equating to approximately 7000 new infections each day, and resulting in 2.1 million AIDS deaths. Since its initial description, an estimated 25 million people have died of AIDS (www.avert.org/worldstats.htm).

Despite the surprise of its emergence, as well as its devastating effect on human populations, HIV is not unique. The virus has a number of relatives—the lentiviruses—that infect horses (equine infectious anaemia virus; EIAV), goats (caprine arthritis encephalitis virus; CAEV), sheep (visna virus), cattle (bovine immunodeficiency virus; BIV), and felids, including domestic cats (feline immunodeficiency virus; FIV). As will be discussed in more detail below, even more closely related viruses infect a wide range of non-human primates (simian immunodeficiency virus; SIV), and are central to understanding the origin of HIV. It is inevitable that a more expansive survey of mammalian species will uncover even more lentivirus infections, particularly given the recent discovery of endogenous copies (Katzourakis et al. 2007; Gifford et al. 2008). The wide diversity of mammals that carry these viruses, as well as the existence of endogenous copies without infectious relatives, not only suggests that they are an ancient viral family, but that there has been a regular birth and death of viral lineages.

7.2.2 The genetic diversity of HIV

Right from the earliest descriptions of genetic variation in HIV-1 it was clear that this virus was remarkably diverse, both within individual hosts (Hahn et al. 1986; Balfe et al. 1990; Holmes et al. 1992) and globally (Korber et al. 1995). As the worldwide sample of HIV-1 began to expand it became clear that viral genetic diversity could often be partitioned into discrete clusters, or clades, on phylogenetic trees, that were eventually christened 'subtypes'. At the time of writing there are nine such subtypes of HIV-1 (denoted A–K, but excluding E and I), as well as a growing database of circulating recombinant forms (CRFs), representing inter-subtype recombinants, some of which are remarkably complex in that they incorporate genetic material from multiple subtypes. Some CRFs have reached relatively high frequencies, with a worrisome example provided by the BF recombinant that circulates widely in Latin America (Carr et al. 2001), and which appears to be spreading more rapidly than its parental subtypes (Aulicino et al. 2007). The frequency with which CRFs are observed clearly reflects how weak a cross-protective immune response is elicited by HIV infection, and highlights the potential for recombination to have a major influence on the genetic structure of HIV. Indeed, inter-subtype recombination is so pervasive in HIV that is arguable that discrete subtypes do not exist at all. The subtypes (and CRFs) of HIV-1 are also notable for their differing geographical distributions: subtype B represents the form of the virus first observed in industrialized nations during the early 1980s and which still dominates in these regions to this day, whereas subtypes A, C, and D are more commonly found in sub-Saharan Africa, with subtype C rising dramatically in frequency, particularly in southern Africa (Fig. 7.7). The other subtypes are found at rather lower frequencies. Phylogenetically defined subtypes of viruses have also been identified in HIV-2, denoted 'epidemic' subtypes A and B and non-epidemic subtypes C–G, although all are usually restricted to West Africa (Lemey et al. 2003).

The identification of lentiviruses in a wide range of non-human primates, particularly chimpanzees (Huet et al. 1990; Santiago et al. 2002, 2003; Nerrienet et al. 2005; Keele et al. 2006), changed the evolutionary context of genetic variation in HIV-1. Specifically, the subtypes and CRFs of HIV-1 described above fall into a single branch on the HIV phylogeny, reflecting a single cross-species transmission from chimpanzees (Fig. 7.8). This cluster of viruses is denoted the M, or Main, group, and contains the vast majority of viruses assigned to HIV-1. Strikingly, two other clusters (groups) of HIV-1 isolates have been identified, although present at far lower frequencies: O, for Outlier, and N for New. That these groups are separated from the M group by SIVcpz from chimpanzees is powerful evidence not only that chimpanzees are the ultimate source for HIV-1, but that species jumps have occurred a number of times (Fig. 7.8; see below). Also of note is that the greatest phylogenetic diversity in HIV (i.e. HIV-2, the M, N, and O groups of HIV-1, and extensive diversity within the M group) is observed in central-west Africa, the most likely birth place for this virus.

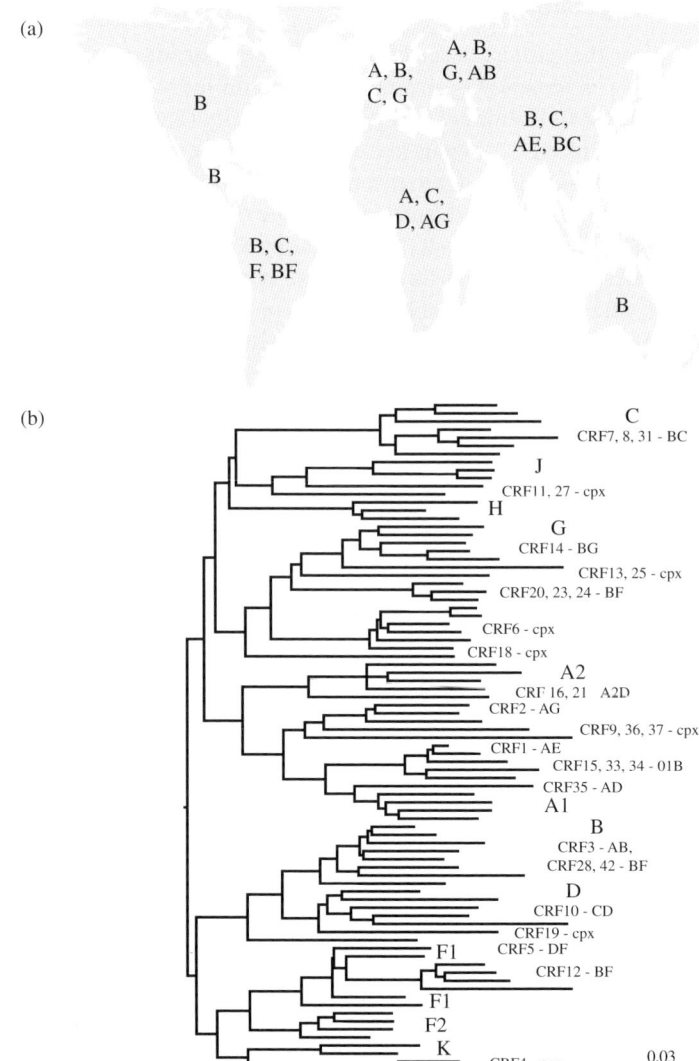

Fig. 7.7 The phylogenetic relationships and global distribution of the subtypes and circulating recombinant forms (CRFs) of HIV-1. The phylogeny was estimating using 83 reference sequences of the *pol* gene (3108 nt) of HIV-1 taken from the Los Alamos database (www.hiv.lanl.gov/content/index). All branch lengths are drawn to a scale of nucleotide substitutions per site, and the tree is mid-point root for purposes of clarity only. Note that because of frequent recombination the subtype structure is no longer clear and different genes will produce different trees.

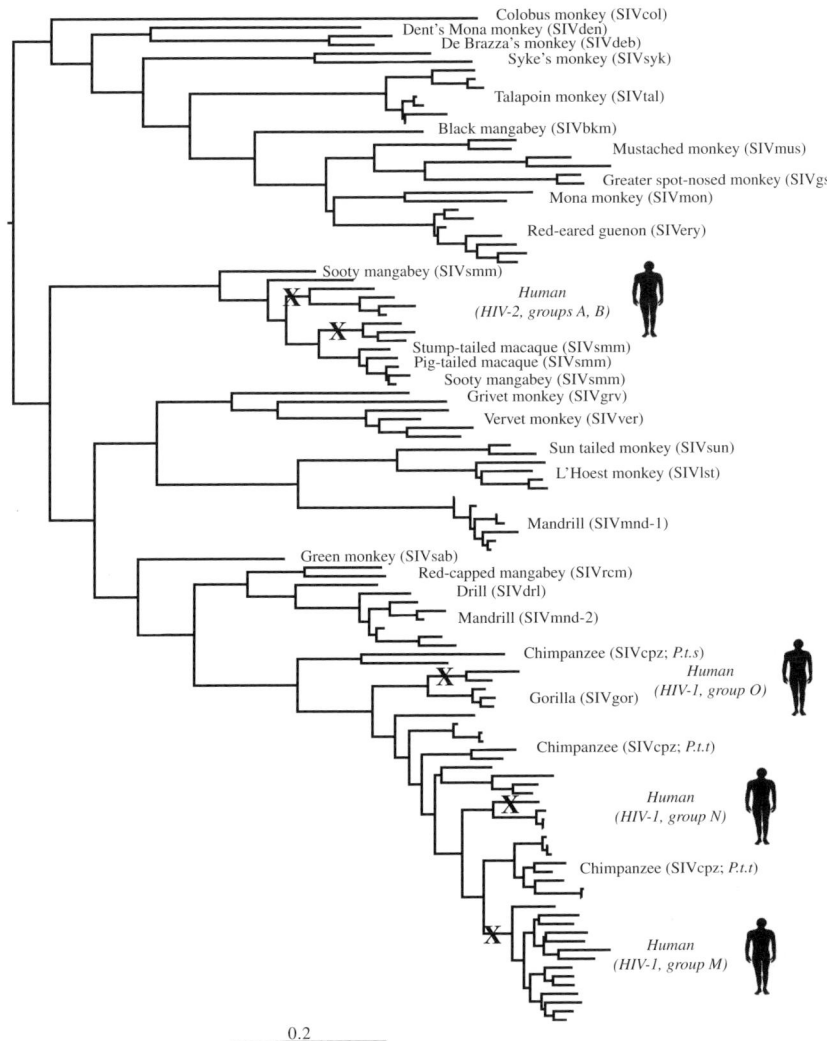

Fig. 7.8 Phylogenetic relationships of the primate lentiviruses showing the cross-species transmission events that led to the emergence of HIV (and marked by the X symbol). *P.t.t* refers to *Pan troglodytes troglodytes* whereas *P.t.s* refers to *Pan troglodytes schweinfurthii*, both subspecies of common chimpanzee. The tree was estimated using the Bayesian method available in the MrBayes package (Ronquist and Huelsenbeck 2003), based on 116 amino acid reference sequences (1108 residues) of the *pol* gene taken from the Los Alamos database (www.hiv.lanl.gov/content/index). All horizontal branch lengths are drawn to a scale of amino acid substitutions per site, and the tree is mid-point-rooted for purposes of clarity only.

7.2.3 What and why are subtypes?

What evolutionary processes are responsible for the population genetic structure of HIV? In particular, could the types, subtypes, and CRFs of HIV-1 differ in fitness, perhaps manifest as differences in virulence (i.e. the time from initial infection to the development of full-blown AIDS), which then explains their differing distributions and prevalences? Before discussing these issues in a detail, it is important to clarify exactly how fitness is measured. For many viruses—and HIV is no exception—it is fairly straightforward to measure fitness *in vitro* through growth assay kinetics. In some cases it is even possible to perform a more powerful 'competition assay', with the strain growing to the highest titre obviously that of highest fitness. While these experiments undoubtedly capture some aspects of viral fitness, particularly the relative abilities of different viruses to infect and replicate in different cell types, they do not speak to the fitness of a virus in nature as there is no consideration of R at the epidemiological scale. In the case of HIV, and perhaps many other viruses as well, there is a fundamental division between fitness as measured in cell culture and fitness as measured through R. For example, a series of elegant *in vitro* experiments have demonstrated that subtype B viruses systematically outcompete subtype C viruses, perhaps due to differences in cell binding and entry, indicating that the former have higher 'fitness' in the cell types assayed (Ball *et al*. 2003; Marozsan *et al*. 2005). While there is no denying the power of these experiments, they cannot be easily translated to the epidemiological scale as subtype C viruses are now globally the most common of all HIV-1 strains and increasing in prevalence.

These complexities notwithstanding, there have been a variety of proposals that the subtypes of HIV-1 differ in fundamental properties that reflect their global prevalence, most notably transmissibility. Hence, subtype C viruses have greater epidemic potential than subtype B viruses. There are, however, a number of reasons to be cautious about such inferences. First, as noted above, it is often difficult to extrapolate from *in vitro* to natural systems, and HIV is no exception. Second, when functional assays are performed on a viral 'subtype' it is inevitably the case that only a handful of isolates are ever surveyed, making it difficult to draw general conclusions, particularly for a virus as variable as HIV. This raises a more generic issue: because viral isolates cluster in a phylogenetic tree does not mean that they are necessary blessed with identical phenotypic properties. Similarly, differences in phenotype are not an automatic consequence of the existence of viral subtypes. Next, the high rate of recombination evident in HIV means that any extrapolation from genotype to phenotype is thwart with uncertainly. Finally, and perhaps of most importance, the current evidence for clade-specific differences in viral fitness is highly debatable. In particular, although it has been claimed that subtype C is more sexually transmissible than other subtypes (Iversen *et al*. 2005; John-Stewart *et al*. 2005), and might be associated with lower virulence (Ariën *et al*. 2007), this has yet to be tested rigorously. As a warning shot, initial suggestions that subtype E (now merged into subtype A) was more readily sexually transmissible than other subtypes, and which may have explained its close association with the explosive HIV epidemic in Thailand (Soto-Ramirez et al. 1996), have not withstood closer scrutiny.

If not fitness, what explains the subtype structure of the HIV-1 M group tree? The most likely explanation is a series of local founder effects (Rambaut et al. 2001). Specifically, viral lineages were by chance exported to other localities from a source population in the Congo region of Africa (Vidal et al. 2000). As these local outbreak strains continued to spread in their new populations they acquired those genetic differences (manifest as relatively long internal branches) that allowed their identification as specific subtypes (Fig. 7.9). Hence, if the subtypes differ fundamentally in phenotype, these traits were acquired in isolation, and not due to any selective pressure caused by inter-subtype competition. Evidence for this essentially 'neutral' model of epidemiological scale evolution is that when a tree is inferred using both the global subtypes of HIV and those viruses sampled from the Congo region of Africa, the distinct subtype structure breaks down (Fig. 7.9). As noted previously, frequent inter-subtype recombination will erode subtype structure even further.

7.2.4 The origins and spread of HIV

While some aspects of HIV research have made little progress, particularly the development of vaccines, the study of the origin and spread of HIV has proven remarkably successful. It would be fair to say that we now know where HIV comes from, with a plausible route of entry into human populations, as well as a rough timescale for these events. In the case of HIV-1 this means that the virus mostly likely emerged in the Congo region of central-west Africa during the first decades of the twentieth century and first entered human populations through exposure to contaminated bush meat from chimpanzees (Gao et al. 1999; Keele et al. 2006; Worobey et al. 2008). A similar picture can be painted for HIV-2, although the place of emergence is likely to be rather further west in Africa, and sooty mangabeys act as the reservoir species (Santiago et al. 2005). In both cases there also appears to have been multiple cross-species transmission events from non-human primates to humans.

As should be apparent from these statements, documenting the diversity of viruses that circulate in a wide variety of non-human primates is critical to understanding the origins of HIV. Although these are routinely referred to as the SIVs, that none seem to cause overt disease in the natural hosts means that the term immunodeficiency is something of a misnomer, so that the primate lentiviruses is a safer phrase. Indeed, SIV in its natural hosts is not associated with a decline in the number of CD4 T cells despite long-term infection and high levels of viraemia, and does not seem to generate an overly strong immune response (Broussard et al. 2001; Silvestri et al. 2003).

Primate lentiviruses are also remarkably abundant. At the time of writing at least 40 of these viruses associated with different primate species have been identified, largely within monkeys of the family *Cercopithecidae* and apes (chimpanzees, gorillas, and humans) of the family *Hominidae* (Hahn et al. 2000; Santiago et al. 2002; Keele et al. 2006; Van Heuverswyn et al. 2006). Crucially, these viruses are only

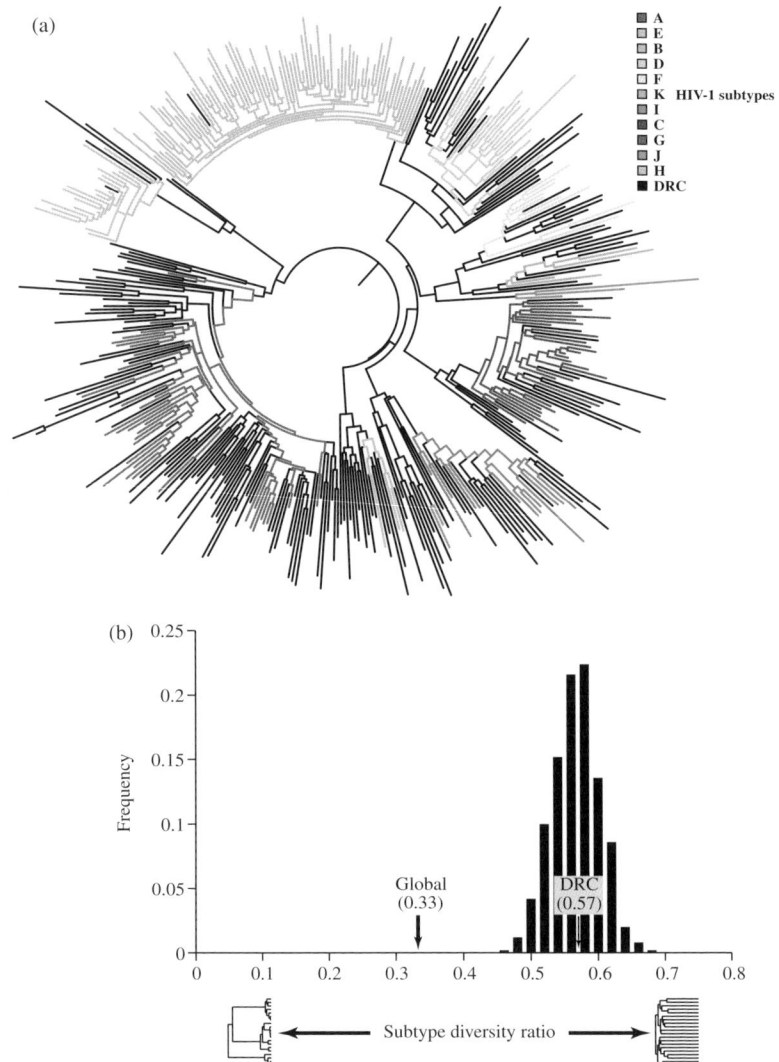

Fig. 7.9 The genesis of the subtype structure of HIV-1. (a) A phylogenetic tree of the V3–V5 region of the *env* gene of HIV-1 showing the different subtypes of HIV-1 (excluding CRFs). The lineages shown in black are from the Democratic Republic of Congo (DRC) and tend to fall in basal locations, suggesting that this population is ancestral. Tree kindly provided by Andrew Rambaut. (b) Analysis of the same tree using the subtype diversity ratio (SDR): the mean path length between tips of a specific subtype divided by the mean path length between tips of different subtypes. There is a far stronger subtype structure (low SDR) in the HIV-1 subtypes excluding the DRC data (Global = 0.33), compared to a tree of the Global and DRC data combined (Congo = 0.57). The latter SDR is also no different from than observed in simulated phylogenies experiencing exponential population growth with no subtype structure (black bars). Adapted from Rambaut *et al.* (2001).

found naturally in animals of African origin. Despite the evidence for the antiquity of the lentiviruses as a whole, this phylogenetic distribution strongly suggests that the *current* lineages of primate lentiviruses were acquired subsequent to the divergence between Old World and New World primates, and that there has been clear species jumping between those viruses that infect Old World monkeys and those that infect chimpanzees, gorillas, and humans (Fig. 7.8).

Despite these notable advances, determining the timescale of primate lentivirus evolution has proven more problematic. Initially, the observation that each species of primate seemed to carry its own unique lentivirus (such that they could be considered 'species-specific'), as well as suggestions that the phylogeny of the viruses matched that of the hosts, argued for long-term co-divergence, perhaps on timescales of millions of years (Allan *et al.* 1991; Beer *et al.* 1999; Bibollet-Ruche *et al.* 2004). However, such an ancient history sits in stark conflict with the timescale of lentivirus evolution inferred from molecular clocks and which is measured in thousands of years at most (Sharp *et al.* 2000; Holmes 2003c). Clearly, these very shallow date estimates are incompatible with a history of co-divergence. In addition, there are a growing number of mismatches between phylogenies of host and virus, indicative of widespread cross-species transmission. For example, the four viruses found in great apes—SIVcpz in chimpanzees, SIVgor in gorillas, and HIV-1 and HIV-2 in humans—are all clearly examples of cross-species transmission, despite how closely these hosts and viruses are related. The origin of SIVcpz is particularly informative, as this virus is a complex multi-species recombinant, with different genomic regions having ancestry in SIVrcm from red-capped mangabeys (*Cercocebus torquatus*) and SIVgsn from greater spot-nosed monkeys (*Cercopithecus nictitans*) (Bailes *et al.* 2003). The explanation for such a complex evolutionary history appears to be that chimpanzees often butcher other primate species for food, a process that provides an obvious route for the virus to jump species boundaries.

The uncertainty over the timescale of lentivirus diversification raises one of the central paradoxes in all of viral evolution: how is it possible to reconcile the observation that viruses can infect a wide range of host species and often at high prevalence, which is suggestive of an ancient ancestry, with very recent divergence times as estimated under molecular clocks? Indeed, if retroviruses (and other RNA viruses) evolve rapidly, then their genome sequences should be unrecognizably divergent after millions of years of evolution, perhaps aside from a few conserved motifs (Holmes 2003c). Yet, while sequence alignment is undoubtedly difficult in some gene regions, the primate lentiviruses are clearly closely related. The simplest explanation is that the lentiviruses that infect cercopithecoid monkeys evolve many orders of magnitude more slowly than other RNA viruses, as proposed for the retrovirus SFV (Switzer *et al.* 2005). However, while rates of nonsynonymous substitution are indeed rather lower in SIVs than HIVs, mostly likely reflecting reduced immune selection pressure, overall substitution rates are similar across the full range of primate lentiviruses, with viruses like SIVsm also showing substantial intra-host genetic diversity (Demma *et al.* 2005). Another possibility is that the models of nucleotide substitution used to estimate genetic distances are so flawed that they hugely over-estimate evolutionary rates on some lineages. While it is undoubtedly the case that these methods have room for improvement, particularly in their common assumption that each nucleotide evolves independently, it is difficult to envisage how an error of the magnitude required to reconcile the very different theories for the origin of primate lentiviruses is possible. For example, while the covarion substitution model (Fitch 1971; Huelsenbeck 2002) may constitute a better descriptor of sequence evolution than models that assume that all sites vary equally across lineages, an equivalent analysis of DENV revealed that it only has a relatively minor impact on estimated divergence times (Dunham and Holmes 2007). Similarly, although differing distributions of the shape parameter (α) of the gamma distribution of among-site rate variation can have a major impact on genetic distances (and phylogenetic accuracy) (Yang 1996), α would need to be unfeasibly low to greatly increase divergence times in the primate lentiviruses (Sharp *et al.* 2000). The most likely explanation for the recent common ancestry of the primate lentiviruses inferred from molecular clocks therefore rests with the general macroevolutionary model outlined in section 5.3: lentiviruses may have been associated with primates for millions of years but experience a continual process of extinction and reinvasion; that is, of lineage birth and death. The contemporary distribution of primate lentiviruses therefore reflects the inter-species spread of recent invasive viral lineages, where all older lineages have suffered extinction (Fig. 5.9).

7.2.5 The intra- and inter-host evolutionary dynamics of HIV

HIV was also the first virus where extensive studies of intra-host diversity and evolution were undertaken, and which have even contributed to theories of HIV pathogenesis (Nowak *et al.* 1991). The most basic observation here is that levels of genetic diversity are often extensive, comprising a turnover of genetic variants through time (Saag *et al.* 1988; Holmes *et al.* 1992; Shankarappa *et al.* 1999), a possible differentiation of virus into tissue-specific types, such as those sampled from blood, brain, and seminal fluid (Sanjuán *et al.* 2004a), and the generation of isolates with increased virulence (Fenyö *et al.* 1989). Patterns of genetic diversity also seem to change according to the time point within the infection cycle. Most notably, levels of genetic diversity are greatly reduced during primary infection (Zhang *et al.* 1993; Zhu *et al.* 1993), then generally increase as the infection proceeds (Shankarappa *et al.* 1999). While some of this initial reduction in genetic diversity is likely due to the population bottleneck that accompanies inter-host transmission, it is less clear what proportion of the genetic variation present in a donor is passed to a recipient at transmission, and whether particular viral variants are favoured by natural selection during the very early days of HIV infection. In the most extreme case, it has been proposed that the majority of HIV transmissions may have been initiated by a single infecting virus (Keele *et al.* 2008; Salazar-Gonzalez *et al.* 2008).

A major complicating factor in studying early HIV evolution is the long-standing confusion between 'transmission' and 'primary infection': just because a viral population is strongly homogeneous during primary infection does not necessarily mean that it was so at transmission some weeks earlier. In particular, it is possible that a diversity of virus particles initiate infection in a new host and that the variant with the highest replicative fitness then outgrows to dominate at primary infection, leading to a substantial reduction in genetic diversity through a selective sweep. In addition, the bottlenecks associated with some forms of transmission (sexual), may be greater than in others (vertical) (Scarlatti 2004). Obviously, estimating genetic diversity at the precise moment of inter-host transmission is extremely difficult to achieve for natural systems, so that most studies of this process are necessarily indirect (Keele *et al.* 2008). The detailed coalescent analysis of a rare case in which sequences were available from both the donor and recipient very close to the point of transmission revealed that over 99% of the genetic diversity in the *env* and *gag* genes was lost as the virus passed between hosts (Edwards *et al.* 2006b). Although this is clearly a substantial population bottleneck, that must have a major impact on evolutionary dynamics, it does not necessarily equate to the transmission of a single virus particle.

7.2.6 The great obsession moves to HIV

Even more controversial is determining what evolutionary processes shape the genetic diversity of HIV following primary infection. Indeed, the debate over whether the intra-host dynamics of HIV are dominated by stochastic (neutral) or deterministic (selected) processes has raged for more than 10 years. Crucially, if stochastic processes dominate HIV evolution, then there will be an inherent randomness in such phenomena as the evolution of drug resistance.

Some authors have suggested that values of d_N/d_S measured at the intra-host level indicate that HIV evolution is essentially a neutral process (Plikat *et al.* 1997; Shriner *et al.* 2004). While it is certainly possible that genetic drift can shape certain aspects of intra-host evolution, particularly during transmission bottlenecks or bottlenecks that follow the initiation of drug treatment, there is far more evidence to suggest that the intra-host evolution of HIV is dominated by natural selection, in both the purifying selection of deleterious mutants (Edwards *et al.* 2006a) and the positive selection of advantageous ones (Bonhoeffer *et al.* 1995; Nielsen and Yang 1998; Williamson 2003). In fact, it is arguable that the intra-host evolution of HIV-1 represents the most powerful force of positive selection recorded for an RNA virus, and perhaps for any organism. It is this capacity for strong natural selection that facilies the rapid evolution of drug resistance, particularly during monotherapy (Kellam *et al.* 1994), and explains why HIV is readily able to evade aspects of the natural immune response, be it innate, T-cell (Price *et al.* 1997) or antibody-based (Wei *et al.* 2003). As such, much of the notion that intra-host evolution is dominated by essentially neutral processes seems to be due to the use of inappropriate analytical methods. In particular, pairwise d_N/d_S methods that average over nucleotide sites experiencing mixed positive and purifying selection are well known to underplay the importance of adaptive evolution (Crandall *et al.* 1999; Zanotto *et al.* 1999).

More contentious are suggestions that genetic drift dominates the intra-host evolutionary dynamics of HIV because effective population sizes are far lower than the census population sizes measured in counts of the numbers of infected cells (i.e. that there is less genetic diversity than expected in HIV given its population size; Leigh Brown 1997). Although population genetic estimates of intra-host N_e in HIV-1 are invariably low—in the range of 10^{-3} to 10^{-4}—and therefore very much smaller than the numbers of infected cells, which may fall in the range 10^7–10^8, these estimates were derived from methods that implicitly assume neutrality (see section 3.3). As a consequence, it is impossible to determine whether these low N_e values—which are better thought of as indicators of restricted genetic diversity—reflect truly stochastic processes, including local bottlenecks following tissue-specific subdivision (Frost et al. 2001), or continual selective sweeps. To my mind, the rapidity with which HIV responds to changing environments strongly suggests that ongoing natural selection is responsible for the reduced levels of intra-host genetic diversity, irrespective of population subdivision (Kouyos et al. 2006). This strongly selective process also in part explains why substitution rates are higher within than among hosts (Lemey et al. 2006). Likewise, the clear similarities in structure between the intra-host phylogenies of HIV-1 and the *inter-host* phylogenies of influenza A virus, both of which exhibit a distinctive ladder-like pattern, strongly argues for the efficient and continual action of natural selection (Grenfell et al. 2004) (Fig. 7.10). In these circumstances, any reported randomness in the evolution of drug resistance in HIV (Leigh Brown and Richman 1997; Nijhuis et al. 1998; Frost et al. 2000) may be due to epistatic interactions, of which little is known.

7.2.7 Epidemiological scale dynamics

While understanding the intra-host evolutionary dynamics of HIV is evidently of great importance, rather less is known about how what occurs within hosts relates to evolution at the epidemiological scale. Perhaps the most interesting question here is whether HIV is adapting to transmission in different human populations, particularly those that exhibit major differences in HLA haplotype. The most provocative idea is that CTL escape mutations that are beneficial within hosts are also favourably transmitted at the population level, leading to epidemiological scale adaptation (Moore et al. 2002; Trachtenberg et al. 2003). Given sufficient time, this process will result in genotypically and phenotypically distinct HIV lineages in human populations that differ in their dominant HLA types, a phenomenon that has recently been demonstrated (Kawashima et al. 2009). However, while it is possible that a proportion of those mutations that are favoured within hosts will also spread efficiently at the epidemiological scale (Edwards et al. 2005; Kosakovsky Pond et al. 2006), detailed analyses have revealed that many of the associations between HIV-1 and specific HLA types are due to shared common ancestry rather than to shared selection pressure (Bhattacharya et al. 2007).

There are also a variety of reasons why it is dangerous to extrapolate from the intra-host to the inter-host evolution of HIV. First, if there is a large-scale (neutral) population bottleneck at transmission, at least via some transmission routes (Edwards et al. 2006b), then this major stochastic event will inhibit the ability of natural selection to favour advantageous mutations in the long term. Similarly, at the epidemiological scale, natural selection works best in large, well-mixed populations.

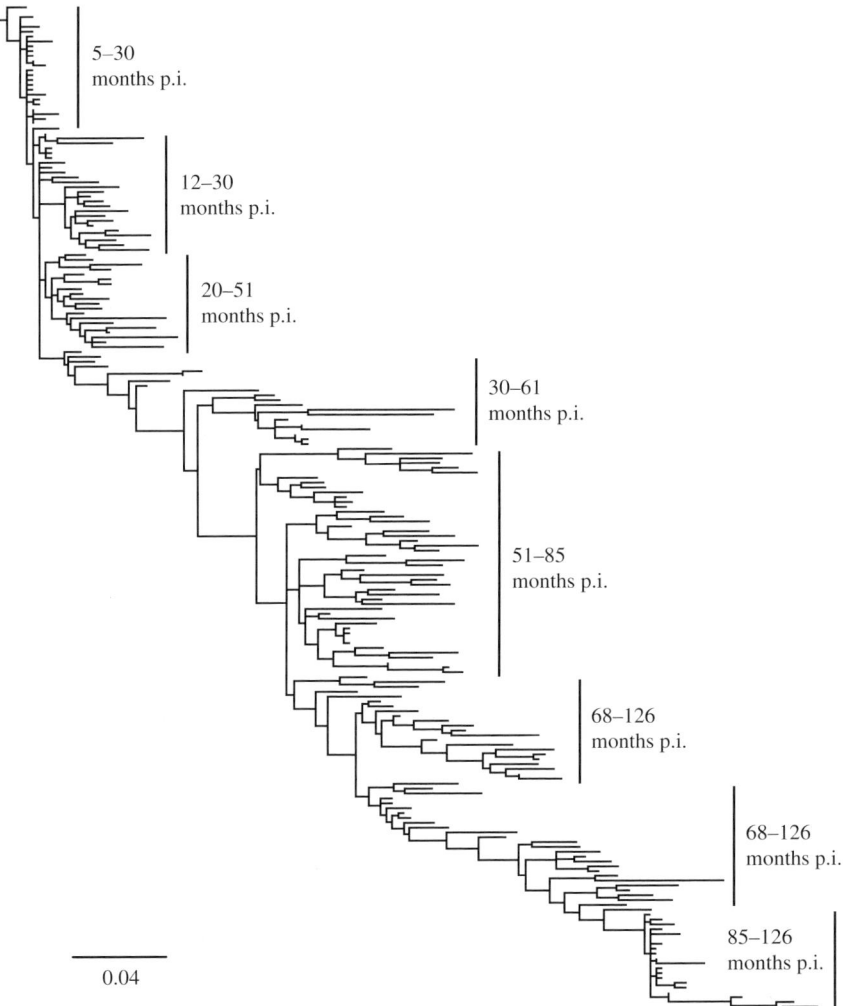

Fig. 7.10 Intra-host phylogenetic tree of HIV-1. Maximum likelihood tree of *env* gene sequences from 'patient 2' of the Shankarappa *et al.* (1999) data set. To show the strength of temporal structuring in this phylogeny, the time of sampling post-infection (p.i.) is shown for a number of arbitrarily chosen groups of sequences. The resemblance to the inter-host tree of the HA1 domain of human influenza A virus is striking (see Fig. 7.3a). All horizontal branches are drawn to a scale of nucleotide substitutions per site, and the tree is rooted on the sequences closest to the point of infection.

While this may be true of some transmission networks—the homosexual and injecting drug-user populations of the late 1970s and early 1980s, and parts of sub-Saharan Africa today—it is unlikely to apply to HIV across its entire geographic range. Indeed, sexual networks of HIV are characterized by enormous variation in rates of partner exchange, with some 'superspreaders' having extremely high numbers of sexual partners and therefore contributing their virus at a disproportionately high frequency to subsequent generations. Such strong heterogeneities in HIV transmission mean that natural selection cannot act with optimal efficiency, as it is possible that a high-fitness variant with emerge in a low-contact network, thereby inhibiting its spread. In short, the global population dynamics of HIV most assuredly do not fit a Wright–Fisher model.

There are two more particular reasons with CTL escape mutations are not guaranteed onward transmission success. The first is that many people may transmit the virus *before* they develop a CTL escape (or other beneficial) mutation. In these circumstances, any intra-host evolution that occurs after the point of inter-host transmission will have no impact on epidemic-scale processes. There are two lines of evidence that this effect is likely to be of importance: that many HIV transmissions occur rapidly following initial infection, and particularly in 'standing networks' where rates of sexual partner or needle exchange, as well as viral loads, are very high (Jacquez *et al.* 1994), and that at least some CTL escape mutations can take many years to develop, in part because these mutations have negative effects on other aspects of viral fitness (Kelleher *et al.* 2001; section 3.3).

The second factor that will inhibit the spread of CTL escape mutations at the population level is that a mutation that is beneficial in an individual with a specific HLA type may be deleterious in another individual with a different HLA type, in which case a reversion mutation may be favoured (Friedrich *et al.* 2004a; Leslie *et al.* 2004; Herbeck *et al.* 2006; Li *et al.* 2007; Matthews *et al.* 2008). Such reversions may also reduce the rate of nucleotide substitution at the inter- compared to the intra-host level (Maljkovic Berry *et al.* 2007). In general, there are four potential fates that await the new CTL escape mutation at the population level: (i) it may be advantageous in all the hosts it enters, which will clearly enhance its spread; (ii) it may be advantageous in hosts with the 'correct' HLA type, but neutral in different HLA backgrounds (Edwards *et al.* 2005); (iii) it may be advantageous in hosts with the correct HLA type, but deleterious in the wrong genetic background, in which case the escape mutation will revert frequently, perhaps resulting in lower virulence in the new host (Chopera *et al.* 2008); and (iv) different HLA types favour different CTL escape mutations at specific nucleotide sites, in which case these positions will exhibit complex patterns of genetic variation. Together, these processes make the inter-host evolution more complex than predicted in simple deterministic models. Indeed, it is striking that the phylogenies of HIV inferred at the global population level are so different from those observed within hosts (Grenfell *et al.* 2004; Holmes 2004), and reflect large-scale, and hence often neutral, epidemiological processes (Lemey *et al.* 2006).

7.3 The evolution of dengue virus

One of the most interesting generalities stemming from research on emerging RNA viruses is that although many spill-over infections are caused by viruses that are transmitted by arthropod vectors (usually mosquitoes), few ever establish sustained transmission networks in humans, indicative of major selective constraints (Woelk and Holmes 2002). As such, exceptions to this pattern—those arboviruses that have successfully established endemic transmission cycles in humans—are of special importance. Of the viruses in this category, it is DENV that merits most attention. DENV is currently the most common arbovirus infection of humans, responsible for perhaps 50 million dengue fever infections in over 100 countries throughout the tropical and sub-tropical world (Gubler 2002; Pan Amercian Health Organization 2002) and some 500 000 hospitalizations (90% of whom are children), although with a relatively low case fatality rate, such that approximately 25 000 dengue deaths are recorded annually (see www.who.int/tdr).

In a minority of cases infection with DENV results in far more serious disease syndromes involving vascular leakage, usually referred to as dengue haemorrhagic fever (DHF) and dengue shock syndrome (DSS), the latter of which is characterized by hypotension and circulatory failure. Severe dengue disease may result in case fatality rates as high as 5% depending on the availability of appropriate clinical management (Gibbons and Vaughn 2002). The first well-documented outbreak of DHF/DSS occurred in Manila during 1953/1954, and was followed by a larger outbreak in Bangkok in 1958 (Halstead 1980). Since this time, DHF/DSS has become endemic in all countries in South-east Asia, where it represents a major paediatric infection and an archetypal emerging disease. Given the increasing size and mobility of the human population, as well as the current lack of an effective vaccine, dengue will doubtless continue to be an important public health problem for the foreseeable future. Global warming, and its knock-on effect on mosquito dispersal, may eventually result in an even wider distribution (Jetten and Focks 1997).

DENV is a member of the genus *Flavivirus* (family *Flaviviridae*) that is transmitted by *Aedes* spp. mosquitoes. It is organized as four serotypes—DENV-1 to DENV-4—that are so phylogenetically distinct (diverging at ≈30% across their polyprotein) that some have argued that they should be classified as different viruses. DENV is also relatively closely related to a number of other important viral (and emerging) infections of humans, notably Japanese encephalitis virus (JEV), tick-borne encephalitis virus (TBE), West Nile virus (WNV), and yellow fever virus (YFV), the latter of which will be discussed shortly. As an interesting historical digression, it is worth noting the role played by Benjamin Rush, physician and signatory of the US Declaration of Independence, in the epidemiology of both DENV and YFV. Rush coined the famous term 'break-bone' fever for dengue infection and produced one of the first and clearest descriptions of epidemic yellow fever, that which occurred in 1793 in Philadelphia and killed some 5000 people. DENV and YFV are also similar in that both viruses exist in an urban (epidemic) cycle, involving human-to-human transmission and anthrophilic *Aedes* mosquitoes, and a sylvatic cycle involving non-human primates, and possibly other mammals (de Thoisy *et al.* 2004), and sylvatic species of mosquito (such as *Aedes furcifer*, *Aedes luteocephalus*, and *Aedes taylori*) (Rudnick 1978; Barrett and Higgs 2007) (Fig. 7.11).

182 • 7 Case studies in evolution and emergence

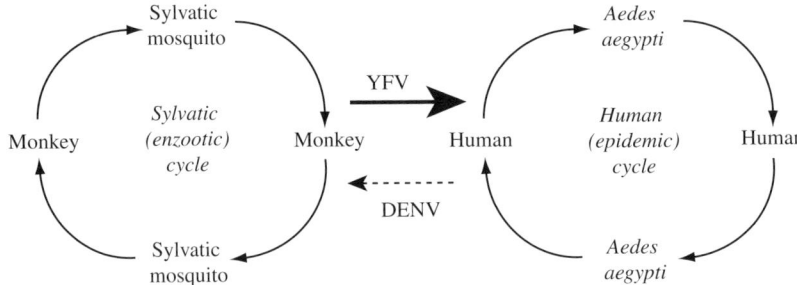

Fig. 7.11 The transmission cycle of dengue (DENV) and yellow fever (YFV) viruses, both of which involve sylvatic (enzootic) and human (epidemic) components. Crucially, while the link to the sylvatic cycle is critical to the emergence of YFV in humans, DENV has effectively broken free of sylvatic transmission, although some spill-back from humans to non-human primates occurs (reflected in the direction and strength of the arrows). Also, it is important to note that while monkeys are likely to be the major sylvatic host for both DENV and YFV, other mammalian species may also play a role in the sylvatic transmission of these viruses.

7.3.1 The origins of DENV

Despite the great interest in DENV, revealing its origins has proven problematic. Perhaps the only strong conclusion that can be drawn at present is that DENV is very unlikely to have originated in the New World as the first recorded disease outbreaks on this continent coincide with the slave trade, in which infected hosts and/or vectors were imported into the Americas for the first time. Consequently, it must be that DENV has its origins in the jungles of either Africa or Asia. However, choosing among these regions has proven difficult. Tentative evidence for an African origin is that (i) the *Aedes* mosquitoes that transmit DENV are themselves of African origin, (ii) early outbreaks of dengue in the Americas are directly related to the slave trade from Africa, and that (iii) people of African origin are seemingly less susceptible to serious dengue disease. However, all these points are debatable at best. In particular, given that DENV evolution is measured on a timescale of a few thousand years (and with a substitution rate that is fairly typical of RNA viruses; Twiddy *et al.* 2003), whereas that of *Aedes* mosquitoes is a good deal longer, the evolutionary history of virus and vector are clearly uncoupled. In addition, although there is strong circumstantial evidence that the global spread of dengue can at least in part be associated with the slave trade from Africa, early outbreaks of DENV were reported in other localities most notably Batavia (now Jakarta, Indonesia). Consequently, it is likely that many aspects of the increased mobility and size of human populations which occurred at this time, with slavery one ugly manifestation, were responsible for the global dissemination of DENV (Zanotto *et al.* 1996b). Further, there is some evidence that DENV might have an Asian origin. In particular, currently the highest prevalence of DENV is found in South-east Asia, and sylvatic strains of DENV isolated from species of Asian primate fall at divergent positions on phylogenetic

7.3 The evolution of dengue virus • 183

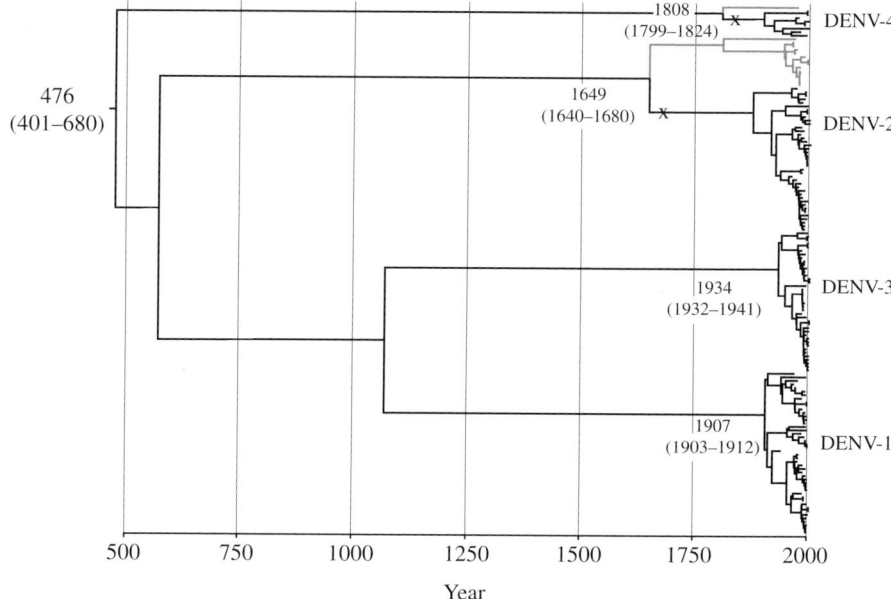

Fig. 7.12 Phylogeny of 150 complete genomes (coding region, 10185 nt) of all four serotypes of DENV, including those sylvatic viruses only found in non-human primates and which are shaded in grey. The branches on which cross-species transmission events from monkeys to humans have occurred are marked with an X symbol. Although complete genomes of clearly sylvatic viruses are currently only available for DENV-2 and DENV-4, it is highly likely that all four serotypes have sylvatic origins. The times to the most recent common ancestor (TMRCAs) (and 95% highest posterior density (HPD) values) for the serotypes are also shown. The phylogeny depicted is the maximum clade credibility tree with a scale given in years and tip times corresponding to sampling times. The tree was estimated using a strict molecular clock model under a Bayesian skyline coalescent prior with the BEAST package (Drummond and Rambaut 2007). More details are available from the author on request.

trees, suggesting that they have been resident in these species for extend time periods (Wang *et al.* 2000; Wolfe *et al.* 2001).

The mention of sylvatic dengue also raises one of the most puzzling aspects of its evolution: that sylvatic strains are found in monkeys of both African and Asian origin and that the divergence between these viruses most likely occurs on a timescale of only hundreds of years (Vasilakis *et al.* 2007a) (Fig. 7.12). Indeed, estimates for the time to the most recent common ancestor (TMRCA) of the four serotypes are usually no more than 2000 years (Twiddy *et al.* 2003; Dunham and Holmes 2007). As monkeys are unable to move large (trans-continental) distances themselves, how can the link between the African and Asian monkeys be explained? At face value there seem to be two possibilities, both of which raise interesting questions about the nature of viral evolution. One explanation is that sylvatic DENVs have been associated with these species since they shared a common ancestor several million years ago. However, this would require a profound slow-down in the tick of the viral molecular clock. Indeed, the genetic diversity within the sylvatic stains of DENV-2 is no greater than that observed with the human viruses, which is clearly measured in timescales of hundreds of years (Holmes and Twiddy 2003; Vasilakis et al. 2007a). The second possibility, more viable by default, is that humans have transmitted the virus to monkey species living on different continents, through the direct movement of infected humans, vectors, or monkeys. In support of this theory are observations of viral infections at high prevalence in Asian temple monkeys that are in regular contact with humans, and which represent 'reverse spill-overs' (Jones-Engel et al. 2006). Further, the timescale of sylvatic DENV evolution concurs with the beginning of global trade in humans (Vasilakis et al. 2007a), which would have also allowed widespread dissemination. Humans may therefore act more regularly as a vector for the transfer of viral infections to other species than previously anticipated (section 6.2).

7.3.2 DENV biodiversity

Typically for an RNA virus, extensive genetic diversity is present within each of the four DENV serotypes. For ease of representation this diversity is often partitioned into phylogenetically discrete clusters of sequences termed genotypes ('subtype' has been used interchangeably), although it is likely that some of these also differ in antigenicity (Holmes and Twiddy 2003). As with all RNA viruses, there is an arbitrary component to genotypic classifications, and a number of different schemes have been proposed. Depending on the scheme, between three and six such genotypes are described within each serotype, with both DENV-2 and DENV-4 containing genotypes that are currently only observed in non-human primates (Figs. 7.12 and 7.13). It is highly likely that more extensive sampling will reveal additional genotypes, and that other genotypes will disappear as 'gaps' in the phylogenetic tree are filled in. These genotypes most likely represent the outcomes of independent evolution following geographical isolation, rather than a selectively driven process. This being said, there have been claims that the genotypes, or clades within genotypes, of DENV do differ in aspects of phenotype, most notably virulence (see below).

Aside from their phylogenetic distinctiveness, the other obvious feature of the genotypes is their differing spatial distributions, with some more widespread than others (Fig. 7.13). This is best documented in DENV-2 where two genotypes are seemingly restricted to South-east Asia, and a third to the Americas. In contrast, a 'cosmopolitan' genotype has a far wider geographical distribution, encompassing almost the full spatial range over which dengue is observed (Twiddy et al. 2002). However, far more extensive epidemiological and genomic surveys will be required to determine whether the widespread distribution of some genotypes is due to enhanced fitness, such that the virus in question has more epidemic potential, or merely chance exportation. Other general conclusions that can be drawn from the phylogenetic analysis of genotype diversity are: (i) that genotypes frequently co-circulate within

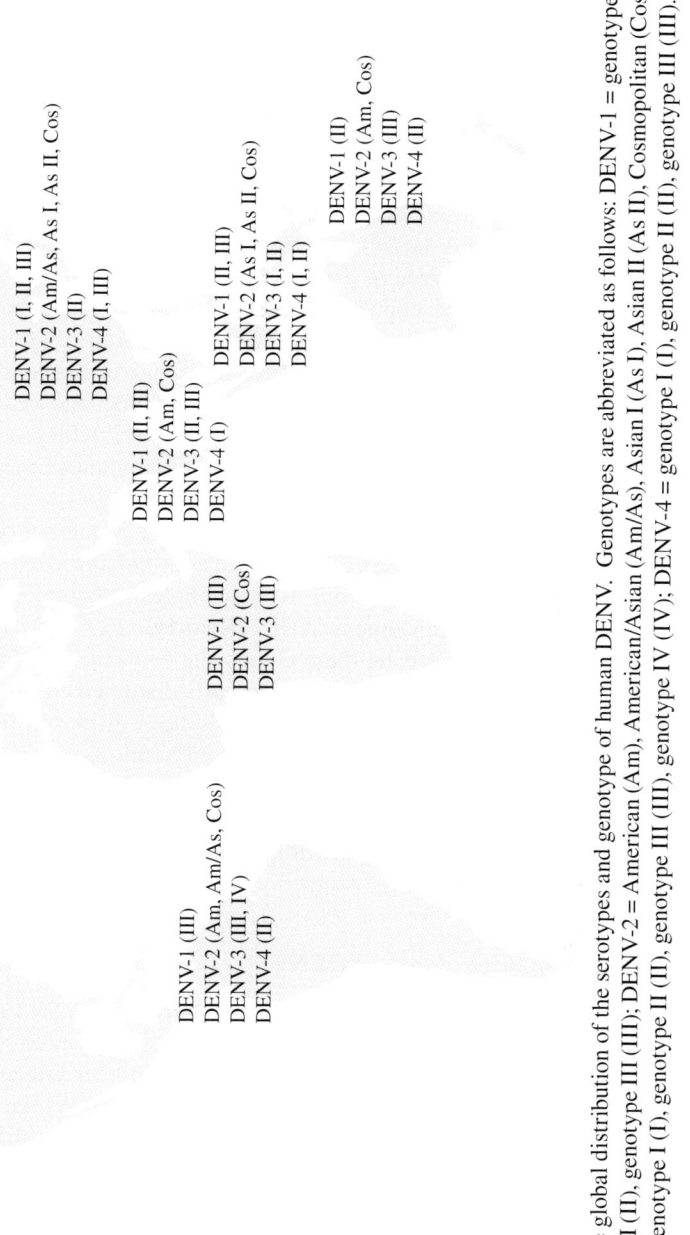

Fig. 7.13 The global distribution of the serotypes and genotype of human DENV. Genotypes are abbreviated as follows: DENV-1 = genotype I (I), genotype II (II), genotype III (III); DENV-2 = American (Am), American/Asian (Am/As), Asian I (As I), Asian II (As II), Cosmopolitan (Cos); DENV-3 = genotype I (I), genotype II (II), genotype III (III), genotype IV (IV); DENV-4 = genotype I (I), genotype II (II), genotype III (III).

the same locality, and particularly in parts of South-east Asia; (ii) that there is a possible distinction between 'endemic' genotypes that have circulated within particular localities for extended time periods, and 'epidemic' genotypes that seem to spread rapidly through populations, with the microevolution of DENV-1 in Pacific islands (endemic) compared to mainland South-east Asia (epidemic) an important example (A-Nuegoonpipat *et al.* 2004); (iii) that there is a partial phylogeographic division between what be regarded as 'northern' (China, Myanmar, Taiwan, Thailand, Vietnam) and 'southern' (East Timor, Indonesia, Malaysia, The Philippines, Singapore) South-east Asia; (iv) that despite this division, South-east Asia in general harbours the greatest degree of genetic diversity, suggesting that it has traditionally acted as a global source population, generating strains that then ignite epidemics elsewhere (although there now appears to be endemic transmission in other localities, including the Americas); and (v) that there is a relatively high rate of clade (including genotype) replacement, so that there is a clear birth and death of viral lineages through time (Sittisombut *et al.* 1997; Wittke *et al.* 2002; Klungthong *et al.* 2004; Thu *et al.* 2004). This process is particularly important for understanding the evolutionary dynamics of DENV and is discussed in more detail below.

Of more clinical relevance are suggestions that the genotypes of DENV also differ in virulence, which in this case can be defined as an enhanced capacity to cause DHF/DSS either in isolation or in the context of a complex immunological reaction termed antibody-dependent enhancement (ADE) (Rico-Hesse *et al.* 1997). Although still a subject of considerable debate, there is growing evidence that DENV genotypes do indeed differ in virulence. This is most famously documented in the case of DENV-2, where an 'American' genotype has seemingly been displaced from its home range in Latin America by a DENV-2 genotype from South-east Asia, the latter of which is also associated with far higher levels of severe dengue disease (Rico-Hesse *et al.* 1997; Rico-Hesse 2003). Similarly, the introduction of genotype III DENV-3 into Sri Lanka in 1989 was associated with the appearance of DHF/DSS in this country (Messer *et al.* 2002, 2003).

7.3.3 Lineage birth-death in DENV

The spread of DENV-2 in Latin America and DENV-3 in Sri Lanka illustrate what has been dubbed 'clade replacement': viral lineages appear, circulate for a period of time, and then die out, usually to be replaced by another lineage, and occasionally resulting in a change of virulence (Sittisombut *et al.* 1997; Wittke *et al.* 2002; Bennett *et al.* 2003; Klungthong *et al.* 2004). A striking, and more localized, example of this process took place in Bangkok, Thailand, involving two clades of viruses assigned to genotype I of DENV-1, and correspondent with a decline in the overall prevalence of DENV-1 and the rise of DENV-4 in this population (Zhang *et al.* 2005). Such major changes in genetic space can be explained by either (i) random population bottlenecks, for example caused by large-scale declines in mosquito numbers during the annual dry season, so that clade-replacement events are driven by neutral epidemiological processes or (ii) because clades differ in fitness so that one is able outcompete another, perhaps by generating higher viral loads which in turn results in greater transmissibility.

Although stochastic processes are likely to play some role in determining the phylogenetic structure of DENV populations, particularly given seasonal fluctuations in vector abundance, which themselves track changes in temperature and precipitation (Hay *et al.* 2000), there is growing evidence that natural selection also plays a major role in determining the dynamics of lineage turnover. For example, that the major clade-replacement event in DENV-1 genotype I in Thailand involved viruses circulating in the same population—a children's hospital in Bangkok—and occurred gradually, so that both clades co-circulated for a number of years, strongly suggests that it was selectively mediated (Zhang *et al.* 2005).

7.3.4 DENV fitness

More difficult is determining exactly how these viruses differ in fitness, and particularly whether these fitness differences relate to virulence. Two general hypotheses have been put forward (Rico-Hesse 2007). The first is that viruses with enhanced virulence produce higher (and perhaps longer) viraemia than viruses of low virulence, which in turn increases their probability of mosquito transmission (Leitmeyer *et al.* 1999). Natural selection by this mechanism was proposed as the explanation for the displacement of American by South-east Asian DENV-2 viruses (Cologna *et al.* 2005). In support of this hypothesis were observations that South-east Asian viruses produced consistently higher virus titres in human dendritic cells (Cologna and Rico-Hesse 2003), and were better able to infect and disseminate in *Aedes aegypti* mosquitoes (Armstrong and Rico-Hesse 2001), than American genotype viruses. Importantly, this implies that there is a direct link between viral titre, transmissibility, fitness, and virulence. In the case of DENV-3 from Sri Lanka, although the indigenous and invading viruses infected a similar proportion of mosquitoes, the latter replicated to higher levels in mosquitoes and disseminated to the head tissue more readily (Hanley *et al.* 2008). Pinpointing exactly which viral mutations are responsible for these differences in fitness has proven more problematic. For example, suggestions that variation in the RNA secondary structure of the 3′ UTR were in part responsible for the differing disease associations among American and Asian DENV-2 viruses (Leitmeyer *et al.* 1999) have yet to be verified (Shurtleff *et al.* 2001; Zhou *et al.* 2006).

The second hypothesis posits that high-virulence (i.e. fitter) DENVs are better able to avoid neutralization by cross-reactive antibodies present in semi-immune hosts (Kochel *et al.* 2002, 2005; Bennett *et al.* 2003). Hence, the selection pressures acting on DENV are strongly immune-mediated, such that the fitness differences between lineages are only manifest given a certain immunological landscape. Specifically, amino acid replacements that are neutral in one immunological landscape—manifest as a particular frequency of the four serotypes—are subject to natural selection when this immunological landscape is altered and the four serotypes change their frequency. This change in fitness occurs because the extent of immunological cross-protection—herd immunity—will change along with the dominant serotype, so that mutations that are cross-protective against one serotype are not so against another. This process will result in a complex relationship between genetic and antigenic evolution (Fig. 7.14). Evidence for such complex *immune-mediated selection* is apparent in DENV-1 from Bangkok: the two clades in question differentiated at a time when this was the dominant serotype, so that the extent of cross-protection against the remaining three serotypes in the population was minimal. The clade replacement event occurred as DENV-1 began to decline in frequency and DENV-4 took over as the dominant serotype, such that one clade was better able to evade cross-protective antibodies generated to DENV-4 than the other (Zhang et al. 2005). In support of this idea, detailed analyses of monthly incidence data from a 20-year period in Bangkok also revealed the DENV-1 and DENV-4 serotypes to be out of phase, with the former dominating when the latter is rare, and vice versa (Nisalak et al. 2003; Adams et al. 2006; Wearing and Rohani 2006).

A similar process of immune-mediated selection occurs in emergent as well as endemic populations. As a case in point, DENV-4 exhibited more rapid rates of population growth in Caribbean populations than DENV-2 after both were imported into this region from South-east Asia (Carrington et al. 2005). The slower epidemiological dynamics of DENV-2 can be attributed to the fact that this serotype had previously spread through the Caribbean, so that some of the host population already carried immunity to it, while DENV-4 represented a virgin soil outbreak, allowing it to spread rapidly through an immunologically naïve population. Hence, the fitness of particular lineages of DENV is determined by what serotypes circulate (and have circulated) in the population and their frequency, and as these serotype distributions change, and the immunological interactions among them are altered, so the fitness of the component viral lineages also changes. Critically, simple laboratory assays of viral fitness will miss this essential epidemiological-scale dynamic.

7.3.5 Comparing dengue and yellow fever

Although it is tempting to consider dengue in isolation, much can be gained by comparing its evolution with that of YFV, to which it shares a number of important features: a close phylogenetic relationship, a transmission cycle that involves primates and various species of *Aedes* mosquitoes, and an epidemiological history that is closely linked to the slave trade (Zanotto et al. 1996b ; Bryant et al. 2007) (Fig. 7.11). However, despite these similarities, YFV and DENV display a number of very important differences that shed light on their emergence and evolution. Most significantly, YFV can still be considered a disease of non-human primates (that is, a sylvatic disease), causing only sporadic outbreaks in humans, which are often little more than spill-over hosts. For example, in 2000–2005 fewer than 5000 cases of human yellow fever were reported in African and South America, although these numbers are likely to be large underestimates (Barrett and Higgs 2007). In addition, YFV has only a single serotype, compared to the four that characterize DENV, and is notoriously absent Asia (and countries of the Pacific), even though both the hosts and

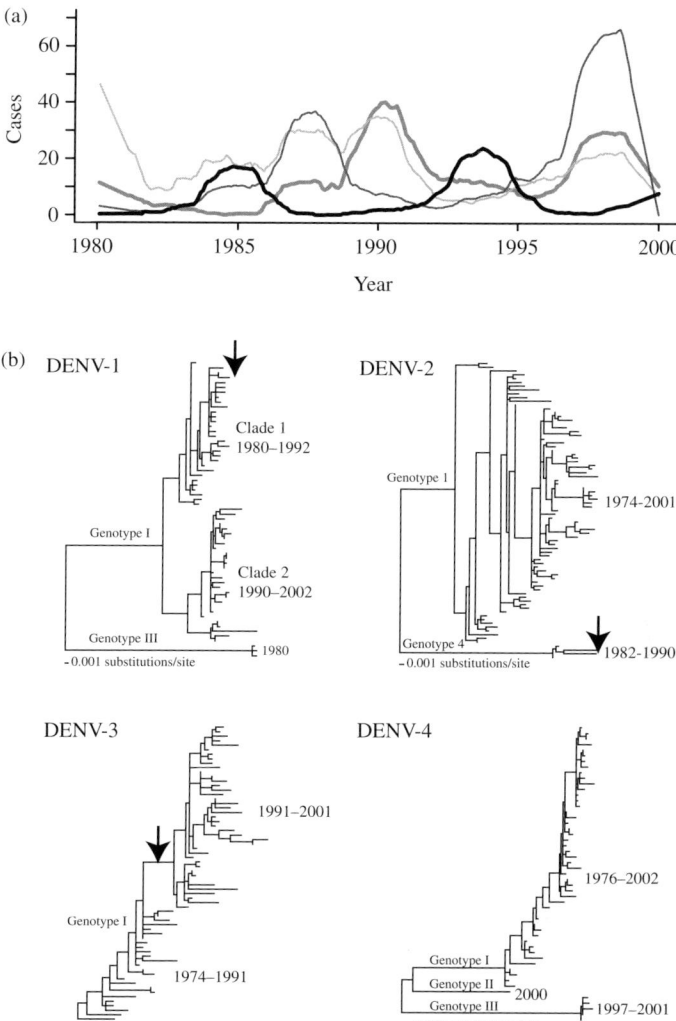

Fig. 7.14 The complex relationship between antigenic and genetic evolution in DENV. (a) Monthly number of cases (smoothed and adjusted for seasonality) of all four DENV serotypes recorded at the Queen Sirikit National Institute of Child Health in Bangkok, Thailand, between 1980 and 2000. DENV-4 (black line) is clearly out of phase with the other three serotypes. (b) Phylogenetic trees of all four DENV serotypes (E gene) sampled between 1973 and 2001 in Bangkok. The major phylogenetic events that coincide with changes in serotype abundance (a) are marked by arrows. The extinction of DENV-1 clade 1 (genotype 1) correspondent with the rise of DENV-4 in the early 1990s is particularly notable. Adapted from Adams *et al.* (2006) with permission.

vectors are present in this region (see below). Finally YFV is far more virulent than DENV, with reported case fatality rates of 25% in Africa.

Given these differences, the most obvious question to address with respect to YFV is why it has not evolved to be an endemic human pathogen? There are a number of possible explanations. It is possible that DENV is some how intrinsically better adapted to replicate in human cells than YFV and therefore better able to spread through human populations. This enhanced replication ability results in a higher and/or longer viraemia and so increases the probability of transmission. Indeed, experimental studies have shown that isolates of sylvatic DENV have little trouble in replicating in human cell types, suggesting that jumping species boundaries in this case was not a major adaptive challenge (Vasilakis *et al.* 2007b). It is also striking that DENV has been able to successfully cross the species barrier from monkeys to humans on at least four occasions. However, this does not explain why YFV is able to cause very major epidemics when the ecological conditions (i.e. density of mosquitoes and hosts) are favourable. As an example, the Philadelphia yellow fever epidemic of 1793 resulted in 17 000 cases and 5000 deaths, some 10% of the total population (a case fatality rate of ≈30%). This suggests that YFV does possess the requisite mutations that allow it to spread efficiently in humans given the right epidemiological conditions. In addition, an increase in the level of viraemia is normally expected to lead to an increase in disease severity in humans, as there is a strong association between viral load and morbidity (Vaughn *et al.* 2000). However, DENV is demonstrably less virulent than YFV, contrary to this prediction.

The major difference in virulence between DENV and YFV may provide an important clue to their differing epidemiological profiles. Specifically, as DENV is less virulent than YFV it requires a smaller critical community size (CCS) to sustain its transmission. Historically, the high virulence of YFV meant that it would quickly burn through human populations, killing hosts before the pool of susceptibles could be replenished. Under these circumstances the virus could not, and still will not, be able to spread efficiently unless it encounters very large human populations, such as those that would have characterized the cities where the symptoms of yellow fever were so famously described. In contrast, DENV, which is characterized by a far lower virulence and where many infections may be asymptomatic, is able to spread through far smaller human populations. It is therefore no surprise that the rise of both viruses is closely tied to the rise of urbanization and the slave trade, both of which would have greatly increased the number of susceptible hosts (Chapter 6).

7.3.6 Why no yellow fever in Asia?

A second interesting puzzle relating to YFV, and one that is intrigued virologists for decades, is why the virus has not spread in Asia, even though the requisite hosts (primates) and vectors are in place. Although Asian strains of *Ae. aegypti* are less competent at transmitting YFV that American strains (Barrett and Higgs 2007), that some transmission does occur indicates that this is not absolute barrier. A rather more compelling (although untested) hypothesis is that YFV has been prevented from gaining a foothold in Asia because of an 'immunological barrier', reflecting strong cross-immunity exerted by another flavivirus that is already at high prevalence in this region. This is analogous to the complex interactions between the four serotypes of DENV. Indeed, as the DENV serotypes are as divergent as many different 'species' of flavivirus, it is possible that their immunological interactions, and particularly cross-protection, extend over wider phylogenetic distances, influencing the distribution of other flaviviruses. Hence, DENV is an obvious candidate for a virus that prevents the establishment of YFV, particularly as it may have originated in South-east Asia and is at very high prevalence in this region. However, other flaviviruses that are more closely related to YFV also exist in South-east Asia, including sepik and wesselsbron, while JEV is also highly prevalent. It is also possible that the cross-immunity is caused by another, as yet undescribed, flavivirus, of which there may be many (Pybus *et al.* 2002). More generally, it is possible that such immunological barriers—reflecting complex patterns of cross-immunity among viruses that are so closely related that they share epitopes (Crill and Chang 2004)—are a major factor shaping the spatial distributions occupied by RNA viruses, and may even lead to viral 'speciation' (section 5.3). Revealing the large-scale interactions among RNA viruses should clearly be a major component of the molecular epidemiology of the future.

7.4 The phylogeography and evolution of rabies virus

Before the emergence of HIV/AIDS, rabies was considered one of the most frightening of human diseases. Without vaccination—which thankfully can occur following exposure—the case fatality rate of rabies infection in humans is close to 100%, with few people ever reported as having recovered after the onset of symptoms. Those who do recover often experience permanent neurological damage. Rabies is also one of the most historically important human infections. The symptoms of rabies, which are gruesomely diagnostic, may have been reported in parts of the Old World before 2300 BC (Steele and Fernandez 1991). Louis Pasteur famously developed an effective vaccine to the virus in 1885, an event considered a major breakthrough in medical science. Today, and despite the effectiveness of vaccination, rabies is still responsible for more than 50 000 deaths globally on an annual basis, most of them in Asia.

Despite the burden of human morbidity and mortality due to rabies, it should not be classed as an endemic human infection. Rather, rabies is a wildlife disease, circulating in a variety of mammalian species, notably dogs, bats, foxes, raccoons, skunks, jackals, etc., which occasionally spill over to infect other species, including humans. Globally, the most common host is the dog, giving rise to what is called 'street rabies' in some parts of the world. The interest in rabies from an evolutionary perspective is that it provides a unique window into the process of viral emergence, as the virus has jumped host species boundaries on a fairly regular basis, and because it exhibits interesting spatial dynamics, particularly the occurrence of transmission waves. To put it another way, the spatial and temporal dynamics of RABV occur on approximately the same scale, so that it represents a powerful case study in viral phylogeography. It is the spatial dynamics of rabies on which I focus here.

7.4.1 The world of lyssaviruses

RABV represents one genotype (genotype 1) of the genus *Lyssavirus*, a group of ssRNA− viruses of the family *Rhabdoviridae* whose name derives from the Greek *rhabdos*, for 'rod', a reference to their distinctive rod- (or bullet-) shaped virion. The rhabdoviruses are particularly interesting for the diversity of host species they infect, including mammals (such as RABV and VSV), fish (such as viral haemorrhagic septicaemia virus), and even plants (including rice yellow stunt virus). The ability to infect both animals and plants is a relatively rare trait among RNA viruses. Further, the rhadoviruses can be both vector-borne or transmitted by other routes, such as infected saliva in the case of RABV (Bourhy *et al.* 2005).

As well as the 'classical' RABVs of genotype 1, other genotypes of lyssaviruses are: Lagos bat virus (LBV; genotype 2); Mokola virus (MOKV; genotype 3); Duvenhage virus (DUVV; genotype 4); European bat lyssavirus type 1 (EBLV-1; genotype 5); European bat lyssavirus type 2 (EBLV-2; genotype 6); and Australian bat lyssavirus (ABLV; genotype 7) (Fig. 7.15). More recently, it has been proposed that LBV should in fact be divided into two distinct genotypes (Delmas *et al.* 2008), and that four putative new genotypes infect bats in Asia (Arai *et al.* 2003). It is therefore highly likely that more intensive sampling will reveal additional lyssaviruses.

All lyssavirus genotypes except MOKV (where the host species are unknown but may involve terrestrial mammals) have bat reservoirs, strongly suggesting that lyssaviruses originated in these mammals, with occasional jumps to other species. This is also true of RABV individually, where there is a major phylogenetic split between those isolates that circulate in bats, and those that infect terrestrial mammals, suggesting that there was a single jump from bats to other mammals (see below; Badrane and Tordo 2001). Also of note in this context is that, to date, the only lyssaviruses found in any host species in the Americas are those assigned to genotype 1. This observation, combined with molecular clock estimates of the timescale of RABV evolution, suggests that RABV may have initially entered the Americas in the post-Colombian period.

There have been a variety of analyses of the global genetic diversity of RABV although, to date, none have considered complete genomes. A variety of important generalities can be drawn from these analyses. First, as noted above, there is fundamental division between those RABV viruses that circulate in bats and those whose hosts are terrestrial mammals. Second, within the RABV that circulate in terrestrial mammals a number of distinct clades can be described (again with a strong arbitrary component and idiosyncratic nomenclature): (i) Africa 2, (ii) Africa 3, (iii) Arctic-related, (iv) Asian, (v) cosmopolitan, and (vi) Indian subcontinent, with a seventh genotype circulating exclusively in bats (Bourhy *et al.* 2008) (Fig. 7.16). As in the case of DENV, these genotypes have different geographical distributions, including the presence of some relatively cosmopolitan clades, although there is no

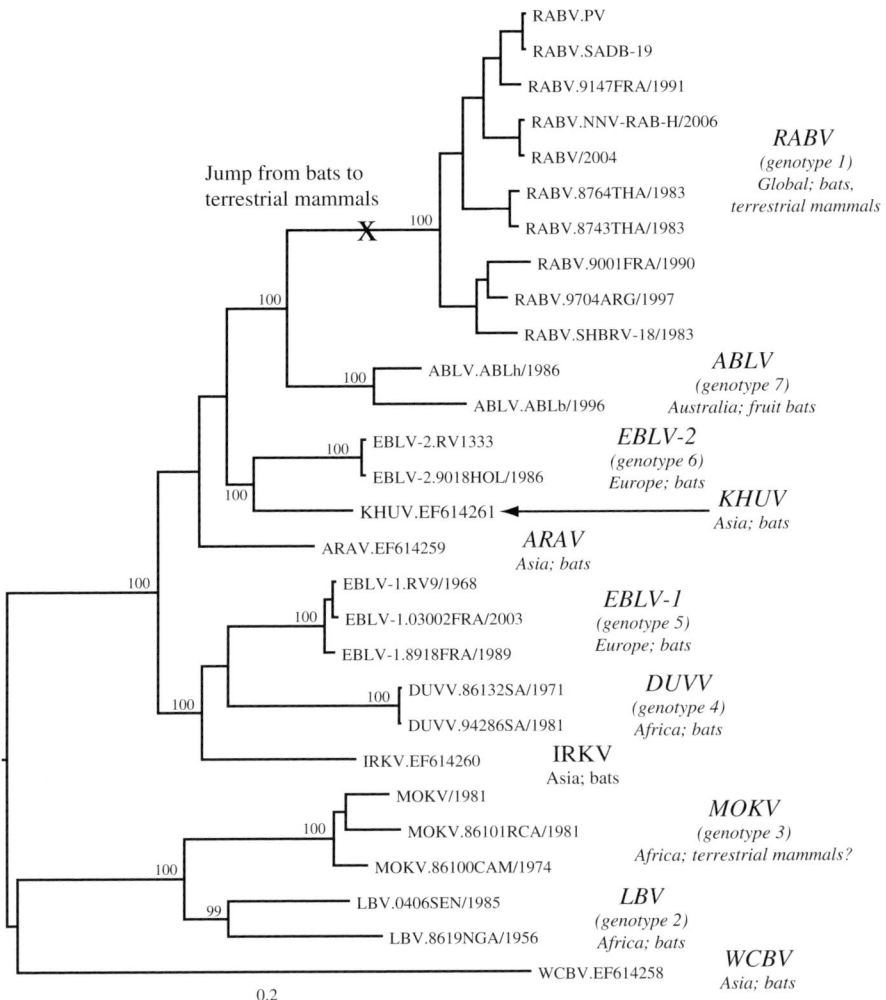

Fig. 7.15 Phylogeny of the lyssaviruses. This maximum likelihood tree (PAUP*; Swofford 2003) was inferred using complete genome sequences with ambiguous regions of alignment removed (total of 10 914 nt). More details are available from the author on request. Seven lyssavirus genotypes have been described to date: RABV (rabies virus), LBV (Lagos bat virus), MOKV (Mokola virus), DUVV (Duvenhage virus), EBLV-1 (European bat lyssavirus type 1), EBLV-2 (European bat lyssavirus type 2), and ABLV (Australian bat lyssavirus). In addition, four new genotypes have been proposed: ARAV (Aravan virus), IRKV (Irkut virus), KHUV (Khujand virus), and WCBV (West Caucasian bat virus). RABV.PV and RABV.SADB-19 are vaccine strains. The probable location of the jump from bats to terrestrial mammals is marked by the X symbol. Bootstrap values are shown for key nodes. All horizontal branch lengths are drawn to a scale of nucleotide substitutions per site.

194 · 7 Case studies in evolution and emergence

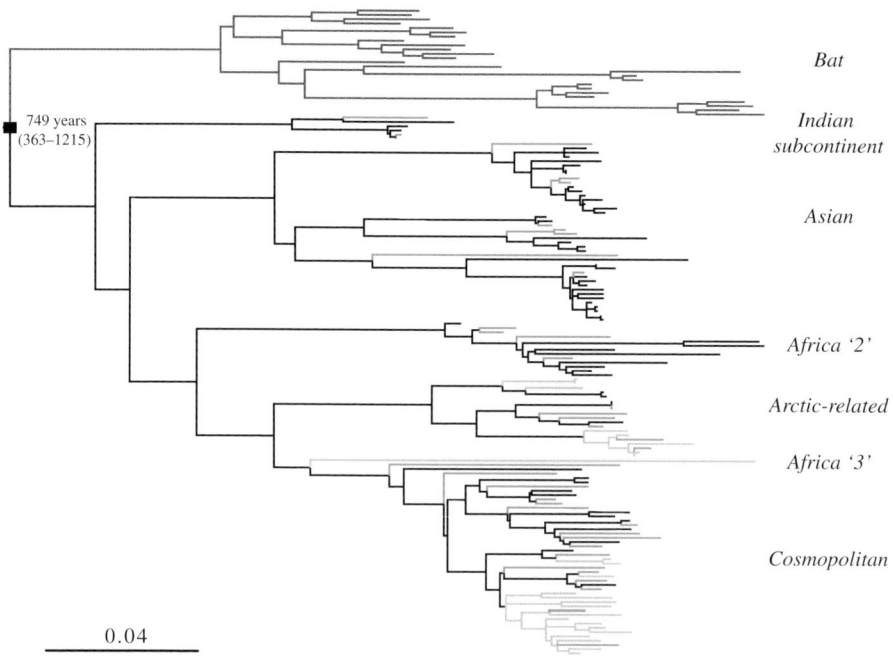

Fig. 7.16 Phylogenetic tree of global RABV inferred from the N gene. The major clades of RABV are indicated, along with the time to the most recent common ancestor (TMRCA) for this sample of viruses, as well as its 95% highest posterior density (HPD) value. Branches are shaded by species group: dogs, black, and bats are dark grey; alternative-reservoir hosts (such as the red fox) are light grey; and spill-over hosts (such as humans and bovines) are mid-grey. Horizontal branches are drawn to a scale of nucleotide substitutions per site. Adapted from Bourhy et al. (2008) with permission.

evidence that these distributions reflect underlying differences in fitness, and all have the dog as the principal host. Third, a more detailed analysis of phylogenetic patterns reveals very little viral gene flow among geographic regions so that, on a global scale, populations of RABV are characterized by strong population subdivision (Bourhy et al. 2008). However, there are also clear, and extremely interesting, instances of spatial diffusion within individual geographical regions, which are discussed in greater detail below.

The marked spatial subdivision of RABV on a global scale is evidently a function of the relatively localized movement of dog populations, at least in comparison to humans. Despite this, occasional long distance movements of viruses are apparent, for example between the Old World and the Americas, and which most likely reflect the long-distance translocation of infected animals by humans. Rather harder to infer from the data currently available is from where (terrestrial mammal) RABV first originated, particularly as there appears to have been a relatively rapid radiation of viral lineages near the base of the RABV tree (Bourhy et al. 2008). In the largest

N gene tree (Fig. 7.16) the most divergent lineages are those sampled from the Indian sub-continent, which would be in accordance with at least some epidemiological records.

As well as global spatial dynamics, it is informative to infer the timescale of this epidemiological history. A variety of analyses of the rates and dates of RABV evolution have been undertaken. These studies have recorded broadly equivalent substitution rates—between 10^{-3} and 10^{-4} subs/site/year (Badrane and Tordo 2001; Holmes *et al.* 2002; Hughes *et al.* 2005; Davis *et al.* 2007; Bourhy *et al.* 2008)— although rather lower rates were estimated in the case of EBLV (Davis *et al.* 2005), and indicate that the current global genetic diversity of RABV appeared within the last 2000 years (Badrane and Tordo 2001; Holmes *et al.* 2002; Bourhy *et al.* 2008). This again supports the view that human translocation events have been critical in the global dissemination of RABV. In some cases, particularly that of fox rabies in Europe, the molecular clock estimates of evolutionary history are also in striking accord with epidemiological records (Bourhy *et al.* 1999). However, if a such a recent timescale of RABV evolution is true, then those suggestions that rabies was present before 2300 BC (Steele and Fernandez 1991) must have involved either lineages of RABV that have since gone extinct or different genotypes of lyssavirus.

7.4.2 The spatiotemporal dynamics of RABV

As noted at the start of this section, RABV has a particular importance in studies of viral evolution and phylogeography as it provides one of the very best illustrations of the link between spatial and evolutionary dynamics. To be more precise, there are now a variety of examples depicting how RABV moves in an invading wave across particular geographical areas, and that the pattern and dynamics of this wave can be determined through phylogenetic analysis. As an interesting aside, that the virus often travels as a wave, and at a specific rate, also means that that its movement is in some sense predictable (Russell *et al.* 2005). Further, because the spread of RABV essentially follows the dispersal patterns of its mammalian hosts, it also provides a beautiful example of how physical barriers, such as rivers and mountains, can inhibit the spread of viral infections. In what follows I will briefly discuss three waves of RABV epidemic spread, although similar spatial dynamics are observed in some of the bat lyssaviruses (Davis *et al.* 2005).

The first wave of RABV transmission I will describe occurred in Europe in a period spanning the 1930s to the present. Before the 1930s, RABV in Europe was most often associated with dogs, and in earlier times wolves. However, in the 1930s the virus established itself in two new host species: the red fox (*Vulpes vulpes*) and the raccoon dog (*Nyctereutes procyonoides*). Although raccoon dogs appear as if they were created using digital imaging software, this interesting group of carnivores was introduced in large numbers into Europe for the purposes of fur farming. The initial jump from dogs to foxes and raccoon dogs seems to have occurred in north-eastern Europe, perhaps near the border between Poland and the former Soviet Union, and then spread both westwards and southwards a rate of 30–60 km/year, reaching

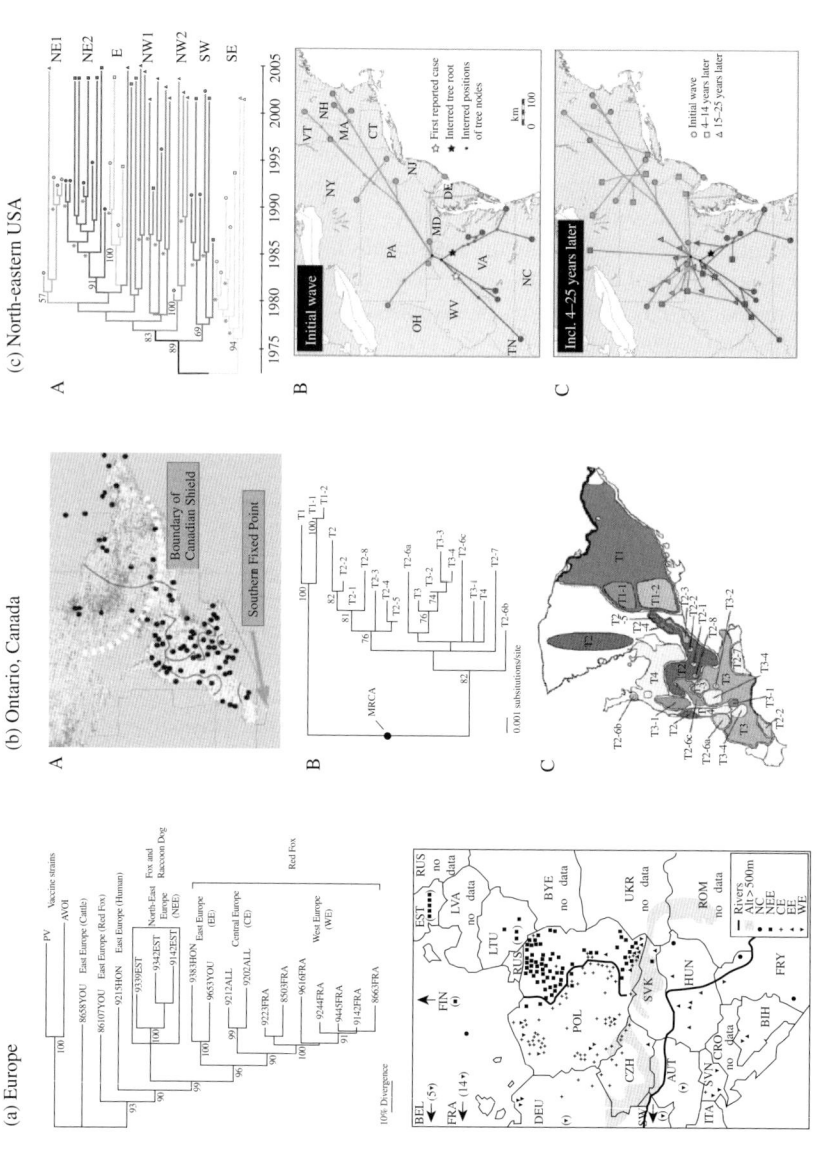

Fig. 7.17 Transmission waves of RABV. (a) Spatial diffusion of RABV in red foxes and raccoon dogs in Europe (adapted from Bourhy *et al.* 1999 with permission). (b) Southerly spread of RABV in red foxes from Ontario, Canada (adapted from Real *et al.* 2005 with permission). (c) Phylogenetic analysis of raccoon RABV in the north-eastern USA and its spatial diffusion (adapted from Biek *et al.* 2007 with permission). In each case the phylogenetic tree on which the inference of spatial dynamics is based is shown.

France in 1968 (Anderson *et al.* 1981; Bourhy *et al.* 1999). Such a spatial dynamic is clearly represented in phylogenetic trees of viral sequence data from isolates across Europe: the phylogeny has a ladder-like structure reflecting the east-to-west spread of the virus (Bourhy *et al.* 1999) (Fig. 7.17). Perhaps of even more interest was that the influence of physical barriers was also apparent in the structure of phylogenetic trees, such as the separation of viruses around the Bohemian and Carpathian mountains, and more dramatically by the Vistula river in Poland which appears to represent a major physical barrier for the virus circulating in raccoon dogs, although not in red foxes (Bourhy *et al.* 1999).

The second spatial wave of RABV described here occurred in red foxes from the Canadian province of Ontario (Nadin-Davis *et al.* 1993, 1999). Not only is the study of this particular epidemic important from the perspective of Canadian wildlife (with the virus in red foxes having jumped from that in arctic foxes), but it was the first to combine both state-of-the-art phylogenetics with spatial epidemiology (Real *et al.* 2005). In this case, the transmission wave moved southwards through Ontario during the 1980s and 1990s, resulting in two virus subpopulations with rather differing geographical distributions and marked isolation by distance (Real *et al.* 2005) (Fig. 7.17).

The final spatial wave of RABV explored here, and the one that has been examined in most detail, occurred in the eastern seaboard of the USA from the late 1970s to the present day. This epidemic involves raccoons (*Procyon lotor*), and has resulted in this species becoming the principal reservoir for RABV in the USA. The epidemic began when a population of raccoons were moved from the southern USA to the border of Virginia and West Virginia for hunting purposes in 1976 (Dobson 2000). The virus then spread in a north-eastwards fashion at a speed of 30–50 km/year, a rate remarkably similar to that recorded for fox rabies in Europe, eventually connecting with other foci of RABV in North America, including the southerly moving Ontario wave (Childs *et al.* 2000). Phylogenetic analyses of sequence data collected from 1977–2005 not only show the northerly advance of the wave-front (Biek *et al.* 2007) but local substructure, as viral lineages become isolated in different geographical regions (Fig. 7.17), in a similar manner to that of RABV in red foxes in Europe. That physical barriers, in this case the Allegheny and Appalachian mountains, have such a profound effect on dispersal highlights their potential for initiating the genetic differentiation of RNA viruses that are not spread by humans.

8

Epilogue

A combination of good theory, rigorous experimentation, and wide-ranging comparative analyses have dramatically improved our knowledge of the patterns and processes of RNA virus evolution. This book, which I hope is at least timely, serves as an attempt to summarize many of the insights that have stemmed from this research programme, albeit with an unashamed bias towards phylogenetic studies. Although I will not repeat my arguments of earlier chapters, it is worth stressing again that so much of the evolutionary biology of RNA viruses, from the number of progeny they produce, to the way their genomes are organized, and even their propensity to jump species barriers, seems to be a function of what I consider to be their defining feature: a remarkably rapid rate of mutation. Of course, this generality hides much of the idiosyncratic biology of individual viruses that we would be foolish to ignore.

Despite this evident progress, it is equally clear that important advances are still needed if we are to come to a complete understanding of the evolutionary biology of these unique and fascinating organisms. In what follows I will briefly outline a number of areas where I think particular attention is required, although I will leave the solutions to others. Thankfully, most of these fall into the category of 'data-limited' problems, which should readily be resolved given sufficient quantities of time, money, and application. Indeed, the remarkable advances in genome sequencing that have occurred in recent years mean that the ability to effectively analyse data is perhaps a greater barrier to future progress than the ability to generate it.

One area where analytical advances are particularly important is in the development of new methods that are able to recover phylogenetic signal in viral genomes that possess no clear similarity in primary sequence. As noted a number of times in this book, those comparative analyses that explore highly divergent gene sequences, either within or among genomes, are particularly compromised as much of the critical phylogenetic signal has been eroded beyond recognition. The situation is especially serious for studies of viral origins, where the inability to infer deep phylogenetic history means that much of the work in this area is understandably little more than speculation. Although there have been some promising recent developments in the recognition and analysis of distant homologies among protein sequences (Chang *et al.* 2008), the study of protein structure still seems the most profitable approach to reconstruct ancient viral evolution, particularly as there is a growing body of evidence that the phylogenetic signal present in patterns of protein folding is more robust than that in primary sequence. Of course, this statement is also a rather glib one because the computational analysis of protein structure evolution is one of the most difficult problems in contemporary biology, with a number of major obstacles to overcome, including: (i) the identification of truly homologous structures in highly divergent proteins (as opposed to structures that have arisen through convergent evolution), (ii) the development of a metric that is able to accurately describe the evolutionary similarities and differences in homologous structures and, perhaps the biggest challenge of all, (iii) the development of a realistic model of protein structure evolution. Without such advances it is difficult to see how the study of viral origins and comparative genomics can move forward in a meaningful manner.

A very different form of phylogenetic limitation currently plagues studies of viral phylogeography. Although studies in this area abound (see Chapter 6) most are limited with respect to the precision of spatial and temporal sampling, so that it is usually only possible to make broad-brush statements on epidemiological dynamics. Hopefully the raise of pyrosequencing (and its descendants) will soon allow genome databases to increase to the required size and detail to allow a more informative molecular epidemiology, as opposed to the rather descriptive studies that are often undertaken at present. In particular, viral genomes must be sampled on at least the same scale as the epidemiological processes under investigation. Of equal importance, it is essential that these expanding databases of viral genomes be combined with other relevant epidemiological and clinical information, such as their precise geographical location, the exact date of sampling and, where it is appropriate, the clinical presentation of the disease (Holmes 2007). If there is one criticism of those sequence-based studies of viral epidemiology undertaken to date it is that they have often been taken out of context of other forms of biological data.

The rise of pyrosequencing may also allow us to close another major data gap in studies of RNA virus evolution: the accumulation and analysis of large-scale 'clonal' sequence data sampled from individual hosts, particularly those generated by acute RNA viruses during natural infections. As I argued in Chapter 3, a detailed description of the extent and structure of intra-host genetic variation is essential for a complete understanding of evolutionary dynamics. In the same way, it is of fundamental importance that these studies of clonal sequence data are able to span epidemiological scales, from individuals to populations, as the impact of many key evolutionary processes may likewise change across scales. To give one simple but extremely important example, the magnitude of the population bottleneck at inter-host transmission, which is central to revealing the respective roles of natural selection and genetic drift in viral evolution, is known in only a very limited number of cases. I wish also to stress the 'natural' aspect of these studies. As a comparative biologist I firmly believe that, where it is reasonably possible, research on viral evolution should recapitulate the natural situation. To be sure, although highly informative, evolution *in vitro* is not necessarily the same as evolution *in natura*.

I suspect those working with bacteriophages may be disappointed with some aspects of this book. Although this is in part due to the fact that my own research interests primarily rest with viruses of humans and other mammals, there is also a marked absence of comparative studies of phage evolution. Indeed, despite their central role in the history of molecular biology, their ubiquity and diversity, and their current popularity in experimental studies of evolution, remarkably little is known about the molecular evolution of bacteriophages. As a simple case in point, there is currently no good estimate of the rate of nucleotide substitution, or of the timescale of evolutionary change, of bacteriophages in nature. Yet, such information would provide a unique insight into whether those measures of evolutionary dynamics inferred from *in vitro* studies can be readily transposed to the natural situation.

Finally, I would argue that the most basic, and therein important, advance required in the study of RNA virus evolution is a far greater understanding of their natural biodiversity. As noted right at the outset of this book it is clear that we are only scratching the surface of the number and spectrum of viruses (or virus-like particles) in nature, such that all generalities about their evolution must necessarily be made with a healthy dose of caution. Simple models of viral macroevolution predict that there are legions of unsampled viruses for any specific viral family (Pybus *et al.* 2002), and which are slowly being discovered (Grard *et al.* 2006; Kapoor *et al.* 2008), as well as a multitude of undescribed families. I suspect that the 'virophage' of mimivirus (La Scola *et al.* 2008) will not be the last new form of virus unearthed. It therefore seems obvious that the intensive exploration of the virosphere should be a research priority in the biological sciences, particularly as it is likely that a number of those viruses newly described will have the potential to spread and cause disease in human populations. Although some may regard this is as essentially a fishing exercise devoid of intellectual challenge, it is paramount if we are to understand how evolution produces such endless forms most beautiful.

References

Aaskov, J., Buzacott, K., Thu, H.M., Lowry, K., and Holmes, E.C. (2006). Long-term transmission of defective RNA viruses in humans and *Aedes* mosquitoes. *Science* **311**, 236–238.
Aaskov, J., Buzacott, K., Field, E., Berlioz-Arthaud, A., and Holmes, E.C. (2007). Multiple recombinant dengue type 1 viruses in an isolate from a dengue patient. *Journal of General Virology* **88**, 3334–3340.
Aaziz, R. and Tepfer, M. (1999). Recombination in RNA viruses and in virus-resistant transgenic plants. *Journal of General Virology* **80**, 1339–1346.
Adams, B., Holmes, E.C., Zhang, C., Mammen, M.P. Jr, Nimmannitya, S., Kalayanarooj, S., and Boots, M. (2006). Cross-protective immunity can account for the alternating epidemic pattern of dengue virus serotypes circulating in Bangkok. *Proceedings of the National Academy of Sciences USA* **103**, 14234–14239.
Adams, M.J. and Antoniw, J.F. (2003). Codon usage bias amongst plant viruses. *Archives of Virology* **149**, 113–135.
Agranovsky, A.A., Boyko, V.P., Karasev, A.V., Koonin, E.V., and Dolja, V.V. (1991). Putative 65 kDa protein of beet yellows closterovirus is a homologue of HSP70 heat shock proteins. *Journal of Molecular Biology* **217**, 603–610.
Akashi, H. (1994). Synonymous codon usage in *Drosophila melanogaster*: natural selection and translational accuracy. *Genetics* **136**, 927–935.
Allan, J.S., Short, M., Taylor, M.E., Su, S., Hirsch, V.M., Johnson, P.R., Shaw, G.M., and Hahn, B.H. (1991). Species-specific diversity among simian immunodeficiency viruses from African green monkeys. *Journal of Virology* **65**, 2816–2828.
Ali, A., Li, H., Schneider, W.L., Sherman, D.J., Gray, S., Smith, D., and Roossinck, M.J. (2006). Analysis of genetic bottlenecks during horizontal transmission of *Cucumber mosaic virus*. *Journal of Virology* **80**, 8345–8350.
Alonso, W.J., Viboud, C., Simonsen, L., Hirano, E.W., Daufenbach, L.Z., and Miller, M.A. (2007). Seasonality of influenza in Brazil: a traveling wave from the Amazon to the subtropics. *American Journal of Epidemiology* **165**, 1434–1442.
Alvarez, D.E., De Lella Ezcurra, A.L., Fucito, S., and Gamarnik, A.V. (2005a). Role of RNA structures present at the 3'UTR of dengue virus on translation, RNA synthesis, and viral replication. *Virology* **339**, 200–212.
Alvarez, D.E., Lodeiro, M.F., Ludueña, S.J., Pietrasanta, L.I., and Gamarnik, A.V. (2005b). Long-range RNA-RNA interactions circularize the dengue virus genome. *Journal of Virology* **79**, 6631–6643.
Anderson, J.P., Daifuku, R., and Loeb, L.A. (2004). Viral error catastrophe by mutagenic nucleosides. *Annual Review of Microbiology* **58**, 183–205.

Anderson, R.M., Jackson, H.C., May, R.M., and Smith, A.M. (1981). Population dynamics of fox rabies in Europe. *Nature* **289**, 765–771.

Andolfatto, P. (2005). Adaptive evolution of non-coding DNA in *Drosophila*. *Nature* **437**, 1149–1152.

Angly, F.E., Felts, B., Breitbart, M., Salamon, P., Edwards, R.A., Carlson, C., Chan, A.M., Haynes, M., Kelley, S., Liu, H. *et al.* (2006). The marine virospheres of four oceanic regions. *PLoS Biology* **4**, e368.

Anishchenko, M., Bowen, R.A., Paessler, S., Austgen, L., Greene, I.P., and Weaver, S.C. (2006). Venezuelan encephalitis emergence mediated by a phylogenetically predicted viral mutation. *Proceedings of the National Academy of Sciences USA* **103**, 4994–4999.

Anisimova, M., Nielsen, R., and Yang, Z. (2003). Effect of recombination on the accuracy of the likelihood ratio method for detecting positive selection at amino acid sites. *Genetics* **164**, 1229–1236.

Antia, R., Regoes, R.R., Koella, J.C., and Bergstrom, C.T. (2003). The role of evolution in the emergence of infectious diseases. *Nature* **426**, 658–661.

Antonovics, J., Hood, M., and Baker, C.H. (2006). Molecular virology: was the 1918 flu avian in origin? *Nature* **440**, E9.

A-Nuegoonpipat, A., Berlioz-Arthaud, A,, Chow, V., Endy, T., Lowry, K., Mai, L.Q., Ninh, T.U., Pyke, A., Reid, M., Reynes, J.-M. *et al.* (2004). Sustained transmission of dengue virus type 1 in the Pacific due to repeated introduction of different Asian genotypes. *Virology* **329**, 505–512.

Appel, N., Herian, U., and Bartenschlager, R. (2005). Efficient rescue of hepatitis C virus RNA replication by trans-complementation with nonstructural protein 5A. *Journal of Virology* **79**, 896–909.

Arai, S., Bennett, S.N., Sumibcay, L., Cook, J.A., Song, J.W., Hope, A., Parmenter, C., Nerurkar, V.R., Yates, T.L., and Yanagihara, R. (2008). Phylogenetically distinct hantaviruses in the masked shrew (*Sorex cinereus*) and dusky shrew (*Sorex monticolus*) in the United States. *American Journal of Tropical Medicine and Hygiene* **78**, 348–351.

Arai, Y.T., Kuzmin, I.V., Kameoka, Y., and Botvinkin, A.D. (2003). New lyssavirus genotype from the Lesser Mouse-eared Bat (*Myotis blythi*), Kyrghyzstan. *Emerging Infectious Diseases* **9**, 333–337.

Arias, A., Ruiz-Jarabo, C.M., Escarmís, C., and Domingo, E. (2004). Fitness increase of memory genomes in a viral quasispecies. *Journal of Molecular Biology* **339**, 405–412.

Archer, A.M. and Rico-Hesse, R. (2002). High genetic divergence and recombination in arenaviruses from the Americas. *Virology* **304**, 274–281.

Ariën, K.K., Vanham, G., and Arts, E.J. (2007). Is HIV-1 evolving to a less virulent form in humans? *Nature Reviews Microbiology* **5**, 141–151.

Arkhipova, I.R., Pyatkov, K.I., Meselson, M., and Evgen'ev, M.B. (2003). Retroelements containing introns in diverse invertebrate taxa. *Nature Genetics* **33**, 123–124.

Armitage, A.E., Katzourakis, A., de Oliveira, T., Welch, J.J., Belshaw, R., Bishop, K.N., Kramer, B., McMichael, A.J., Rambaut, A., and Iversen, A.K. (2008). Conserved footprints of APOBEC3G on hypermutated HIV-1 and HERV-K(HML2) sequences. *Journal of Virology* **82**, 8743–8761.

Armstrong, P.M. and Rico-Hesse, R. (2001). Differential susceptibility of *Aedes aegypti* to infection by the American and Southeast Asian genotypes of dengue type 2 virus. *Vector-Borne and Zoonotic Diseases* **1**, 159–168.

Arnaud, F., Caporale, M., Varela, M., Biek, R., Chessa, B., Alberti, A., Golder, M., Mura, M., Zhang, Y.-P., Yu, L. *et al.* (2007). A paradigm for virus-host coevolution: sequential counter-adaptations between endogenous and exogenous retroviruses. *PLoS Pathogens* **3**, e170.

Asquith, B. and Bangham, C.R. (2008). How does HTLV-I persist despite a strong cell-mediated immune response? *Trends in Immunology* **29**, 4–11.

Au, K.S., Mattion, N.M., and Estes, M.K. (1993). A subviral particle binding domain on the rotavirus nonstructural glycoprotein NS28. *Virology* **194**, 665–673.

Aulicino, P.C., Holmes, E.C., Rocco, C., Mangano, A.M., and Sen, L. (2007). Extremely rapid spread of HIV-1 BF recombinants in Argentina. *Journal of Virology* **81**, 427–429.

Azevedo, R.B., Lohaus, R., Srinivasan, S., Dang, K.K., and Burch, C.L. (2006). Sexual reproduction selects for robustness and negative epistasis in artificial gene networks. *Nature* **440**, 87–90.

Badrane, H. and Tordo, N. (2001). Host switching in *Lyssavirus* history from the Chiroptera to the Carnivora orders. *Journal of Virology* **75**, 8096–8104.

Baigent, S.J. and McCauley, J.W. (2003). Influenza type A in humans, mammals and birds: determinants of virus virulence, host-range and interspecies transmission. *Bioessays* **25**, 657–671.

Bailes, E., Gao, F., Bibollet-Ruche, F., Courgnaud, V., Peeters, M., Marx, P.A., Hahn, B.H., and Sharp, P.M. (2003). Hybrid origin of SIV in chimpanzees. *Science* **300**, 1713.

Balfe, P., Simmonds, P., Ludlam, C.A., Bishop, J.O., and Leigh Brown, A.J. (1990). Concurrent evolution of human immunodeficiency virus type 1 in patients infected from the same source: rate of sequence change and low frequency of inactivating mutations. *Journal of Virology* **64**, 6221–6233.

Ball, S.C., Abraha, A., Collins, K.R., Marozsan, A.J., Baird, H., Quiñones-Mateu, M.E., Penn-Nicholson, A., Murray, M., Richard, N., Lobritz, M. *et al.* (2003). Comparing the *ex vivo* fitness of CCR5-tropic human immunodeficiency virus type 1 isolates of subtypes B and C. *Journal of Virology* **77**, 1021–1038.

Baltimore, D. (1971). Expression of animal virus genomes. *Bacteriological Reviews* **35**, 235–241.

Baltimore, D. (1980). Evolution of RNA viruses. *Annals of the New York Academy of Sciences* **354**, 492–497.

Bamford, D.H., Grimes, J.M., and Stuart, D.I. (2005). What does structure tell us about viral evolution? *Current Opinion in Structural Biology* **15**, 1–9.

Bândea, C.I. (1983). A new theory on the origin and the nature of viruses. *Journal of Theoretical Biology* **105**, 591–602.

Bangham, C.R.M. and Kirkwood, T.B.L. (1990). Defective interfering particles: effects on modulating virus growth and persistence. *Virology* **179**, 821–826.

Baranowski, E., Ruiz-Jarabo, C.M., and Domingo, E. (2001). Evolution of cell recognition by viruses. *Science* **292**, 1102–1105.

Baroth, M., Orlich, M., Thiel, H.J., and Becher, P. (2000). Insertion of cellular NEDD8 coding sequences in a pestivirus. *Virology* **278**, 456–466.

Barré-Sinoussi, F., Chermann, J.C., Rey, F., Nugeytre, M.T., Chamaret, S., Gruest, J., Dauguet, C., Axler-Blue, C., Vezinet-Brun, F., Rouzioux, C. *et al.* (1983). Isolation of a T-lymphotropic retrovirus from a patient at risk for acquired immune deficiency syndrome (AIDS). *Science* **220**, 868–871.

Barrett, A.D.T. and Higgs, S. (2007). Yellow fever: a disease that has yet to be conquered. *Annual Review of Entomology* **52**, 209–229.

Bartlett, M.S. (1957). Measles periodicity and community size. *Journal of the Royal Statistical Society A* **120**, 48–70.

Barton, N.H. and Turelli, M. (1989). Evolutionary quantitative genetics: how little do we know? *Annual Review of Genetics* **23**, 337–370.

Beer, B.E., Bailes, E., Goeken, R., Dapolito, G., Coulibaly, C., Norley, S.G., Kurth, B., Gautier, J.-P., Gautier-Hion, A., Vallet, D. *et al.* (1999). Simian immunodefiency virus (SIV) from sun-tailed monkeys (*Ceropithecus solatus*): evidence for host-dependent evolution of SIV within the *C. lhoesti* superspecies. *Journal of Virology* **73**, 7734–7744.

Bell, P.J. (2001). Viral eukaryogenesis: was the ancestor of the nucleus a complex DNA virus? *Journal of Molecular Evolution* **53**, 251–256.

Belshaw, R., Pybus, O.G., and Rambaut, A. (2007). The evolution of genome compression in RNA viruses. *Genome Research* **17**, 1496–1504.

Belshaw, R., Gardner, A., Rambaut, A., and Pybus, O.G. (2008). Pacing a small cage: mutation and RNA viruses. *Trends in Ecology and Evolution* **23**, 188–193.

Bennett, S.N., Holmes, E.C., Chirivella, M., Rodriguez, D.M., Beltran, M., Vorndam, V., Gubler, D.J., and McMillan, W.O. (2003). Selection-driven evolution of emergent dengue virus. *Molecular Biology and Evolution* **20**, 1650–1658.

Benson, S.D., Bamford, J.K.H., Bamford, D.H., and Burnett, R.M. (2004). Does common architecture reveal a viral lineage spanning all three domains of life. *Molecular Cell* **16**, 673–685.

Bergstrom, C.T., McElhany, P., and Real, L.A. (1999). Transmission bottlenecks as determinants of virulence in rapidly evolving pathogens. *Proceedings of the National Academy of Sciences USA* **96**, 5095–5100.

Berkhoff, E.G., de Wit, E., Geelhoed-Mieras, M.M., Boon, A.C., Symons, J., Fouchier, R.A.M., Osterhaus, A.D.M.E., and Rimmelzwaan, G.F. (2005). Functional constraints of influenza A virus epitopes limit escape from cytotoxic T lymphocytes. *Journal of Virology* **79**, 11239–11246.

Bernard, H.-U. (1994). Coevolution of papillomaviruses with human populations. *Trends in Microbiology* **2**, 140–143.

Bertolotti, L., Kitron, U.D., Walker, E.D., Ruiz, M.O., Brawn, J.D., Loss, S.R., Hamer, G.L., and Goldberg, T.L. (2008). Fine-scale genetic variation and evolution of West Nile Virus in a transmission "hot spot" in suburban Chicago, USA. *Virology* **374**, 381–389.

Bhattacharya, T., Daniels, M., Heckerman, D., Foley, B., Frahm, N., Kadie, C., Carlson, J., Yusim, K., McMahon, B., Gaschen, B. *et al.* (2007). Founder effects in the assessment of HIV polymorphisms and HLA allele associations. *Science* **315**, 1583–1586.

Bibollet-Ruche, F., Bailes, E., Gao, F., Pourrut, X., Barlow, K.L., Clewley, J.P., Mwenda, J.M., Langat, D.K., Chege, G.K., McClure, H.M. *et al.* (2004). New simian immunodeficiency virus infecting De Brazza's monkeys (*Cercopithecus neglectus*): evidence for a cercopithecus monkey virus clade. *Journal of Virology* **78**, 7748–7762.

Biebricher, C.K. and Eigen, M. (2005). The error threshold. *Virus Research* **107**, 117–127.

Biek, R., Walsh, P.D., Leroy, E.M., and Real, L.A. (2006). Recent common ancestry of Ebola Zaire virus found in a bat reservoir. *PLoS Pathogens* **2**, e90.

Biek, R., Henderson, J.C., Waller, L.A., Rupprecht, C.E., and Real, L.A. (2007). A high-resolution genetic signature of demographic and spatial expansion in epizootic rabies virus. *Proceedings of the National Academy of Sciences USA* **104**, 7993–7998.

Black, F.L. (1975). Infectious diseases in primitive societies. *Science* **187**, 515–518.

Boerlijst, M.C., Bonhoeffer, S., and Nowak, M.A. (1996). Viral quasi-species and recombination. *Proceedings of the Royal Society of London Series B Biological Sciences* **263**, 1577–1584.

Boivin, G., Goyette, N., and Bernatchez, H. (2002). Prolonged excretion of amantadine-resistant influenza a virus quasi species after cessation of antiviral therapy in an immunocompromised patient. *Clinical Infectious Diseases* **34**, E23–E25.

Bonhoeffer, S., Holmes, E.C., and Nowak, M.A. (1995). Causes of HIV diversity. *Nature* **376**, 125.

Bonhoeffer, S., Chappey, C., Parkin, N.T., Whitcomb, J.M., and Petropoulos, C.J. (2004). Evidence for positive epistasis in HIV-1. *Science* **306**, 1547–1550.

Boni, M.F., Gog, J.R., Andreasen, V., and Feldman, M.W. (2006). Epidemic dynamics and antigenic evolution in a single season of influenza A. *Proceedings of the Royal Society of London Series B Biological Sciences* **273**, 1307–1316.

Boni, M.F., Zhou, Y., Taubenberger, J.K., and Holmes, E.C. (2008). Homologous recombination is very rare or absent in human influenza A virus. *Journal of Virology* **82**, 4807–4811.

Borucki, M.K., Chandler, L.J., Parker, B.M., Blair, C.D., and Beaty, B.J. (1999). Bunyavirus superinfection and segment reassortment in transovarially infected mosquitoes. *Journal of General Virology* **80**, 3173–3179.

Botstein, D. (1980). A theory of modular evolution for bacteriophages. *Annals of the New York Academy of Sciences* **354**, 484–490.

Botten, J., Mirowsky, K., Kusewitt, D., Ye, C., Gottlieb, K., Prescott, J., and Hjelle, B. (2003). Persistent Sin Nombre virus infection in the deer mouse (*Peromyscus maniculatus*) model: sites of replication and strand-specific expression. *Journal of Virology* **77**, 1540–1550.

Bourhy, H., Kissi, B., Audry, L., Smreczak, M., Sadkowska-Todys, M., Kulonen, K., Tordo, N., Zmudzinski, J.F., and Holmes, E.C. (1999). Ecology and evolution of rabies virus in Europe. *Journal of General Virology* **80**, 2545–2558.

Bourhy, H., Cowley, J.A., Larrous, F., Holmes, E.C., and Walker, P. (2005). Phylogenetic relationships among the rhabdoviruses inferred using the L polymerase gene. *Journal of General Virology* **86**, 2849–2858.

Bourhy, H., Reynes, J.-M., Dunham, E.J., Dacheux, L., Larouss, F., Hong, V.T.Q., Xu, G., Yan, J., Miranda, M.E.G., and Holmes, E.C. (2008). The origin and phylogeography of dog rabies virus. *Journal of General Virology* **89**, 2673–2681.

Boyko, V.P., Karasev, A.V., Agranovsky, A.A., Koonin, E.V., and Dolja, V.V. (1992). Coat protein gene duplication in a filamentous RNA virus of plants. *Proceedings of the National Academy of Sciences USA* **89**, 9156–9160.

Bracho, M.A., Moya, A., and Barrio, E. (1998). Contribution of *Taq* polymerase-induced errors to the estimation of RNA virus diversity. *Journal of General Virology* **79**, 2921–2928.

Bragstad, K., Nielsen, L.P., and Fomsgaard, A. (2008). The evolution of human influenza A viruses from 1999 to 2006: a complete genome study. *Virology Journal* **5**, 40.

Brault, A.C., Powers, A.M., Holmes, E.C., Woelk, C.H., and Weaver, S.C. (2002). Positively charged amino acid substitutions in the E2 envelope glycoprotein are associated with the emergence of Venezuelan equine encephalitis virus. *Journal of Virology* **76**, 1718–1730.

Brault, A.C., Huang, C.Y.-H., Langevin, S.A., Kinney, R.M., Bowen, R.A., Ramey, W.N., Panella, N.A., Holmes, E.C., Powers, A.M., and Miller, B.R. (2007). A single positively selected West Nile viral mutation confers increased avian virogenesis in American crows. *Nature Genetics* **39**, 1162–1166.

Bright, R.A., Medina, M.J., Xu, X., Perez-Oronoz, G., Wallis, T.R., Davis, X.M., Povinelli, L., Cox, N.J., and Klimov, A.I. (2005). Incidence of adamantane resistance among influenza A (H3N2) viruses isolated worldwide from 1994 to 2005: a cause for concern. *Lancet* **366**, 1175–1181.

Bright, R.A., Shay, D.K., Shu, B., Cox, N.J., and Klimov, A.I. (2006). Adamantane resistance among influenza A viruses isolated early during the 2005–2006 influenza season in the United States. *Journal of the American Medical Association* **295**, 891–894.

Briones, C., Domingo, E., and Molina-París, C. (2003). Memory in retroviral quasispecies: experimental evidence and theoretical model for human immunodeficiency virus. *Journal of Molecular Biology* **331**, 213–229.

Broussard, S.R., Staprans, S.I., White, R., Whitehead, E.M., Feinberg, M.B., and Allan, J.S. (2001). Simian immunodeficiency virus replicates to high levels in naturally infected African green monkeys without inducing immunologic or neurologic disease. *Journal of Virology* **75**, 2262–2275.

Brown, J.R. (2003). Ancient horizontal gene transfer. *Nature Reviews Genetics* **4**, 121–132.

Bruenn, J.A. (1991). Relationships among the positive strand and double-strand RNA viruses as viewed through their RNA-dependent RNA polymerases. *Nucleic Acids Research* **19**, 217–226.

Bryant, J.E., Holmes, E.C., and Barrett, A.D.T. (2007). Out of Africa: a molecular perspective on the introduction of Yellow Fever Virus into the Americas. *PLoS Pathogens* **3**, e75.

Bull, J.J., Badgett, M.R., Wichman, H.A., Huelsenbeck, J.P., Hillis, D.M., Gulati, A., Ho, C., and Molineux, I.J. (1997). Exceptional convergent evolution in a virus. *Genetics* **147**, 1497–1507.

Bull, J.J., Meyers, L.A., and Lachmann, M. (2005). Quasispecies made simple. *PLoS Computational Biology* **1**, e61.

Bull, J.J., Sanjuán, R., and Wilke, C.O. (2007). Theory of lethal mutagenesis for viruses. *Journal of Virology* **81**, 2930–2939.

Bulmer, M. (1987). Coevolution of codon usage and transfer RNA abundance. *Nature* **325**, 728–730.

Buonagurio, D.A., Nakada, S., Desselberger, U., Krystal, M., Palese, P., and Fitch, W.M. (1985). Noncumulative sequence changes in the hemagglutinin genes of influenza C virus isolates. *Virology* **146**, 221–232.

Burch, C.L. and Chao, L. (2000). Evolvability of an RNA virus is determined by its mutational neighbourhood. *Nature* **406**, 625–528.

Burch, C.L. and Chao, L. (2004). Epistasis and its relationship to canalization in the RNA virus ϕ6. *Genetics* **167**, 559–567.

Burch, C.L., Turner, P.E., and Hanley, K.A. (2003). Patterns of epistasis in RNA viruses: a review of the evidence from vaccine design. *Journal of Evolutionary Biology* **16**, 1223–1235.

Burch, C.L., Guyader, S., Samarov, D., and Shen, H. (2007). Experimental estimate of the abundance and effects of nearly neutral mutations in the RNA virus ϕ6. *Genetics* **176**, 467–476.

Burke, D.S. (1998). Evolvability of emerging viruses. In A.M. Nelson and C.R. Horsburgh Jr, eds, *Pathology of Emerging Infections 2*, pp. 1–12. American Society for Microbiology, Washington DC.

Bush, R.M., Fitch, W.M., Bender, C.A., and Cox, N.J. (1999). Positive selection on the H3 hemagglutinin gene of human influenza virus A. *Molecular Biology and Evolution* **16**, 1457–1465.

Calvignac, S., Terme, J.M., Hensley, S.M., Jalinot, P., Greenwood, A.D., and Hänni, C. (2008). Ancient DNA identification of early 20th century simian T-cell leukemia virus type 1. *Molecular Biology and Evolution* **25**, 1093–1098.

Cao, D., Barro, M. and Hoshino, Y. (2008). Porcine rotavirus bearing an aberrant gene stemming from an intergenic recombination of the NSP2 and NSP5 genes is defective and interfering. *Journal of Virology* **82**, 6073–6077.

Carr, J.K., Avila, M., Gomez Carrillo, M., Salomon, H., Hierholzer, J., Watanaveeradej, V., Pando, M.A., Negrete, M., Russell, K.L., Sanchez, J. *et al.* (2001). Diverse BF recombinants have spread widely since the introduction of HIV-1 into South America. *AIDS* **15**, F41–F47.

Carrillo, C., Lu, Z., Borca, M.V., Vagnozzi, A., Kutish, G.F., and Rock, D.L. (2007). Genetic and phenotypic variation of foot-and-mouth disease virus during serial passages in a natural host. *Journal of Virology* **81**, 11341–11351.

Carrington, C.V.F., Foster, J.E., Pybus, O.G., Bennett, S.N., and Holmes, E.C. (2005). Invasion and maintenance of dengue virus type 2 and type 4 in the Americas. *Journal of Virology* **79**, 14680–14687.

Carroll, S.M., Higa, H.H., and Paulson, J.C. (1981). Different cell-surface receptor determinants of antigenically similar influenza virus hemagglutinins. *Journal of Biological Chemistry* **256**, 8357–8363.

Cech, T.R. (1987). The chemistry of self-splicing RNA and RNA enzymes. *Science* **236**, 1532–1539.

Chamary, J.-V. and Hurst, L.D. (2004). Similar rates but different modes of sequence evolution in introns and at exonic silent sites in rodents: evidence for selectively driven codon usage. *Molecular Biology and Evolution* **21**, 1014–1023.

Chang, G.S., Hoon, Y., Ko, K.D., Bhardwaj, G., Holmes, E.C., Patterson, R.L., and van Rossum, D.B. (2008). Phylogenetic profiles reveal evolutionary relationships within the 'twilight zone' of sequence similarity. *Proceedings of the National Academy of Sciences USA* **105**, 13474–13479.

Chantawannakul, P. and Cutler, R.W. (2008). Convergent host-parasite codon usage between honeybee and bee associated viral genomes. *Journal of Invertebrate Pathology* **98**, 206–210.

Chao, D.-Y., King, C.-C., Wang, W.-K., Chen, W.-J., Wu, H.-L., and Chang, G.-J. (2005). Strategically examining the full-genome of dengue virus type 3 in clinical isolates reveals its mutation spectra. *Virology Journal* **2**, 72.

Chao, L. (1988). Evolution of sex in RNA viruses. *Journal of Theoretical Biology* **133**, 99–112.

Chao, L. (1990). Fitness of RNA virus decreased by Muller's ratchet. *Nature* **348**, 454–455.

Chao, L. (1991). Levels of selection, evolution of sex in RNA viruses, and the origin of life. *Journal of Theoretical Biology* **153**, 229–246.

Chao, L. (1994). Evolution of genetic exchange in RNA viruses. In S.S. Morse, ed., *The Evolutionary Biology of Viruses*, pp. 233–250. Raven Press, New York.

Chao, L., Tran, T.R., and Matthews, C. (1992). Muller's ratchet and the advantage of sex in the RNA virus φ6. *Evolution* **46**, 289–299.

Chao, L., Tran, T.T., and Tran, T.T. (1997). The advantage of sex in the RNA virus φ6. *Genetics* **147**, 953–959.

Chao, L., Rang, C.U., and Wong, L.E. (2002). Distribution of spontaneous mutants and inferences about the replication mode of the RNA bacteriophage φ6. *Journal of Virology* **76**, 3276–3281.

Chare, E.R. and Holmes, E.C. (2004). Selection pressures in the capsid genes of plant RNA viruses reflect mode of transmission. *Journal of General Virology* **85**, 3149–3157.

Chare, E.R., Gould, E.A., and Holmes, E.C. (2003). Phylogenetic analysis reveals a low rate of homologous recombination in negative-sense RNA viruses. *Journal of General Virology* **84**, 2691–2703.

Charleston, M.A. and Robertson, D.L. (2002). Preferential host switching by primate lentiviruses can account for phylogenetic similarity with the primate phylogeny. *Systematic Biology* **51**, 528–535.

Charini, W.A., Todd, S., Gutman, G.A., and Semler, B.L. (1994). Transduction of a human RNA sequence by poliovirus. *Journal of Virology* **68**, 6547–6552.

Charrel, R.N., de Micco, P., and de Lamballerie, X. (1999). Phylogenetic analysis of GB viruses A and C: evidence for cospeciation between virus isolates and their primate hosts. *Journal of General Virology* **80**, 2329–2335.

Charrel, R.N., Feldmann, H., Fulhorst, C.F., Khelifa, R., de Chesse, R. and de Lamballerie, X. (2002). Phylogeny of New World arenaviruses based on the complete coding sequences of the small genomic segment identified an evolutionary lineage produced by intrasegmental recombination. *Biochemical and Biophysical Research Communications* **296**, 1118–1124.

Chen, H., Smith, G.J., Zhang, S.Y., Qin, K., Wang, J., Li, K.S., Webster, R.G., Peiris, J.S., and Guan, Y. (2005). Avian flu: H5N1 virus outbreak in migratory waterfowl. *Nature* **436**, 191–192.

Chen, R. and Holmes, E.C. (2006). Avian influenza virus exhibits rapid evolutionary dynamics. *Molecular Biology and Evolution* **23**, 2336–2341.

Chetverin, A.B., Chetverina, H.V., Demidenko, A.A., and Ugarov, V.I. (1997). Nonhomologous RNA recombination in a cell-free system: evidence for a transesterification mechanism guided by secondary structure. *Cell* **88**, 503–513.

Childs, J.E., Curns, A.T., Dey, M.E., Real, L.A., Feinstein, L., Bjørnstad, O.N., and Krebs, J.W. (2000). Predicting the local dynamics of epizootic rabies among raccoons in the United States. *Proceedings of the National Academy of Sciences USA* **97**, 13666–13671.

Chopera, D.R., Woodman, Z., Mlisana, K., Mlotshwa, M., Martin, D.P., Seoighe, C., Treurnicht, F., de Rosa, D.A., Hide, W., Karim, S.A., Gray, C.M., Williamson, C. and CAPRISA 002 Study Team (2008). Transmission of HIV-1 CTL escape variants provides HLA-mismatched recipients with a survival advantage. *PLoS Pathogens* **4**, e1000033.

Claverie, J.-M. (2006). Viruses take center stage in cellular evolution. *Genome Biology* **7**, 110.

Codoñer, F.M., Daròs, J.A., Sole, R.V., and Elena, S.F. (2006). The fittest versus the flattest: Experimental confirmation of the quasispecies effect with subviral pathogens. *PLoS Pathogens* **2**, e136.

Coen, D.M. and Ramig, R.F. (1996). Viral genetics. In B.N. Fields, D.M. Knipe, and P.M. Howley, eds, *Fundamental Virology*, 3rd edn, pp. 101–139. Lippincott Williams and Wilkins, Philadelphia, PA.

Coffey, L.L., Vasilakis, N., Brault, A.C, Powers, A.M., Tripet, F., and Weaver, S.C. (2008). Arbovirus evolution *in vivo* is constrained by host alternation. *Proceedings of the National Academy of Sciences USA* **105**, 6970–6975.

Cole, C.N. and Baltimore, D. (1973). Defective interfering particles of poliovirus. III. Inference and enrichment. *Journal of Molecular Biology* **76**, 345–361.

Cole, T.E., Hong, Y., Brasier, C.M., and Buck, K.W. (2000). Detection of an RNA-dependent RNA polymerase in mitochondria from a mitovirus-infected isolate of the Dutch Elm disease fungus, Ophiostoma novo-ulmi. *Virology* **268**, 239–423.

Coleman, J.R., Papamichail, D., Skiena, S., Futcher, B., Wimmer, E., and Mueller, S. (2008). Virus attenuation by genome-scale changes in codon pair bias. *Science* **320**, 1784–1787.

Cologna, R. and Rico-Hesse, R. (2003). American genotype structures decrease dengue virus output from human monocytes and dendritic cells. *Journal of Virology* **77**, 3929–3938.

Cologna, R., Armstrong, P.M., and Rico-Hesse, R. (2005). Selection for virulent dengue viruses occurs in humans and mosquitoes. *Journal of Virology* **79**, 853–859.

Comas, I., Moya, A., and Gonzalez-Candelas, F. (2005). Validating viral quasispecies with digital organisms: A re-examination of the critical mutation rate. *BMC Evolutionary Biology* **5**, 5.

Cooper, A. and Poinar, H.N. (2000). Ancient DNA: do it right or not at all. *Science* **289**, 1139.

Cottam, E.M., Haydon, D.T., Paton, D.J., Gloster, J., Wilesmith, J.W., Ferris, N.P., Hutchings, G.H., and King, D.P. (2006). Molecular epidemiology of the foot-and-mouth disease virus outbreak in the United Kingdom in 2001. *Journal of Virology* **80**, 11274–11282.

Cottam, E.M., Wadsworth, J., Shaw, A.E., Rowlands, R.J., Goatley, L., Maan, S., Maan, N.S., Mertens, P.P., Ebert, K., Li, Y. *et al.* (2008). Transmission pathways of foot-and-mouth disease virus in the United Kingdom in 2007. *PLoS Pathogens* **4**, e1000050.

Coulibaly, F., Chevalier, C., Gutsche, I., Pous, J., Navaza, J., Bressanelli, S., Delmas, B., and Rey, F.A. (2005). The birnavirus crystal structure reveals structural relationships among icosahedral viruses. *Cell* **120**, 761–772.

Cox, A.L., Mosbruger, T., Mao, Q., Liu, Z., Wang, X.-H., Yang, H.-C., Sidney, J., Sette, A., Pardoll, D., Thomas, D.L., and Ray, S.C. (2005). Cellular immune selection with hepatitis C virus persistence in humans. *Journal of Experimental Medicine* **201**, 1741–1752.

Cox-Foster, D.L., Conlan, S., Holmes, E.C., Palacios, G., Evans, J.D., Moran, N.A., Quan, P.-L., Briese, T., Geiser, D.M., Martinson, V.*et al.* (2007). A metagenomic survey of microbes in honey bee colony collapse disorder. *Science* **318**, 283–287.

Coyne, K.P., Gaskell, R.M., Dawson, S., Porter, C.J., and Radford, A.D. (2007). Evolutionary mechanisms of persistence and diversification of a calicivirus within endemically infected natural host populations. *Journal of Virology* **81**, 1961–1971.

Craig, S., Thu, H.M., Lowry, K., Wang, X.-F., Holmes, E.C., and Aaskov, J. (2003). Diverse dengue type 2 virus populations contain recombinant and both parental viruses in a single mosquito host. *Journal of Virology* **77**, 4463–4467.

Crandall, K.A., Kelsey, C.R., Imamichi, H., Lane, H.C., and Salzman, N.P. (1999). Parallel evolution of drug resistance in HIV: failure of nonsynonymous/synonymous substitution rate ratio to detect selection. *Molecular Biology and Evolution* **16**, 372–382.

Crawford, P.C., Dubovi, E.J., Castleman, W.L., Stephenson, I., Gibbs, E.P., Chen, L., Smith, C., Hill, R.C., Ferro, P., Pompey, J. *et al.* (2005). Transmission of equine influenza virus to dogs. *Science* **310**, 482–485.

Crill, W.D. and Chang, G.J. (2004). Localization and characterization of flavivirus envelope glycoprotein cross-reactive epitopes. *Journal of Virology* **78**, 13975–13986.

Crochu, S., Cook, S., Attoui, H., Charrel, R.N., De Chesse, R., Belhouchet, M., Lemasson, J.J., De Micco, P., and De Lamballerie, X. (2004). Sequences of flavivirus-related RNA viruses persist in DNA form integrated in the genome of *Aedes* spp. mosquitoes. *Journal of General Virology* **85**, 1971–1980.

Crotty, S., Maag, D., Arnold, J.J., Zhong, W., Lau, J.Y., Hong, Z., Andino, R., and Cameron, C.E. (2000). The broad-spectrum antiviral ribonucleoside ribavirin is an RNA virus mutagen. *Nature Medicine* **6**, 1375–1379.

Crotty, S., Cameron, C.E., and Andino, R. (2001). RNA virus error catastrophe: direct test by using ribavirin. *Proceedings of the National Academy of Sciences USA* **98**, 6895–6900.

Cubit, B., Ly, C. and de la Torre, J.C. (2001). Identification and characterization of a new intron in Borna disease virus. *Journal of General Virology* **82**, 641–646.

Cuevas, J.M., Elena, S.F., and Moya, A. (2002). Molecular basis of adaptive convergence in experimental populations of RNA viruses. *Genetics* **162**, 533–542.

Culley, A.I., Lang, A.S., and Suttle, C.A. (2003). High diversity of unknown picorna-like viruses in the sea. *Nature* **424**, 1054–1057.

Culley, A.I., Lang, A.S., and Suttle, C.A. (2006). Metagenomic analysis of coastal RNA virus communities. *Science* **312**, 1795–1798.

Daly, J.M., Lai, A.C., Binns, M.M., Chambers, T.M., Barrandeguy, M., and Mumford, J.A. (1996). Antigenic and genetic evolution of equine H3N8 influenza A viruses. *Journal of General Virology* **77**, 661–671.

Daròs, J.A., Elena, S.F., and Flores, R. (2006). Viroids: an Ariadne's thread into the RNA labyrinth. *EMBO Reports* **7**, 593–598.

Davies, T.J. and Pedersen, A.B. (2008). Phylogeny and geography predict pathogen community similarity in wild primates and humans. *Proceedings of the Royal Society of London Series B Biological Sciences* **275**, 1695–1701.

Davis, P.L., Holmes, E.C., Larrous, F., Van de Poel, W.H.M., Tjørnehøj, K., Alonso, W.J., and Bourhy, H. (2005). The phylogeography, population dynamics, and molecular evolution of European bat lyssaviruses. *Journal of Virology* **79**, 10487–10497.

Davis, P.L., Rambaut, A., Bourhy, H., and Holmes, E.C. (2007). The evolutionary dynamics of canid and mongoose rabies virus in Southern Africa. *Archives of Virology* **152**, 1251–1258.

DeFilippis, V.R. and Villarreal, L.P. (2000). An introduction to the evolutionary ecology of viruses. In C.J. Hurst, ed., *Viral Ecology*, pp. 126–208. Academic Press, New York.

de Jong, J.C., Beyer, W.E.P., Palache, A.M., Rimmelzwaan, G.F., and Osterhaus, A.D.M.E. (2000). Mismatch between the 1997/1998 influenza vaccine and the major epidemic A(H3N2) virus strain as the cause of an inadequate vaccine-induced antibody response to this strain in the elderly. *Journal of Medical Virology* **61**, 94–99.

de la Iglesia, F. and Elena, S.F. (2007). Fitness declines in Tobacco etch virus upon serial bottleneck transfers. *Journal of Virology* **81**, 4941–4947.

de la Torre, J.C. and Holland, J.J. (1990). RNA virus quasispecies populations can suppress vastly superior mutant progeny. *Journal of Virology* **64**, 6278–6281.

Delmas, O., Holmes, E.C., Talbi, C., Larrous, F., Dacheux, L., Bouchier, C., and Bourhy, H. (2008). Genomic diversity and evolution of the lyssaviruses. *PLoS ONE* **3**, e2057.

Demma, L.J., Logsdon, J.M., Jr, Vanderford, T.H., Feinberg, M.B., and Staprans, S.I. (2005). SIVsm quasispecies adaptation to a new simian host. *PLoS Pathogens* **1**, e3.

DePristo, M.A., Weinreich, D.M., and Hartl, D.L. (2005). Missense meanderings in sequence space: a biophysical view of protein evolution. *Nature Reviews Genetics* **6**, 678–687.

de Silva, A. and Messer, W. (2004). Arguments for live flavivirus vaccines. *Lancet* **364**, 500.

de Thoisy, B., Dussart, P., and Kazanji, M. (2004). Wild terrestrial rainforest mammals as potential reservoirs for flaviviruses (yellow fever, dengue 2 and St Louis encephalitis viruses) in French Guiana. *Transactions of the Royal Society of Tropical Medicine and Hygiene* **98**, 409–412.

de Visser, J.A., Hermisson, J., Wagner, G.P., Ancel Meyers, L., Bagheri-Chaichian, H., Blanchard, J.L., Chao, L., Cheverud, J.M., Elena, S.F., Fontana, W. *et al.* (2003). Evolution and detection of genetic robustness. *Evolution* **57**, 1959–1972.

Diamond, J. (2002). Evolution, consequences and future of plant and animal domestication. *Nature* **418**, 700–707.

Dobson, A. (2000). Raccoon rabies in time and space. *Proceedings of the National Academy of Sciences USA* **97**, 14041–14043.

Dobson, A.P. and Carper, E.R. (1996). Infectious diseases and human population history. *Bioscience* **46**, 115–126.

Domingo, E. (1992). Genetic variation and quasi-species. *Current Opinion in Genetics and Development* **2**, 61–63.
Domingo, E. (2002). Quasispecies theory in virology. *Journal of Virology* **76**, 463–465.
Domingo, E. and Holland, J.J. (1997). RNA virus mutations for fitness and survival. *Annual Review of Microbiology* **51**, 151–178.
Domingo, E., Sabo, D., Taniguchi, T., and Weissman, C. (1978). Nucleotide sequence heterogeneity of an RNA phage population. *Cell* **13**, 735–744.
Domingo, E., Martinez-Salas, E., Sobrino, F., de la Torre, J.C., Portela, A., Ortin, J., Lopez-Galindez, C., Perez-Brena, P., Villanueva, N., Najera, R. *et al.* (1985). The quasispecies (extremely heterogeneous) nature of viral RNA genome populations: biological relevance - a review. *Gene* **40**, 1–8.
Domingo, E., Escarmís, C., Menéndez-Arias, L., and Holland, J.J. (1999). Viral quasispecies and fitness variations. In E. Domingo, R. Webster, and J.J. Holland, eds, *Origin and Evolution of Viruses*, pp. 141–161. Academic Press, London.
Domingo, E., Biebricher, C., Eigen, M., and Holland, J.J., eds. (2001). *Quasispecies and RNA Virus Evolution: Principles and Consequences*. Landes Bioscience, Austin, TX.
Domingo, E., Escarmís, C., Lázaro, E., and Manrubia, S.C. (2005). Quasispecies dynamics and RNA virus extinction. *Virus Research* **107**, 129–139.
Doolittle, R.F. (1986). *Of Urfs and Orfs: a primer on how to analyze derived amino acid sequences*. University Science Books, Mill Valley, CA.
Dopazo, J., Dress, A., and von Haeseler, A. (1993). Split decomposition: a technique to analyse viral evolution. *Proceedings of the National Academy of Sciences USA* **90**, 10320–10324.
Dopazo, J., Sobrino, F., Palma, E.L., Domingo, E., and Moya, A. (1988). Gene encoding capsid protein VP1 of foot-and-mouth disease virus: a quasispecies model of molecular evolution. *Proceedings of the National Academy of Sciences USA* **85**, 6811–6815.
Drake, J.W. (1993). Rates of spontaneous mutation among RNA viruses. *Proceedings of the National Academy of Sciences USA* **90**, 4171–4175.
Drake, J.W. (2007). Too many mutants with multiple mutations. *Critical Reviews in Biochemistry and Molecular Biology* **42**, 247–258.
Drake, J.W. and Holland, J.J. (1999). Mutation rates among RNA viruses. *Proceedings of the National Academy of Sciences USA* **96**, 13910–13913.
Drake, J.W. and Hwang, C.B.C. (2005). On the mutation rate of herpes simplex virus type 1. *Genetics* **170**, 969–970.
Drake, J.W., Charlesworth, B., Charlesworth, D., and Crow, J.F. (1998). Rates of spontaneous mutation. *Genetics* **148**, 1667–1686.
Drake, J.W., Bebenek, A., Kissling, G.E., and Peddada, S. (2005). Clusters of mutations from transient hypermutability. *Proceedings of the National Academy of Sciences USA* **102**, 12849–12854.
Drummond, A.J. and Rambaut, A. (2007). BEAST: Bayesian evolutionary analysis by sampling trees. *BMC Evolutionary Biology* **7**, 214.
Drummond, A.J., Pybus, O.G., and Rambaut, A. (2003a). Inference of viral evolutionary rates from molecular sequences. *Advances in Parasitology* **54**, 331–358.
Drummond, A.J., Pybus, O.G., Rambaut, A., Forsberg, R., and Rodrigo, A.G. (2003b). Measurably evolving populations. *Trends in Ecology and Evolution* **18**, 481–488.
Drummond, A.J., Rambaut, A., Shapiro, B., and Pybus, O.G. (2005). Bayesian coalescent inference of past population dynamics from molecular sequences. *Molecular Biology and Evolution* **22**, 1185–1192.

Drummond, A.J., Ho, S.Y.W., Phillips, M.J., and Rambaut, A. (2006). Relaxed phylogenetics and dating with confidence. *PLoS Biology* **4**, e88.

Duarte, E., Clarke, D., Moya, A., Domingo, E., and Holland, J.J. (1992). Rapid fitness losses in mammalian RNA virus clones due to Muller's ratchet. *Proceedings of the National Academy of Sciences USA* **89**, 6015–6019.

Duffy, S. and Holmes, E.C. (2008). Phylogenetic evidence for rapid rates of molecular evolution in the single-stranded DNA begomovirus *Tomato Yellow Leaf Curl Virus*. *Journal of Virology* **82**, 957–965.

Duffy, S., Turner, P.E., and Burch, C.L. (2006). Pleiotropic costs of niche expansion in the RNA bacteriophage ɸ6. *Genetics* **172**, 751–757.

Duffy, S., Burch, C.L., and Turner, P.E. (2007). Evolution of host specificity drives reproductive isolation among RNA viruses. *Evolution* **61**, 2614–2622.

Duffy, S., Shackelton, L.A., and Holmes, E.C. (2008). Rates of evolutionary change in viruses: Patterns and determinants. *Nature Reviews Genetics* **9**, 267–276.

Duhaut, S.D. and McCauley, J.W. (1996). Defective RNAs inhibit the assembly of influenza virus genome segments in a segment-specific manner. *Virology* **216**, 326–337.

Dugan, V.G., Chen, R., Spiro, D.J., Sengamalay, N., Zaborsky, J., Ghedin, E., Nolting, J., Swayne, D.E., Runstadler, J.A., Happ, G.M. et al. (2008). The evolutionary genetics and emergence of avian influenza viruses in wild birds. *PLoS Pathogens* **4**, e1000076.

Dunham, E.J. and Holmes, E.C. (2007). Inferring the time-scale of dengue virus evolution under realistic models of DNA substitution. *Journal of Molecular Evolution* **64**, 656–661.

Dutta, R.N., Rouzine, I.M., Smith, S.D., Wilke, C.O., and Novella, I.S. (2008). Rapid adaptive amplification of preexisting variation in an RNA virus. *Journal of Virology* **82**, 4354–4362.

Edwards, C.T.T., Pfafferot, K.J., Goulder, P.G.R., Phillips, R.E., and Holmes, E.C. (2005). Intrapatient escape in the A*0201-restricted epitope SLYNTVATL drives evolution of human immunodeficiency virus type 1 at the population level. *Journal of Virology* **79**, 9363–9366.

Edwards, C.T.T., Holmes, E.C., Pybus, O.G., Wilson, D.J., Viscidi, R.P., Abrams, E.J., Phillips, R.E., and Drummond, A.J. (2006a). Evolution of the HIV-1 envelope is dominated by purifying selection. *Genetics* **174**, 1441–1453.

Edwards, C.T.T., Holmes, E.C., Wilson, D.J., Viscidi, R.P., Abrams, E.J., Phillips, R.E., and Drummond, A.J. (2006b). Population genetic estimation of the loss of genetic diversity during horizontal transmission of HIV-1. *BMC Evolutionary Biology* **6**, 28.

Eickbush, T.H. (1994). Origin and evolutionary relationships of retroelements. In S.S. Morse, ed., *The Evolutionary Biology of Viruses*, pp. 121–157. Raven Press, New York.

Eickbush, T.H. (1997). Telomerase and retrotransposons: which came first? *Science* **277**, 911–912.

Eigen, M. (1971). Self-organization of matter and the evolution of biological macromolecules. *Naturwissenschaften* **58**, 465–523.

Eigen, M. (1992). *Steps Towards Life*. Oxford University Press, New York.

Eigen, M. (1996). On the nature of viral quasispecies. *Trends in Microbiology* **4**, 216–218.

Eigen, M. and Schuster, P. (1977). The hypercycle, a principle of natural self-organization. Part A: emergence of the hypercycle. *Naturwissenschaften* **64**, 541–565.

Elena, S.F. (2002). Restrictions to RNA virus adaptation: an experimental approach. *Antonie van Leeuwenhoek* **81**, 135–142.

Elena, S.F. and Moya, A. (1999). Rate of deleterious mutation and the distribution of its effects on fitness in vesicular stomatitis virus. *Journal of Evolutionary Biology* **12**, 1078–1088.

Elena, S.F. and Lenski, R.E. (2003). Evolution experiments with microorganisms: the dynamics and genetic bases of adaptation. *Nature Reviews Genetics* **4**, 457–470.
Elena, S.F. and Sanjuán, R. (2005). Adaptive value of high mutation rates of RNA viruses: separating causes from consequences. *Journal of Virology* **79**, 11555–11558.
Elena, S.F., Dopazo, J., Flores, R., Diener, T.O., and Moya, A. (1991). Phylogeny of viroids, viroidlike satellite RNAs, and the viroidlike domain of hepatitis δ virus RNA. *Proceedings of the National Academy of Sciences USA* **88**, 5631–5634.
Elena, S.F., Sanjuán, R., Bordería, A.V., and Turner, P.E. (2001). Transmission bottlenecks and the evolution of fitness in rapidly evolving RNA viruses. *Infection, Genetics and Evolution* **1**, 41–48.
Elena, S.F., Carrasco, P., Daròs, J.A., and Sanjuán, R. (2006). Mechanisms of genetic robustness in RNA viruses. *EMBO Reports* **7**, 168–173.
Enami, M., Sharma, G., Benham, C., and Palese, P. (1991). An influenza virus containing nine different RNA segments. *Virology* **185**, 291–298.
Evans, K.D. and Kline, M.K. (1995). Prolonged influenza A infection responsive to rimantadine therapy in a human immunodeficiency virus-infected child. *The Pediatric Infectious Disease Journal* **14**, 332–334.
Eyre-Walker, A. (1999). Evidence of selection on silent site base composition in mammals: potential implications for the evolution of isochores and junk DNA. *Genetics* **152**, 675–683.
Eyre-Walker, A. and Keightley, P.D. (2007). The distribution of fitness effects of new mutations. *Nature Reviews Genetics* **8**, 610–618.
Fares, M.A., Ruiz-González, M.X., Moya, A., Elena, S.F., and Barrio, E. (2002). Endosymbiotic bacteria: groEL buffers against deleterious mutations. *Nature* **417**, 39.
Fargette, D., Pinel, A., Rakotomalala, M., Sangu, E., Traoré, O., Sérémé, D., Sorho, F., Issaka, S., Hébrard, E., Séré, Y. *et al.* (2008a). Rice yellow mottle virus, an RNA plant virus, evolves as rapidly as most RNA animal viruses. *Journal of Virology* **82**, 3584–3589.
Fargette, D., Pinel-Galzi, A., Sérémé, D., Lacombe, S., Hébrard, E., Traoré, O., and Konaté, G. (2008b). Diversification of rice yellow mottle virus and related viruses spans the history of agriculture from the neolithic to the present. *PLoS Pathogens* **4**, e1000125.
Fauquet, C.M., Mayo, M.A., Maniloff, J., Desselberger, U., and Ball, L.A. (2005). *Virus Taxonomy: VIIIth Report of the International Committee on Taxonomy of Viruses*. Elsevier Academic Press, San Diego, CA.
Felsenstein, J. (1974). The evolutionary advantage of recombination. *Genetics* **78**, 737–756.
Fenyö, E.M., Albert, J., and Asjö, B. (1989). Replicative capacity, cytopathic effect and cell tropism of HIV. *AIDS* **3**, S5–12.
Ferber, D. (2000). Human diseases threaten great apes. *Science* **289**, 1277–1278.
Ferguson, N.M., Galvani, A.P., and Bush, R.M. (2003). Ecological and immunological determinants of influenza evolution. *Nature* **422**, 428–433.
Figueroa, F., Günther, E., and Klein, J. (1988). MHC polymorphism pre-dating speciation. *Nature* **355**, 265–267.
Finkbeiner, S.R., Allred, A.F., Tarr, P.I., Klein, E.J., Kirkwood, C.D., and Wang, D. (2008). Metagenomic analysis of human diarrhea: viral detection and discovery. *PLoS Pathogens* **4**, e1000011.
Finkelman, B.S., Viboud, C., Koelle, K., Ferrari, M.J., Bharti, N., and Grenfell, B.T. (2007). Global patterns in seasonal activity of influenza A/H3N2, A/H1N1, and B from 1997 to 2005: viral coexistence and latitudinal gradients. *PLoS ONE* **2**, e1296.
Finkelstein, D.B, Mukatira, S., Mehta, P.K., Obenauer, J.C., Su, X., Webster, R.G., and Naeve, C.W. (2007). Persistent host markers in pandemic and H5N1 influenza viruses. *Journal of Virology* **81**, 10292–10299.

Fisher, R.A. (1930). *The Genetical Theory of Natural Selection*. Clarendon Press, Oxford.

Fitch, W.M. (1971). Rate of change of concomitantly variable codons. *Journal of Molecular Evolution* **1**, 84–96.

Fitch, W.M., Leiter, J.M.E., Li, X., and Palese, P. (1991). Positive Darwinian evolution in human influenza A viruses. *Proceedings of the National Academy of Sciences USA* **88**, 4270–4274.

Fitch, W.M., Bush, R.M., Bender, C.A., and Cox, N.J. (1997). Long term trends in the evolution of H(3) HA1 human influenza type A. *Proceedings of the National Academy of Sciences USA* **94**, 7712–7718.

Forrester, N.L., Moss, S.R., Turner, S.L., Schirrmeier, H., and Gould, E.A. (2008). Recombination in rabbit haemorrhagic disease virus: possible impact on evolution and epidemiology. *Virology* **376**, 390–396.

Forss, S. and Schaller, H. (1982). A tandem repeat gene in a picornavirus. *Nucleic Acids Research* **10**, 6441–6450.

Forterre, P. (2005). The two ages of the RNA world, and the transition to the DNA world: a story of viruses and cells. *Biochimie* **87**, 793–803.

Fouchier, R.A.M., Schneeberger, P.M., Rozendaal, F.W., Broekman, J.M., Kemink, S.A., Munster, V., Kuiken, T., Rimmelzwaan, G.F., Schutten, M., Van Doornum, G.J. *et al.* (2004). Avian influenza A virus (H7N7) associated with human conjunctivitis and a fatal case of acute respiratory distress syndrome. *Proceedings of the National Academy of Sciences USA* **101**, 1356–1361.

Fouchier, R.A.M., Munster, V., Wallensten, A., Bestebroer, T.M., Herfst, S., Smith, D., Rimmelzwaan, G.F., Olsen, B., and Osterhaus, A.D.M.E. (2005). Characterization of a novel influenza A virus hemagglutinin subtype (H16) obtained from black-headed gulls. *Journal of Virology* **79**, 2814–2822.

Fraile, A., Escriu, F., Aranda, M.A., Malpica, J.M., Gibbs, A.J., and García-Arenal, F. (1997). A century of tobamovirus evolution in an Australian population of *Nicotiana glauca*. *Journal of Virology* **71**, 8316–8320.

French, R. and Stenger, D.C. (2003). Evolution of Wheat streak mosaic virus: dynamics of population growth within plants may explain limited variation. *Annual Review of Phytopathology* **41**, 199–214.

Frey, T.K. and Youngner, J.S. (1984). Further studies of the RNA synthesis phenotype selected during persistent infection with vesicular stomatitis virus. *Virology* **136**, 211–220.

Friedrich, T.C., Dodds, E.J., Yant, L.J., Vojnov, L., Rudersdorf, R., Cullen, C., Evans, D.T., Desrosiers, R.C., Mothe, B.R., Sidney, J. *et al.* (2004a). Reversion of CTL escape-variant immunodeficiency viruses *in vivo*. *Nature Medicine* **10**, 275–281.

Friedrich, T.C., Frye, C.A., Yant, L.J., O'Connor, D.H., Kriewaldt, N.A., Benson, M., Vojnov, L., Dodds, E.J., Cullen, C., Rudersdorf, R. *et al.* (2004b). Extraepitopic compensatory substitutions partially restore fitness to simian immunodeficiency virus variants that escape from an immunodominant cytotoxic-T-lymphocyte response. *Journal of Virology* **78**, 2581–2585.

Froissart, R., Michalakis, Y., and Blanc, S. (2002). Helper component-transcomplementation in the vector transmission of plant viruses. *Phytopathology* **92**, 576–579.

Froissart, R., Wilke, C.O., Montville, R., Remold, S.K., Chao, L., and Turner, P.E. (2004). Co-infection weakens selection again epistatic mutations in RNA viruses. *Genetics* **168**, 9–19.

Froissart, R., Rose, D., Uzest, M., Galibert, L., Blanc, S., and Michalakis, Y. (2005). Recombination every day: Abundant recombination in a virus during a single multi-cellular host infection. *PLoS Biology* **3**, e89.

Frost, S.D.W., Nijhuis, M., Schuurman, R., Boucher, C.A., and Leigh Brown, A.J. (2000). Evolution of lamivudine resistance in human immunodeficiency virus type 1-infected individuals: the relative roles of drift and selection. *Journal of Virology* **74**, 6262–6268.

Frost, S.D.W., Dumaurier, M.-J., Wain-Hobson, S., and Leigh Brown, A.J. (2001). Genetic drift and within-host metapopulation dynamics of HIV-1 infection. *Proceedings of the National Academy of Sciences USA* **98**, 6975–6980.

Furió, V., Moya, A., and Sanjuán, R. (2005). The cost of replication fidelity in an RNA virus. *Proceedings of the National Academy of Sciences USA* **102**, 10233–10237.

Furió, V., Moya, A., and Sanjuán, R. (2007). The cost of replication fidelity in human immunodeficiency virus type 1. *Proceedings of the Royal Society of London Series B Biological Sciences* **274**, 225–230.

Gago, S., Elena, S.F., Flores, R., and Sanjuán, R. (2009). Extremely high mutation rate of a hammerhead viroid. *Science* **323**, 1308.

Gao, F., Bailes, E., Robertson, D.L., Chen, Y., Rodenburg, C.M., Michael, S.F., Cummins, L.B., Arthur, L.O., Peeters, M., Shaw, G.M. *et al.* (1999). Origin of HIV-1 in the chimpanzee *Pan troglodytes troglodytes*. *Nature* **397**, 436–441.

Gao, F., Chen, Y., Levy, D.N., Conway, J.A., Kepler, T.B., and Hui, H. (2004). Unselected mutations in the human immunodeficiency virus type 1 genome are mostly nonsynonymous and often deleterious. *Journal of Virology* **78**, 2426–2433.

García-Arenal, F., Fraile, A., and Malpica, J.M. (2001). Variability and genetic structure of plant virus populations. *Annual Review of Phytopathology* **39**, 157–186.

García-Arenal, F., Fraile, A., and Malpica, J.M. (2003). Variation and evolution of plant virus populations. *International Microbiology* **6**, 225–232.

García-Arriaza, J., Manrubia, S.C., Toja, M., Domingo, E., and Escarmís, C. (2004). Evolutionary transition toward defective RNAs that are infectious by complementation. *Journal of Virology* **78**, 11678–11685.

Garcia-Diaz, M. and Bebenek, K. (2007). Multiple functions of DNA polymerases. *Critical Reviews in Plant Sciences* **26**, 105–122.

Garriga, D., Navarro, A., Querol-Audí, J., Abaitua, F., Rodríguez, J.F., and Verdaguer, N. (2007). Activation mechanism of a noncanonical RNA-dependent RNA polymerase. *Proceedings of the National Academy of Sciences USA* **104**, 20540–20545.

Gaunt, M.W., Sall, A.A., de Lamballerie, X., Falconar, A.K.I., Dzhivanian, T.I., and Gould, E.A. (2001). Phylogenetic relationships of flaviviruses correlate with their epidemiology, disease association and biogeography. *Journal of General Virology* **82**, 1867–1876.

Ge, L.M., Zhang, J.T., Zhou, X.P., and Li, H.Y. (2007). Genetic structure and population variability of Tomato yellow leaf curl China virus. *Journal of Virology* **81**, 5902–5907.

Geigenmüller-Gnirke, U., Weiss, B., Wright, R., and Schlesinger, S. (1991). Complementation between Sindbis viral RNAs produces infectious particles with a bipartite genome. *Proceedings of the National Academy of Sciences USA* **88**, 3253–3257.

Geleziunas, R., Bour, S., and Wainberg, M.A. (1994). Cell surface down-modulation of CD4 after infection by HIV-1. *FASEB Journal* **8**, 593–600.

Ghedin, E., Fitch, A., Boyne, A., DePasse, J., Bera, J., Halpin, R.A., Griesemer, S., Smit, M., Jennings, L., St. George, K., *et al.* Mixed infection and the genesis of influenza diversity. Submitted to *Journal of Virology*.

Ghedin, E., Sengamalay, N.A., Shumway, M., Zaborsky, J., Feldblyum, T., Subbu, V., Spiro, D.J., Sitz, J., Koo, H., Bolotov, P. *et al.* (2005). Large-scale sequencing of human influenza reveals the dynamic nature of viral genome evolution. *Nature* **437**, 1162–1166.

Gibbons, R.V. and Vaughn, D.W. (2002). Dengue: an escalating problem. *British Medical Journal* **324**, 1563–1566.

Gibbs, A.J. (1987). Molecular evolution of viruses; 'trees', 'clocks' and 'modules'. *Journal of Cell Science Supplement* **7**, 319–337.

Gibbs, A.J. and Gibbs, M.J. (2006a). A broader definition of 'the virus species'. *Archives of Virology* **151**, 1419–1422.

Gibbs, A.J., Gibbs, M.J., and Armstrong, J.S. (2004). The phylogeny of SARS coronavirus. *Archives of Virology* **149**, 621–624.

Gibbs, A.J., Gibbs, M.J., Ohshima, K., and García-Arenal, F. (2008a). More about plant virus evolution: past, present and future. In E. Domingo, C.R. Parrish, and J.J. Holland, eds, *Origin and Evolution of Viruses*, 2nd edn, pp. 229–250. Academic Press, London.

Gibbs, A.J., Ohshima, K., Phillips, M.J., and Gibbs, M.J. (2008b). The prehistory of potyviruses: their initial radiation was during the dawn of agriculture. *PLoS ONE* **3**, e2523.

Gibbs, M.J. and Weiller, G.F. (1999). Evidence that a plant virus switched hosts to infect a vertebrate and then recombined with a vertebrate-infecting virus. *Proceedings of the National Academy of Sciences USA*. **96**, 8022–8027.

Gibbs, M.J. and Gibbs, A. (2006b). Molecular virology: was the 1918 pandemic caused by a bird flu? *Nature* **440**, E8.

Gibbs, M.J., Armstrong, J.S., and Gibbs, A.J (2001). Recombination in the hemagglutinin gene of the 1918 "Spanish flu" *Science* **293**, 1842–1845.

Gifford, R., Kabat, P., Martin, J., Lynch, C., and Tristem, M. (2005). Evolution and distribution of class II-related endogenous retroviruses. *Journal of Virology* **79**, 6478–6486.

Gifford, R.J., Katzourakis, A., Tristem, M., Pybus, O.G., Winters, M., and Shafer, R.W. (2008). A transitional endogenous lentivirus from the genome of a basal primate and implications for lentivirus evolution. *Proceedings of the National Academy of Sciences USA* **105**, 20362–20367.

Gilbert, M.T., Rambaut, A., Wlasiuk, G., Spira, T.J., Pitchenik, A.E., and Worobey, M. (2007). The emergence of HIV/AIDS in the Americas and beyond. *Proceedings of the National Academy of Sciences USA* **104**, 18566–18570.

Gilbert, W. (1986). The RNA world. *Nature* **319**, 618.

Gillespie, J.H. (1991). *The Causes of Molecular Evolution*. Oxford University Press, Oxford.

Gillespie, J.H. (1998). *Population Genetics: a Concise Course*. Johns Hopkins University Press, Baltimore, MD.

Gog, J.R., Rimmelzwaan, G.F., Osterhaus, A.D.M.E., and Grenfell, B.T. (2003). Population dynamics of rapid fixation in cytotoxic T lymphocyte escape mutants of influenza A. *Proceedings of the National Academy of Sciences USA* **100**, 11143–11147.

Gojobori, T., Moriyama, E.N., and Kimura, M. (1990). Molecular clock of viral evolution, and the neutral theory. *Proceedings of the National Academy of Sciences USA* **87**, 10015–10018.

Goldbach, R. and de Haan, P. (1994). RNA viral supergroups and evolution of RNA viruses. In S.S. Morse, ed., *The Evolutionary Biology of Viruses*, pp. 105–119. Raven Press, New York.

Goncalvez, A.P., Escalante, A.A., Pujol, F.H., Ludert, J.E., Tovar, D., Salas, R.A., and Liprandi, F. (2002). Diversity and evolution of the envelope gene of dengue virus type 1. *Virology* **303**, 110–119.

Gorbalenya, A.E. and Koonin, E.V. (1989). Viral proteins containing the purine NTP-binding sequence pattern. *Nucleic Acids Research* **17**, 8413–8440.

Gorbalenya, A.E., Pringle, F.M., Zeddam, J.L., Luke, B.T., Cameron, C.E., Kalmakoff, J., Hanzlik, T.N., Gordon, K.H., and Ward, V.K. (2002). The palm subdomain-based active

site is internally permuted in viral RNA-dependent RNA polymerases of an ancient lineage. *Journal of Molecular Biology* **324**, 47–62.
Gorbalenya, A.E., Enjuanes, L., Ziebuhr, J., and Snijder, E.J. (2006). *Nidovirales*: evolving the largest RNA virus genome. *Virus Research* **117**, 17–37.
Gottlieb, M.S., Schroff, R., Schanker, H.M., Weisman, J.D., Fan, P.T., Wolf, R.A., and Saxon, A. (1981). *Pneumocystis carinii* pneumonia and mucosal candidiasis in previously healthy homosexual men: evidence of a new acquired cellular immunodeficiency. *New England Journal of Medicine* **305**, 1425–1431.
Goudsmit, J., De Ronde, A., Ho, D.D., and Perelson, A.S. (1996). Human immunodeficiency virus fitness in vivo: calculations based on a single zidovudine resistance mutation at codon 215 of reverse transcriptase. *Journal of Virology* **70**, 5662–5664.
Gough, J. (2005). Convergent evolution of domain architectures (is rare). *Bioinformatics* **21**, 1464–1471.
Grakoui, A., Shoukry, N.H., Woollard, D.J., Han, J.H., Hanson, H.L., Ghrayeb, J., Murthy, K.K., Rice, C.M., and Walker, C.M. (2003). HCV persistence and immune evasion in the absence of memory T cell help. *Science* **302**, 659–662.
Grande-Pérez, A., Sierra, S., Castro, M.G., Domingo, E., and Lowenstein, P.R. (2002). Molecular indetermination in the transition to error catastrophe: systematic elimination of lymphocytic choriomeningitis virus through mutagenesis does not correlate linearly with large increases in mutant spectrum complexity. *Proceedings of the National Academy of Sciences USA* **99**, 12938–12943.
Grande-Pérez, A., Gómez-Mariano, G., Lowenstein, P.R., and Domingo, E. (2005). Mutagenesis-induced, large fitness variations with an invariant arenavirus consensus genomic nucleotide sequence. *Journal of Virology* **79**, 10451–10459.
Grard, G., Lemasson, J.-J., Sylla, M., Dudot, A., Cook, S., Molez, J.-F., Charrel, R.N., Gonzalez, J.-P., Munderloh, U., Holmes, E.C., and de Lamballerie, X. (2006). Ngoye virus: a novel evolutionary lineage within the genus *Flavivirus*. *Journal of General Virology* **87**, 3273–3277.
Greenbaum, B.D., Levine, A.J., Bhanot, G., and Rabadan, R. (2008). Patterns of evolution and host gene mimicry in influenza and other RNA viruses. *PLoS Pathogens* **4**, e1000079.
Greene, I.P., Wang, E., Deardorff, E.R., Milleron, R., Domingo, E., and Weaver, S.C. (2005). Effect of alternating passage on adaptation of sindbis virus to vertebrate and invertebrate cells. *Journal of Virology* **79**, 14253–14260.
Grenfell, B.T., Bjørnstad, O.N., and Kappey, J. (2001). Travelling waves and spatial hierarchies in measles epidemics. *Nature* **414**, 716–723.
Grenfell, B.T., Pybus, O.G., Gog, J.R., Wood, J.L.N., Daly, J.M., Mumford, J.A., and Holmes, E.C. (2004). Unifying the epidemiological and evolutionary dynamics of pathogens. *Science* **303**, 327–332.
Gritsun, T.S. and Gould, E.A. (2006). The 3' untranslated regions of Kamiti River virus and Cell fusing agent virus originated by self-duplication. *Journal of General Virology* **87**, 2615–2619.
Gubler, D.J. (2002). Epidemic dengue/dengue hemorrhagic fever as a public health, social and economic problem in the 21st century. *Trends in Microbiology* **10**, 100–103.
Haag-Liautard, C., Dorris, M., Maside, X., Macaskill, S., Halligan, D.L., Houle, D., Charlesworth, B., and Keightley, P.D. (2007). Direct estimation of per nucleotide and genomic deleterious mutation rates in *Drosophila*. *Nature* **445**, 82–85.
Hahn, B.H., Shaw, G.M., Taylor, M.E., Redfield, R.R., Markham, P.D., Salahuddin, S.Z., Wong-Staal, F., Gallo, R.C., Parks, E.S., and Parks, W.P. (1986). Genetic variation in

HTLV-III/LAV over time in patients with AIDS or at risk for AIDS. *Science* **232**, 1548–1553.

Hahn, B.H., Shaw, G.M., de Cock K.M., and Sharp, P.M. (2000). AIDS as a zoonosis: scientific and public health implications. *Science* **287**, 607–614.

Halstead, S.B. (1980). Dengue haemorrhagic fever – a public health problem and a field for research. *Bulletin of the World Health Organization* **58**, 1–21.

Hamburger, Z.A., Brown, M.S., Isberg, R.R., and Bjorkman, P.J. (1999). Crystal structure of invasin: a bacterial integrin-binding protein. *Science* **286**, 291–295.

Hanada, K., Suzuki, Y., and Gojobori, T. (2004). A large variation in the rates of synonymous substitution for RNA viruses and its relationship to a diversity of viral infection and transmission modes. *Molecular Biology and Evolution* **21**, 1074–1080.

Hanger, J.J., Bromham, L.D., McKee, J.J., O'Brien, T.M., and Robinson, W.F. (2000). The nucleotide sequence of koala (*Phascolarctos cinereus*) retrovirus: a novel type C endogenous virus related to Gibbon ape leukemia virus. *Journal of Virology* **74**, 4264–4272.

Hanley, K.A., Nelson, J.T., Schirtzinger, E.E., Whitehead, S.S., and Hanson, C.T. (2008). Superior infectivity for mosquito vectors contributes to competitive displacement among strains of dengue virus. *BMC Ecology* **8**, 1.

Harvey, P.H. and Pagel, M.D (1991). *The Comparative Method in Evolutionary Biology*. Oxford University Press, Oxford.

Harvey, P.H., Holmes, E.C., and Nee, S. (1994a). Model phylogenies to explain the real world. *BioEssays* **16**, 767–770.

Harvey, P.H., May, R.M., and Nee, S. (1994b). Phylogenies without fossils. *Evolution* **48**, 523–529.

Haseloff, J., Goelet, P., Zimmern, D., Ahlquist, P., Dasgupta, R., and Kaesberg, P. (1984). Striking similarities in amino acid sequence among nonstructural proteins encoded by RNA viruses that have dissimilar genomic organization. *Proceedings of the National Academy of Sciences USA* **81**, 4358–4362.

Hatta, M., Hatta, Y., Kim, J.H., Watanabe, S., Shinya, K., Nguyen, T., Lien, P.S., Le, Q.M., and Kawaoka, Y. (2007). Growth of H5N1 influenza A viruses in the upper respiratory tracts of mice. *PLoS Pathogens* **3**, 1374–1379.

Hatwell, J.N. and Sharp, P.M. (2000). Evolution of human polyomavirus JC. *Journal of General Virology* **81**, 1191–1200.

Hay, S.I., Myers, M.F., Burke, D.S., Vaughn, D.W., Endy, T., Ananda, N., Shanks, G.D., Snow, R.W., and Rogers, D.J. (2000). Etiology of interepidemic periods of mosquito-borne disease. *Proceedings of the National Academy of Sciences USA* **97**, 9335–9339.

Hayden, F.G. (2006). Antiviral resistance in influenza viruses – implications for management and pandemic response. *New England Journal of Medicine* **354**, 785–788.

Heeney, J.L., Dalgleish, A.G., and Weiss, R.A. (2006). Origins of HIV and the evolution of resistance to AIDS. *Science* **313**, 462–466.

Hein, J. and Støvlbæk, J. (1995). A maximum-likelihood approach to analyzing nonoverlapping and overlapping reading frames. *Journal of Molecular Evolution* **40**, 181–189.

Heldwein, E.E., Lou, H., Bender, F.C., Cohen, G.H., Eisenberg, R.J., and Harrison, S.C. (2006). Crystal structure of glycoprotein B from herpes simplex virus 1. *Science* **313**, 217–220.

Hellen, C.U.T. and de Breyne, S. (2007). A distinct group of hepacivirus/pestivirus-like internal ribosomal entry sites in members of diverse picornavirus genera: evidence for modular exchange of functional noncoding RNA elements by recombination. *Journal of Virology* **81**, 5850–5863.

Hendrix, R.W., Smith, M.C.M., Burns, R.N., Ford, M.E., and Hatfull, G.F. (1999). Evolutionary relationships among diverse bacteriophage and prophages: All the world's a phage. *Proceedings of the National Academy of Sciences USA* **96**, 2192–2197.

Hendrix, R.W., Lawrence, J.G., Hatfull, G.F., and Casjens, S. (2000). The origins and ongoing evolution of viruses. *Trends in Microbiology* **8**, 504–508.

Herbeck, J.T., Nickle, D.C., Learn, G.H., Gottlieb, G.S., Curlin, M.E., Heath, L., and Mullins, J.I. (2006). Human immunodeficiency virus type 1 *env* evolves toward ancestral states upon transmission to a new host. *Journal of Virology* **80**, 1637–1644.

Herniou, E.A., Olszewski, J.A., O'Reilly, D.R., and Cory, J.S. (2004). Ancient coevolution of baculoviruses and their insect hosts. *Journal of Virology* **78**, 3244–3251.

Higgs, P.G. (1998). Compensatory neutral mutations and the evolution of RNA. *Genetica* **102/103**, 91–101.

Hirano, A. (1992). Subacute sclerosing panencephalitis virus dominantly interferes with replication of wild-type measles virus in a mixed infection: implication for viral persistence. *Journal of Virology* **66**, 1891–1898.

Ho, S.Y., Shapiro, B., Phillips, M.J., Cooper, A., and Drummond, A.J. (2007). Evidence for time dependency of molecular rate estimates. *Systematic Biology* **56**, 515–522.

Hofmann, H., Pyrc, K., van der Hoek, L., Geier, M., Berkhout, B., and Pöhlmann, S. (2005). Human coronavirus NL63 employs the severe acute respiratory syndrome coronavirus receptor for cellular entry. *Proceedings of the National Academy of Sciences USA* **102**, 7988–7893.

Holland, J.J. and Villarreal, L.P. (1975). Purification of defective interfering T particles of vesicular stomatitis and rabies viruses generated *in vivo* in brains of newborn mice. *Virology* **67**, 438–449.

Holmes, E.C. (2003a). Patterns of intra- and inter-host nonsynonymous variation reveal strong purifying selection in dengue virus. *Journal of Virology* **77**, 11296–11298.

Holmes, E.C. (2003b). Error thresholds and the constraints to RNA virus evolution. *Trends in Microbiology* **11**, 543–546.

Holmes, E.C. (2003c). Molecular clocks and the puzzle of RNA virus origins. *Journal of Virology* **77**, 3893–3897.

Holmes, E.C. (2004). The phylogeography of human viruses. *Molecular Ecology* **13**, 745–756.

Holmes, E.C. (2005). On being the right size. *Nature Genetics* **37**, 923–924.

Holmes, E.C. (2007). Viral evolution in the genomic age. *PLoS Biology* **5**, e278.

Holmes. E.C. (2008). The evolutionary history and phylogeography of human viruses. *Annual Review of Microbiology* **62**, 307–328.

Holmes, E.C. and Moya, A. (2002). Is the quasispecies concept relevant to RNA viruses? *Journal of Virology* **76**, 460–462.

Holmes, E.C. and Twiddy, S.S. (2003). The origin, emergence and evolutionary genetics of dengue virus. *Infection, Genetics and Evolution* **3**, 19–28.

Holmes, E.C. and Rambaut, A. (2004). Viral evolution and the emergence of SARS coronavirus. *Philosophical Transactions of the Royal Society of London Series B Biological Sciences* **359**, 1059–1065.

Holmes, E.C., Zhang, L.Q., Simmonds, P., Ludlam, C.A., and Leigh Brown, A.J. (1992). Convergent and divergent sequence evolution in the surface envelope glycoprotein of human immunodeficiency virus type 1 within a single infected patient. *Proceedings of the National Academy of Sciences USA* **89**, 4835–4839.

Holmes, E.C., Woelk, C.H., Kassis, R., and Bourhy, H. (2002). Genetic constraints and the adaptive evolution of rabies virus. *Virology* **292**, 247–257.

Holmes, E.C,. Ghedin, E., Miller, N., Taylor, J., Bao, Y., St. George, K., Grenfell, B.T., Salzberg, S.L., Fraser, C.M., Lipman, D.J., and Taubenberger, J.K. (2005). Whole genome analysis of human influenza A virus reveals multiple persistent lineages and reassortment among recent H3N2 viruses. *PLoS Biology* **3**, e300.

Holmes, E.C,. Lipman, D.J., Zamarin, D., and Yewdell, J.W. (2006). Comment on 'Large-scale analysis of avian influenza isolates'. *Science* **313**, 1573b.

Hon, C.-C., Lam, T.-Y., Shi, Z.-L., Drummond, A.J., Yip, C.-W., Zeng, F., Lam, P.-Y., and Leung, F.-C. (2008). Evidence of the recombinant origin of a bat severe acute respiratory syndrome (SARS)-like coronavirus and its implications on the direct ancestor of SARS coronavirus. *Journal of Virology* **82**, 1819–1826.

Horimoto, T. and Kawaoka, Y. (2005). Influenza: lessons from past pandemics, warnings from current incidents. *Nature Reviews Microbiology* **3**, 591–600.

Huang, A.S. and Baltimore, D. (1970). Defective viral particles and viral disease processes. *Nature* **226**, 325–327.

Huang, I.-C., Li, W., Sui, J., Marasco, W., Choe, H., and Farzan, M. (2008). Influenza A virus neuraminidase limits viral superinfection. *Journal of Virology* **82**, 4834–4843.

Hueffer, K., Parker, J.S., Weichert, W.S., Geisel, R.E., Sgro, J.Y., and Parrish, C.R. (2003). The natural host range shift and subsequent evolution of canine parvovirus resulted from virus-specific binding to the canine transferrin receptor. *Journal of Virology* **77**, 1718–1726.

Huelsenbeck, J.P. (2002). Testing a covariotide model of DNA substitution. *Molecular Biology and Evolution* **19**, 698–707.

Huet, T., Cheynier, R., Meyerhans, A., Roelants, G., and Wain-Hobson, S. (1990). Genetic organisation of a chimpanzee lentivirus related to HIV-1. *Nature* **345**, 356–359.

Hughes, A.L. and Friedman, R. (2000). Evolutionary diversification of protein-coding genes of hantaviruses. *Molecular Biology and Evolution* **17**, 1558–1568.

Hughes, A.L. and Friedman, R. (2005). Poxvirus genome evolution by gene gain and loss. *Molecular Phylogenetics and Evolution* **35**, 186–195.

Hughes, G.J., Orciari, L.A., and Rupprecht, C.E. (2005). Evolutionary timescale of rabies virus adaptation to North American bats inferred from the substitution rate of the nucleoprotein gene. *Journal of General Virology* **86**, 1467–1474.

Hull, R., Covey, S., and Dale, P. (2000). Genetically modified plants and the 35S promoter: assessing the risks and enhancing the debate. *Microbial Ecology in Health and Disease* **12**, 1–5.

Huson, D.H. and Bryant, D. (2006). Application of phylogenetic networks in evolutionary studies. *Molecular Biology and Evolution* **23**, 254–267.

Ikemura, T. (1985). Codon usage and tRNA content in unicellular and multicellular organisms. *Molecular Biology and Evolution* **2**, 13–34.

Imbert, I., Guillemot, J.C., Bourhis, J.M., Bussetta, C., Coutard, B., Egloff, M.P., Ferron, F., Gorbalenya, A.E., and Canard, B. (2006). A second, non-canonical RNA-dependent RNA polymerase in SARS coronavirus. *EMBO Journal* **25**, 4933–4942.

Ina, Y. and Gojobori, T. (1994). Statistical analysis of nucleotide sequences of the hemagglutinin gene of human influenza A viruses. *Proceedings of the National Academy of Sciences USA* **91**, 8388–8392.

Isfort, R., Jones, D., Kost, R., Witter, R., and Kung, H.J. (1992). Retrovirus insertion into herpesvirus *in vitro* and *in vivo*. *Proceedings of the National Academy of Sciences USA* **89**, 991–995.

Isnard, M., Granier, M., Frutos, R., Reynaud, B., and Peterschmitt, M. (1998). Quasispecies nature of three Maize streak virus isolates obtained through different modes of selection from a population used to assess response to infection of maize cultivars. *Journal of General Virology* **79**, 3091–3099.

Itoh, H. and Melnick, J.L. (1959). Double infections of single cells with ECHO 7 and Coxsackie A9 viruses. *Journal of Experimental Medicine* **109**, 393–406.

Iversen, A.K., Learn, G.H., Skinhøj, P., Mullins, J.I., McMichael, A.J., and Rambaut, A. (2005). Preferential detection of HIV subtype C' over subtype A in cervical cells from a dually infected woman. *AIDS* **19**, 990–993.

Jackson, A.P. and Charleston, M.A. (2004). A cophylogenetic perspective of RNA-virus evolution. *Molecular Biology and Evolution* **21**, 45–57.

Jacquez, J.A., Koopman, J.S., Simon, C.P., and Longini, I.M., Jr (1994). Role of the primary infection in epidemics of HIV infection in gay cohorts. *Journal of Acquired Immune Deficiency Syndromes* **7**, 1169–1184.

Jalasvuori, M. and Bamford, J.K. (2008). Structural co-evolution of viruses and cells in the primordial world. *Origins of Life and Evolution of the Biosphere* **38**, 165–181.

Jaspars, E.M.J. (1974). Plant viruses with multipartite genomes. *Advances in Virus Research* **19**, 37–149.

Jeffers, S.A., Tusell, S.M., Gillim-Ross, L., Hemmila, E.M., Achenbach, J.E., Babcock, G.J., Thomas, W.D., Jr, Thackray, L.B., Young, M.D., Mason, R.J. *et al.* (2004). CD209L (L-SIGN) is a receptor for severe acute respiratory syndrome coronavirus. *Proceedings of the National Academy of Sciences USA* **101**, 15748–15753.

Jenkins, G.M. and Holmes, E.C. (2003). The extent of codon usage bias in human RNA viruses and its evolutionary origin. *Virus Research* **92**, 1–7.

Jenkins, G.M., Woelk, C., Rambaut, A., and Holmes, E.C. (2000). Testing the extent of sequence similarity among viroids, satellite RNAs and HDV. *Journal of Molecular Evolution* **50**, 98–102.

Jenkins, G.M., Pagel, M., Gould, E.A., Zanotto, P.M. de A., and Holmes, E.C. (2001a). Evolution of base composition and codon usage bias in the genus *Flavivirus*. *Journal of Molecular Evolution* **52**, 383–390.

Jenkins, G.M., Worobey, M., Woelk, C.H., and Holmes, E.C. (2001b). Evidence for the non-quasispecies evolution of RNA viruses. *Molecular Biology and Evolution* **18**, 987–994.

Jenkins, G.M., Rambaut, A., Pybus, O.G., and Holmes, E.C. (2002). Rates of molecular evolution in RNA viruses: a quantitative phylogenetic analysis. *Journal of Molecular Evolution* **54**, 152–161.

Jerzak, G., Bernard, K.A., Kramer, L.D., and Ebel, G.D. (2005). Genetic variation in West Nile virus from naturally infected mosquitoes and birds suggests quasispecies structure and strong purifying selection. *Journal of General Virology* **86**, 2175–2183.

Jetten, T.H. and Focks, D.A. (1997). Potential changes in the distribution of dengue transmission under climate warming. *American Journal of Tropical Medicine and Hygiene* **57**, 285–297.

Jetzt, A.E., Yu, H. Klarmann, G.J., Ron, Y., Preston, B.D., and Dougherty, J.P. (2000). High rate of recombination throughout the human immunodeficiency virus type 1 genome. *Journal of Virology* **74**, 1234–1240.

Johne, R., Fernández-de-Luco, D., Höfle, U., and Müller, H. (2006). Genome of a novel circovirus of starlings, amplified by multiply primed rolling-circle amplification. *Journal of General Virology* **87**, 1189–1195.

Johnson, N.P.A.S. and Mueller, J. (2002). Updating the accounts: global mortality of the 1918–20 "Spanish" influenza epidemic. *Bulletin of the History of Medicine* **76**, 105–115.

John-Stewart, G.C., Nduati, R.W., Rousseau, C.M., Mbori-Ngacha, D.A., Richardson, B.A., Rainwater, S., Panteleeff, D.D., and Overbaugh, J. (2005). Subtype C Is associated with increased vaginal shedding of HIV-1. *Journal of Infectious Diseases* **192**, 492–496.

Jones, K.E., Patel, N.G., Levy, M.A., Storeygard, A., Balk, D., Gittleman, J.L., and Daszak, P. (2008). Global trends in emerging infectious diseases. *Nature* **451**, 990–993.

Jones-Engel, L., Engel, G.A., Heidrich, J., Chalise, M., Poudel, N., Viscidi, R., Barry, P.A., Allan, J.S., Grant, R., and Kyes, R. (2006). Temple monkeys and health implications of commensalism, Kathmandu, Nepal. *Emerging Infectious Diseases* **12**, 900–906.

Joyce, G.F. (2002). The antiquity of RNA-based evolution. *Nature* **418**, 214–221.

Jung, A., Maier, R., Vartanian, J.P., Bocharov, G., Jung, V., Fischer, U., Meese, E., Wain-Hobson, S., and Meyerhans, A. (2002). Recombination – multiply infected spleen cells in HIV patients. *Nature* **418**, 144.

Kalinina, O., Norder, H., Mukomolov, S., and Magnius, L.O. (2002). A natural intergenotypic recombinant of hepatitis C virus identified in St. Petersburg. *Journal of Virology* **76**, 4034–4043.

Kan, B., Wang, M., Jing, H., Xu, H., Jiang, X., Yan, M., Liang, W., Zheng, H., Wan, K., Liu, Q. *et al.* (2005). Molecular evolution analysis and geographic investigation of severe acute respiratory syndrome coronavirus-like virus in palm civets at an animal market and on farms. *Journal of Virology* **79**, 11892–11900.

Kandun, I.N., Wibisono, H., Sedyaningsih, E.R., Yusharmen, Hadisoedarsuno, W., Purba, W., Santoso, H., Septiawati, C., Tresnaningsih, E., Heriyanto, B. *et al.* (2006). Three Indonesian clusters of H5N1 virus infection in 2005. *New England Journal of Medicine* **355**, 2186–2194.

Kanegae, Y., Sugita, S., Endo, A., Ishida, M., Senya, S., Osako, K., Nerome, K., and Oya, A. (1990). Evolutionary pattern of the hemagglutinin gene of influenza B viruses isolated in Japan: cocirculating lineages in the same epidemic season. *Journal of Virology* **64**, 2860–2865.

Kapoor, A., Victoria, J., Simmonds, P., Wang, C., Shafer, R.W., Nims, R., Nielsen, O., and Delwart, E. (2008). A highly divergent picornavirus in a marine mammal. *Journal of Virology* **82**, 311–320.

Karasev, A.V., Boyko, V.P., Gowda, S., Nikolaeva, O.V., Hilf, M.E., Koonin, E.V., Niblett, C.L., Cline, K., Gumpf, D.J., Lee, R.F. *et al.* (1995). Complete sequence of the citrus tristeza virus RNA genome. *Virology* **208**, 511–520.

Katzourakis, A. and Tristem, M. (2005). Phylogeny of human endogenous and exogenous retroviruses. In E.D. Sverdlov, ed., *Retroviruses and Primate Genome Evolution*, pp. 186–203. Landes Bioscience, Austin, TX.

Katzourakis, A., Rambaut, A., and Pybus, O.G. (2005). The evolutionary dynamics of endogenous retroviruses. *Trends in Microbiology* **13**, 463–468.

Katzourakis, A., Tristem, M., Pybus, O.G., and Gifford, R.J. (2007). Discovery and analysis of the first endogenous lentivirus. *Proceedings of the National Academy of Sciences USA* **104**, 6261–6265.

Kawashima, Y., Pfafferott, K., Frater, J., Matthews, P., Payne, R., Addo, M., Gatanaga, H., Fujiwara, M., Hachiya, A., Koizumi, H., *et al.* (2009). Adaptation of HIV-1 to human leukocyte antigen class I. *Nature* **458**, 641–645.

Keele, B.F., Van Heuverswyn, F., Li, Y., Bailes, E., Takehisa, J., Santiago, M.L., Bibollet-Ruche, F., Chen, Y., Wain, L.V., Liegeois, F. *et al.* (2006). Chimpanzee reservoirs of pandemic and nonpandemic HIV-1. *Science* **313**, 523–526.

Keele, B.F., Giorgi, E.E., Salazar-Gonzalez, J.F., Decker, J.M., Pham, K.T., Salazar, M.G., Sun, C., Grayson, T., Wang, S., Li, H. *et al.* (2008). Identification and characterization of transmitted and early founder virus envelopes in primary HIV-1 infection. *Proceedings of the National Academy of Sciences USA* **105**, 7552–7557.

Keese, P.K. and Gibbs, A. (1992). Origins of genes: "big bang" or continuous creation? *Proceedings of the National Academy of Sciences USA* **89**, 9489–9493.

Keightley, P.D. and Eyre-Walker, A. (2000). Deleterious mutations and the evolution of sex. *Science* **290**, 331–333.

Kellam, P., Boucher, C.A.B., Tijnagel, J.M.G.H., and Larder, B.A. (1994). Zidovudine treatment results in the selection of human immunodeficiency virus type 1 variants whose genotypes confer increasing levels of drug resistance. *Journal of General Virology* **75**, 341–351.

Kelleher, A.D., Long, C., Holmes, E.C., Allen, R.L., Wilson, J., Conlon, C., Workman, C., Shaunak, S., Wulfestieg, K., Goulder, P. *et al.* (2001). Clustered mutations in HIV-1 gag are consistently required for escape from HLA-B27 restricted CTL responses. *Journal of Experimental Medicine* **193**, 375–385.

Kew, O.M., Sutter, R.W., Nottay, B.K., McDonough, M.J., Prevots, D.R., Quick, L., and Pallansch, M.A. (1998). Prolonged replication of a type 1 vaccine-derived poliovirus in an immunodeficient patient. *Journal of Clinical Microbiology* **36**, 2893–2899.

Kew, O.M., Sutter, R.W., de Gourville, E.M., Dowdle, W.R., and Pallansch, M.A. (2005). Vaccine-derived polioviruses and the endgame strategy for global polio eradication. *Annual Review of Microbiology* **59**, 587–635.

Khatchikian, D., Orlich, M., and Rott, R. (1989). Increased viral pathogenicity after insertion of a 28S ribosomal RNA sequence into the haemagglutinin gene of an influenza virus. *Nature* **340**, 156–157.

Khudyakov, Y.E., Cong, M.E., Nichols, B., Reed, D., Dou, X.G., Viazov, S.O., Chang, J., Fried, M.W., Williams, I., Bower, W. *et al.* (2000). Sequence heterogeneity of TT virus and closely related viruses. *Journal of Virology* **74**, 2990–3000.

Kilbourne, E.D., Taylor, A.H., Whitaker, C.W., Sahai, R., and Caton, A.J. (1988). Hemagglutinin polymorphism as the basis for low- and high-yield phenotypes of swine influenza virus. *Proceedings of the National Academy of Sciences USA* **85**, 7782–7785.

Kilpatrick, A.M., Daszak, P., Goodman, S.J., Rogg, H., Kramer, L.D., Cedeño, V., and Cunningham, A.A. (2006). Predicting pathogen introduction: West Nile virus spread to Galápagos. *Conservation Biology* **20**, 1224–1231.

King, J.L. and Jukes, T.H. (1969). Non-Darwinian evolution: random fixation of selectively neutral mutations. *Science* **164**, 788–798.

Kingman, J.F.C. (1982). On the genealogy of large populations. *Journal of Applied Probability* **19A**: 27–43.

Klimov, A.I., Rocha, E., Hayden, F.G., Shult, P.A., Roumillat, L.F., and Cox, N.J. (1995). Prolonged shedding of amantadine-resistant influenzae A viruses by immunodeficient patients: detection by polymerase chain reaction-restriction analysis. *Journal of Infectious Diseases* **172**, 1352–1355.

Klungthong, C., Zhang, C., Mammen, M.P., Jr, Ubol, S., and Holmes, E.C. (2004). The molecular epidemiology of dengue virus serotype 4 in Bangkok, Thailand. *Virology* **329**, 168–179.

Kochel, T.J., Watts, D.M., Halstead, S.B., Hayes, C.G., Espinoza, A., Felices, V., Caceda, R., Bautista, C.T., Montoya, Y., Douglas, S., and Russell, K.L. (2002). Effect of dengue-1 antibodies on American dengue-2 viral infection and dengue haemorrhagic fever. *Lancet* **360**, 310–312.

Kochel, T.J., Watts, D.M., Gozalo, A.S., Ewing, D.F., Porter, K.R., and Russell, K.L. (2005). Cross-serotype neutralization of dengue virus in *Aotus nancymae* monkeys. *Journal of Infectious Diseases* **191**, 1000–1004.

Koelle, K., Cobey, S., Grenfell, B., and Pascual, M. (2006). Epochal evolution shapes the phylodynamics of interpandemic influenza A (H3N2) in humans. *Science* **314**, 1898–1903.

Köndgen, S., Kühl, H., N'Goran, P.K., Walsh, P.D., Schenk, S., Ernst, N., Biek, R., Formenty, P., Mätz-Rensing, K., Schweiger, B. *et al.* (2008). Pandemic human viruses cause decline of endangered great apes. *Current Biology* **18**, 260–264.

Kondrashov, A.S. (1988). Deleterious mutations and the evolution of sexual reproduction. *Nature* **336**, 435–440.

Koonin, E.V. (1991). The phylogeny of RNA-dependent RNA polymerases of positive-strand RNA viruses. *Journal of General Virology* **72**, 2197–2206.

Koonin, E.V. and Dolja, V.V. (1993). Evolution and taxonomy of positive-strand RNA viruses: implications of comparative analysis of amino acid sequences. *Critical Reviews in Biochemistry and Molecular Biology* **28**, 375–430.

Koonin, E.V., Senkevich, T.G., and Dolja, V.V. (2006). The ancient virus world and evolution of cells. *Biology Direct* **1**, 29.

Korber, B.T., Allen, E.E., Farmer, A.D., and Myers, G.L. (1995). Heterogeneity of HIV-1 and HIV-2. *AIDS* **9**, S5–S18.

Kosakovsky Pond, S.L. and Frost, S.D.W. (2005). Not so different after all: a comparison of methods for detecting amino-acid sites under selection. *Molecular Biology and Evolution* **22**, 1208–1222.

Kosakovsky Pond, S.L., Frost, S.D.W., Grossman, Z., Gravenor, M.B., Richman, D.D., and Leigh Brown, A.J. (2006). Adaptation to different human populations by HIV-1 revealed by codon-based analyses. *PLoS Computational Biology* **23**, e62.

Kouyos, R.D., Althaus, C.L., and Bonhoeffer, S. (2006). Stochastic or deterministic: what is the effective population size of HIV-1? *Trends in Microbiology* **14**, 507–511.

Krakauer, D.C. and Plotkin, J.B. (2002). Redundancy, antiredundancy, and the robustness of genomes. *Proceedings of the National Academy of Sciences USA* **99**, 1405–1409.

Krakauer, D.C. and Komarova, N.L. (2003). Levels of selection in positive-strand virus dynamics. *Journal of Evolutionary Biology* **16**, 64–73.

Kremer, J.R., Brown, K.E., Jin, L., Santibanez, S., Shulga, S.V., Aboudy, Y., Demchyshyna, I.V., Djemileva, S., Echevarria, J.E., Featherstone, D.F. *et al.* (2008). High genetic diversity of measles virus, World Health Organization European Region, 2005–2006. *Emerging Infectious Diseases* **14**, 107–114.

Kruger, K., Grabowski, P.J., Zaug, A.J., Sands, J., Gottschling, D.E., and Cech, T.R. (1982). Self-splicing RNA: autoexcision and autocyclization of the ribosomal RNA intervening sequence of *Tetrahymena*. *Cell* **31**, 147–157.

Kryazhimskiy, S., Bazykin, G.A., and Dushoff, J. (2008). Natural selection for nucleotide usage at synonymous and nonsynonymous sites in influenza A virus genes. *Journal of Virology* **82**, 4938–4945.

Kuiken, T., Holmes, E.C., McCauley, J., Rimmelzwaan, G.F., Williams, C.S., and Grenfell, B.T. (2006). Host species barriers to influenza virus infections. *Science* **312**, 394–397.

Kuipers, E.J., Israel, D.A., Kusters, J.G., Gerrits, M.M., Weel, J., van Der Ende, A., van Der Hulst, R.W., Wirth, H.P., Höök-Nikanne, J., Thompson, S.A., and Blaser, M.J. (2000). Quasispecies development of *Helicobacter pylori* observed in paired isolates obtained years apart from the same host. *Journal of Infectious Diseases* **181**, 273–282.

Kun, A., Santos, M. and Szathmáry, E. (2005). Real ribozymes suggest a relaxed error threshold. *Nature Genetics* **37**, 1008–1011.

Lai, M.M.C. (1992). RNA recombination in animal and plant viruses. *Microbiological Reviews* **56**, 61–79.

Lai, M.M.C. (1996). Recombination in large RNA viruses: coronaviruses. *Seminars in Virology* **7**, 381–388.

Lai, M.M.C. (2005). RNA replication without RNA-dependent RNA polymerase: surprises from hepatitis delta virus. *Journal of Virology* **79**, 7951–7958.

Laín, S., Riechmann, J.L., Martín, M.T., García, J.A. (1989). Homologous potyvirus and flavivirus proteins belonging to a superfamily of helicase-like proteins. *Gene* **82**, 357–362.

LaPierre, L.A., Holzschu, D.L., Bowser, P.R., and Casey, J.W. (1999). Sequence and transcriptional analyses of the fish retroviruses walleye epidermal hyperplasia virus types 1 and 2: evidence for a gene duplication. *Journal of Virology* **73**, 9393–9403.

Larder, B.A. and Kemp, S.D. (1989). Multiple mutations in HIV-1 reverse transcriptase confer high-level resistance to Zidovudine (AZT). *Science* **246**, 1155–1158.

La Scola, B., Audic, S., Robert, C., Jungang, L., de Lamballerie, X., Drancourt, M., Birtles, R., Claverie, J.-M., and Raoult, D. (2003). A giant virus in amoebae. *Science* **299**, 2033.

La Scola, B., Desnues, C., Pagnier, I., Robert, C., Barrassi, L., Fournous, G., Merchat, M., Suzan-Monti, M., Forterre, P., Koonin, E., and Raoult, D. (2008). The virophage as a unique parasite of the giant mimivirus. *Nature* **455**, 100–104.

Lavenu, A., Leruez-Ville, M., Chaix, M.L., Boelle, P.Y., Rogez, S., Freymuth, F., Hay, A., Rouzioux, C., and Carrat, F. (2006). Detailed analysis of the genetic evolution of influenza virus during the course of an epidemic. *Epidemiology and Infection* **134**, 514–520.

Lázaro, E., Escarmís, C., Pérez-Mercader, J., Manrubia, S.C., and Domingo, E. (2003). Resistance of virus to extinction on bottleneck passages: study of a decaying and fluctuating pattern of fitness loss. *Proceedings of the National Academy of Sciences USA* **100**, 10830–10835.

Ledinko, N. and Hirst, G.K. (1961). Mixed infection of cells with poliovirus types 1 and 2. *Virology* **14**, 207–219.

Lefeuvre, P., Lett, J.M., Reynaud, B., and Martin, D.P. (2007). Avoidance of protein fold disruption in natural virus recombinants. *PLoS Pathogens* **3**, e181.

Le Gall, O., Christian, P., Fauquet, C.M., King, A.M., Knowles, N.J., Nakashima, N., Stanway, G., and Gorbalenya, A.E. (2008). Picornavirales, a proposed order of positive-sense single-stranded RNA viruses with a pseudo-T = 3 virion architecture. *Archives of Virology* **153**, 715–727.

Lehmann, E., Brueckner, F., and Cramer, P. (2007). Molecular basis of RNA-dependent RNA polymerase II activity. *Nature* **450**, 445–459.

Leigh Brown, A.J. (1997). Analysis of HIV-1 *env* gene sequences reveals evidence for a low effective number in the viral population. *Proceedings of the National Academy of Sciences USA* **94**, 1862–1865.

Leigh Brown, A.J. and Richman, D.D. (1997). HIV-1: Gambling on the evolution of drug resistance? *Nature Medicine* **3**, 268–271.

Leitmeyer, K.C., Vaughn, D.W., Watts, D.M., Salas, R., Villalobos de Chacon, I., Ramos, C., and Rico-Hesse, R. (1999). Dengue virus structural differences that correlate with pathogenesis. *Journal of Virology* **73**, 4738–4747.

Lemey, P., Pybus, O.G., Wang, B., Saksena, N.K., Salemi, M., and Vandamme, A.M. (2003). Tracing the origin and history of the HIV-2 epidemic. *Proceedings of the National Academy of Sciences USA* **100**, 6588–6592.

Lemey, P., Pybus, O.G., Van Dooren, S., and Vandamme, A.M. (2005). A Bayesian statistical analysis of human T-cell lymphotropic virus evolutionary rates. *Infection, Genetics and Evolution* **5**, 291–298.

Lemey, P., Rambaut, A., and Pybus, O.G. (2006). HIV dynamics within and among hosts. *AIDS Reviews* **8**, 125–140.

Lenski, R.E., Barrick, J.E., and Ofria, C. (2006). Balancing robustness and evolvability. *PLoS Biology* **4**, e428.

Leroy, E.M., Kumulungui, B., Pourrut, X., Rouquet, P., Hassanin, A., Yaba, P., Delicat, A., Paweska, J.T., Gonzalez, J.P., and Swanepoel, R. (2005). Fruit bats as reservoirs of Ebola virus. *Nature* **438**, 575–566.

Leslie, A.J., Pfafferott, K.J., Chetty, P., Draenert, R., Addo, M.M., Feeney, M., Holmes, E.C., Allen, T., Prado, J.G., Altfeld, M. *et al*. (2004). HIV evolution: CTL escape mutation and reversion following transmission. *Nature Medicine* **10**, 282–289.

Levine, A.J. (1996). The origins of virology. In B.N. Fields, D.M. Knipe, and P.M. Howley, eds, *Fundamental Virology*, 3rd edn, pp. 1–14. Lippincott Williams and Wilkins, Philadelphia, PA.

Li, B., Gladden, A.D., Altfeld, M., Kaldor, J.M., Cooper, D.A., Kelleher, A.D., and Allen, T.M. (2007). Rapid reversion of sequence polymorphisms dominates early human immunodeficiency virus type 1 evolution. *Journal of Virology* **81**, 193–201.

Li, H.C., Fujiyoshi, T., Lou, H., Yashiki, S., Sonoda, S., Cartier, L., Nunez, L., Munoz, I., Horai, S., and Tajima, K. (1999). The presence of ancient human T-cell lymphotropic virus type I provirus DNA in an Andean mummy. *Nature Medicine* **5**, 1428–1432.

Li, H.Y. and Roossinck, M.J. (2004). Genetic bottlenecks reduce population variation in an experimental RNA virus population. *Journal of Virology* **78**, 10582–10587.

Li, W., Moore, M.J., Vasilieva, N., Sui, J., Wong, S.K., Berne, M.A., Somasundaran, M., Sullivan, J.L., Luzuriaga, K., Greenough, T.C. *et al*. (2003). Angiotensin-converting enzyme 2 is a functional receptor for the SARS coronavirus. *Nature* **426**, 450–454.

Li, W., Shi, Z., Yu, M., Ren, W., Smith, C., Epstein, J.H., Wang, H., Crameri, G., Hu, Z., Zhang, H. *et al*. (2005). Bats are natural reservoirs of SARS-like coronaviruses. *Science* **310**, 676–679.

Li, W., Wong, S.K., Li, F., Kuhn, J.H., Huang, I.C., Choe, H., and Farzan, M. (2006). Animal origins of the severe acute respiratory syndrome coronavirus: insight from ACE2-S-protein interactions. *Journal of Virology* **80**, 4211–4219.

Liljas, L., Tate, J., Lin, T., Christian, P., and Johnson, J.E. (2002). Evolutionary and taxonomic implications of conserved structural motifs between picornaviruses and insect picorna-like viruses. *Archives of Virology* **147**, 59–84.

Lin, S.-R., Hsieh, S.-C., Yueh, Y.-Y., Lin, T.-H., Chao, D.-Y., Chen, W.-J., King, C.-C., and Wang, W.-K. (2004). Study of sequence variation of dengue type 3 virus in naturally infected mosquitoes and human hosts: implications for transmission and evolution. *Journal of Virology* **78**, 12717–12721.

Liu, W., Worobey, M., Li, Y., Keele, B.F., Bibollet-Ruche, F., Guo, Y., Goepfert, P.A., Santiago, M.L., Ndjango, J.B., Neel, C. *et al*. (2008). Molecular ecology and natural history of simian foamy virus infection in wild-living chimpanzees. *PLoS Pathogens* **4**, e1000097.

Liu, Y., McNevin, J., Zhao, H., Tebit, D.M., Troyer, R.M., McSweyn, M., Ghosh, A.K., Shriner, D., Arts, E.J., McElrath, M.J., and Mullins, J.I. (2007). Evolution of human

immunodeficiency virus type 1 cytotoxic T-lymphocyte epitopes: fitness-balanced escape. *Journal of Virology* **81**, 12179–12188.
Lloyd-Smith, J.O., Schreiber, S.J., Kopp, P.E., and Getz, W.M. (2005). Superspreading and the effect of individual variation on disease emergence. *Nature* **438**, 355–359.
Loeb, L.A., Essigmann, J.M., Kazazi, F., Zhang, J., Rose, K.D., and Mullins, J.I. (1999). Lethal mutagenesis of HIV with mutagenic nucleoside analogs. *Proceedings of the National Academy of Sciences USA* **96**, 1492–1497.
Lopez-Bueno, A., Villarreal, L.P., and Almendral, J.M. (2006). Parvovirus variation for disease: A difference with RNA viruses? *Current Topics in Microbiology and Immunology* **299**, 349–370.
Lozupone, C.A., Knight, R.D., and Landweber, L.F. (2001). The molecular basis of nuclear genetic code change in ciliates. *Current Biology* **11**, 65–74.
Lucks, J.B., Nelson, D.R., Kudla, G.R., and Plotkin, J.B. (2008). Genome landscapes and bacteriophage codon usage. *PLoS Computational Biology* **4**, e1000001.
Luytjes, W., Bredenbeek, P.J., Noten, A.F., Horzinek, M.C., and Spaan, W.J. (1988). Sequence of mouse hepatitis virus A59 mRNA 2: indications for RNA recombination between coronaviruses and influenza C virus. *Virology* **166**, 415–422.
Lynch, M. (2007). The frailty of adaptive hypotheses for the origins of organismal complexity. *Proceedings of the National Academy of Sciences USA* **104**, 8597–8604.
Lynch, M. and Conery, J.S. (2000). The evolutionary fate and consequences of duplicate genes. *Science* **290**, 1151–1155.
Lynch, M. and Conery, J.S. (2003). The origins of genome complexity. *Science* **302**, 1401–1404.
Ma, W., Vincent, A.L., Gramer, M.R., Brockwell, C.B., Lager, K.M., Janke, B.H., Gauger, P.C., Patnayak, D.P., Webby, R.J., and Richt, J.A. (2007). Identification of H2N3 influenza A viruses from swine in the United States. *Proceedings of the National Academy of Sciences USA* **104**, 20949–20954.
Maan, S., Maan, N.S., Samuel, A.R., Rao, S., Attoui, H., and Mertens, P.P. (2007). Analysis and phylogenetic comparisons of full-length VP2 genes of the 24 bluetongue virus serotypes. *Journal of General Virology* **88**, 621–630.
Maaty, W.S., Ortmann, A.C., Dlakić, M., Schulstad, K., Hilmer, J.K., Liepold, L., Weidenheft, B., Khayat, R., Douglas, T., Young, M.J., and Bothner, B. (2006). Characterization of the archaeal thermophile *Sulfolobus* turreted icosahedral virus validates an evolutionary link among double-stranded DNA viruses from all domains of life. *Journal of Virology* **80**, 7625–7635.
Maljkovic Berry, I., Ribeiro, R., Kothari, M., Athreya, G., Daniels, M., Lee, H.Y., Bruno, W., and Leitner, T. (2007). Unequal evolutionary rates in the human immunodeficiency virus type 1 (HIV-1) pandemic: the evolutionary rate of HIV-1 slows down when the epidemic rate increases. *Journal of Virology* **81**, 10625–10635.
Malik, H.S., Burke, W.D., and Eickbush, T.H. (1999). The age and evolution of non-LTR retrotransposable elements. *Molecular Biology and Evolution* **16**, 793–805.
Malik, H.S., Henikoff, S., and Eickbush, T.H. (2000). Poised for contagion: evolutionary origins of the infectious abilities of invertebrate retroviruses. *Genome Research* **10**, 1307–1318.
Malpica, J.M., Fraile, A., Moreno, I., Obies, C.I., Drake, J.W., and García-Arenal, F. (2002). The rate and character of spontaneous mutation in an RNA virus. *Genetics* **162**, 1505–1511.
Mangeat, B., Turelli, P., Caron, G., Friedli, M., Perrin, L., and Trono, D. (2003). Broad antiretroviral defence by human APOBEC3G through lethal editing of nascent reverse transcripts. *Nature* **424**, 99–103.

Manrubia, S.C., Escarmís, C., Domingo, E., and Lázaro, E. (2005). High mutation rates, bottlenecks, and robustness of RNA viral quasispecies. *Gene* **347**, 273–282.

Mansky, L.M. (1998). Retrovirus mutation rates and their role in genetic variation. *Journal of General Virology* **79**, 1337–1345.

Mansky, L.M. and Temin, H.M. (1995). Lower *in vivo* mutation rate of human immunodeficiency virus type 1 than that predicted from the fidelity of purified reverse transcriptase. *Journal of Virology* **69**, 5087–5094.

Mansky, L.M. and Bernard, L.C. (2000). 3'-Azido-3'-deoxythymidine (AZT) and AZT-resistant reverse transcriptase can increase the in vivo mutation rate of human immunodeficiency virus type 1. *Journal of Virology* **74**, 9532–9539.

Mansky, L.M. and Cunningham, K.S. (2000). Virus mutators and antimutators: roles in evolution, pathogenesis and emergence. *Trends in Genetics* **16**, 512–517.

Mansky, L.M., Durand, D.P., and Hill, J.H. (1995). Evidence for complementation of plant potyvirus pathogenic strains in mixed infection. *Journal of Phytopathology* **143**, 247–250.

Marco, C.F. and Aranda, M.A. (2005). Genetic diversity of a natural population of Cucurbit yellow stunting disorder virus. *Journal of General Virology* **86**, 815–822.

Marozsan, A.J., Moore, D.M., Lobritz, M.A, Fraundorf, E., Abraha, A., Reeves, J.D., and Arts, E.J. (2005). Differences in the fitness of two diverse wild-type human immunodeficiency virus type 1 isolates are related to the efficiency of cell binding and entry. *Journal of Virology* **79**, 7121–7134.

Marques, J.T. and Carthew, R.W. (2007). A call to arms: coevolution of animal viruses and host innate immune responses. *Trends in Genetics* **23**, 359–364.

Marsh, G.A., Rabadán, R., Levine, A.J., and Palese, P. (2008). Highly conserved regions of influenza a virus polymerase gene segments are critical for efficient viral RNA packaging. *Journal of Virology* **82**, 2295–2304.

Martin, D.P., van der Walt, E., Posada, D., and Rybicki, E.P. (2005). The evolutionary value of recombination is constrained by genome modularity. *PLoS Genetics* **51**, 475–479.

Martín, J., Dunn, G., Hull, R., Patel, V., and Minor, P.D. (2000). Evolution of the Sabin strain of type 3 poliovirus in an immunodeficient patient during the entire 637-day period of virus excretion. *Journal of Virology* **74**, 3001–3010.

Martin, G., Elena, S.F., and Lenormand, T. (2007). Distributions of epistasis in microbes fit predictions from a fitness landscape model. *Nature Genetics* **39**, 555–560.

Martin, V., Grande-Perez, A., and Domingo, E. (2008). No evidence of selection for mutational robustness during lethal mutagenesis of lymphocytic choriomeningitis virus. *Virology* **378**, 185–192.

Matrosovich, M.N., Gambaryan, A.S., Teneberg, S., Piskarev, V.E., Yamnikova, S.S., Lvov, D.K., Robertson, J.S., and Karlsson, K.A. (1997). Avian influenza A viruses differ from human viruses by recognition of sialyloligosaccharides and gangliosides and by a higher conservation of the HA receptor-binding site. *Virology* **233**, 224–234.

Matthews, P.C., Prendergast, A., Leslie, A., Crawford, H., Payne, R., Rousseau, C., Rolland, M., Honeyborne, I., Carlson, J., Kadie, C. *et al.* (2008). Central role of reverting mutations in HLA associations with human immunodeficiency virus set point. *Journal of Virology* **82**, 8548–8559.

Matthijnssens, J., Ciarlet, M., Heiman, E., Arijs, I., Delbeke, T., McDonald, S.M., Palombo, E.A., Iturriza-Gómara, M., Maes, P., Patton, J.T. *et al.* (2008). Full genome-based classification of rotaviruses reveals a common origin between human Wa-Like and porcine rotavirus strains and human DS-1-like and bovine rotavirus strains. *Journal of Virology* **82**, 204–219.

Maynard Smith, J. and Szathmáry, E. (1995). *The Major Transitions of Evolution.* W.H. Freeman and Co., Oxford.
Mayo, M.A. and Jolly, C.A. (1991). The 5′-terminal sequence of potato leafroll virus RNA: evidence of recombination between virus and host RNA. *Journal of General Virology* **72**, 2591–2595.
McGarvey, M.J., Iqbal, M., Nastos, T., and Karayiannis, P. (2008). Restricted quasispecies variation following infection with the GB virus B. *Virus Research* **135**, 181–186.
McGeoch, D.J. and Gatherer, D. (2005). Integrating reptilian herpesviruses into the family Herpesviridae. *Journal of Virology* **79**, 725–731.
McLysaght, A., Baldi, P.F., and Gaut, B.S. (2003). Extensive gene gain associated with adaptive evolution of poxviruses. *Proceedings of the National Academy of Sciences USA* **100**, 14960–14965.
McMullan, L.K., Grakoui, A., Evans, M.J., Mihalik, K., Puig, M., Branch, A.D., Feinstone, S.M., and Rice, C.M. (2007). Evidence for a functional RNA element in the hepatitis C virus core gene. *Proceedings of the National Academy of Sciences USA* **104**, 2879–2884.
McVean, G., Awadalla, P., and Fearnhead, P. (2002). A coalescent-based method for detecting and estimating recombination from gene sequences. *Genetics* **160**, 1231–1241.
Meiering, C.D. and Linial, M.L. (2001). Historical perspective of foamy virus epidemiology and infection. *Clinical Microbiology Reviews* **14**, 165–176.
Messer, W.B., Vitarana, U.T., Sivananthan, K., Elvtigala, J., Preethimala, L.D., Ramesh, R., Withana, N., Gubler, D.J., and De Silva, A.M. (2002). Epidemiology of dengue in Sri Lanka before and after the emergence of epidemic dengue hemorrhagic fever. *American Journal of Tropical Medicine and Hygiene* **66**, 765–773.
Messer, W.B., Gubler, D.J., Harris, E., Sivananthan, K. and de Silva, A.M. (2003). Emergence and global spread of a dengue serotype 3, subtype III virus. *Emerging Infectious Diseases* **9**, 800–809.
Meyers, G., Rumenapf, T., and Thiel, H.J. (1989). Ubiquitin in a togavirus. *Nature* **341**, 491.
Michalakis, Y. and Roze, D. (2004). Epistasis in RNA viruses. *Science* **306**, 1492–1493.
Michod, R.E. and Levin, B.R. (1988). *The Evolution of Sex: an Examination of Current Ideas.* Sinauer Associates, Sunderland, MA.
Michod, R.E., Bernstein, H., and Nedelcu, A.M. (2008). Adaptive value of sex in microbial pathogens. *Infection, Genetics and Evolution* **8**, 267–285.
Mills, D.R., Peterson, R.L., and Spiegelman, S. (1967). An extracellular Darwinian experiment with a self-duplicating nucleic acid molecule. *Proceedings of the National Academy of Sciences USA* **58**, 217–224.
Minskaia, E., Hertzig, T., Gorbalenya, A.E., Campanacci, V., Cambillau, C., Canard, B., and Ziebuhr, J. (2006). Discovery of an RNA virus 3′->5′ exoribonuclease that is critically involved in coronavirus RNA synthesis. *Proceedings of the National Academy of Sciences USA* **103**, 5108–5113.
Miralles, R., Gerrish, P.J., Moya, A., and Elena, S.F. (1999). Clonal interference and the evolution of RNA viruses. *Science* **285**, 1745–1747.
Moncayo, A.C., Fernandez, Z., Ortiz, D., Diallo, M., Sall, A., Hartman, S., Davis, C.T., Coffey, L., Mathiot, C.C., Tesh, R.B., and Weaver, S.C. (2004). Dengue emergence and adaptation to peridomestic mosquitoes. *Emerging Infectious Diseases* **10**, 1790–1796.
Montville, R., Froissart, R., Remold, S.K., Tenaillon, O., and Turner, P.E. (2005). Evolution of mutational robustness in an RNA virus. *PLoS Biology* **3**, e381.

Moore, C.B., John, M., James, I.R., Christiansen, F.T., Witt, C.S., and Mallal, S.A. (2002). Evidence for HIV-1 adaptation to HLA-restricted immune responses at the population level. *Science* **296**, 1439–1443.

Moran, N.A. (1996). Accelerated evolution and Muller's rachet in endosymbiotic bacteria. *Proceedings of the National Academy of Sciences USA* **93**, 2873–2878.

Moran, N.A. (2002). Microbial minimalism: genome reduction in bacterial pathogens. *Cell* **108**, 583–586.

Moran, N.A. and Mira, A. (2001). The process of genome shrinkage in the obligate symbiont *Buchnera aphidicola*. *Genome Biology* **2**, 0054.

Moreno, I.M., Malpica, J.M., Rodríguez-Cerezo, E., and García-Arenal, F. (1997). A mutation in tomato aspermy cucumovirus that abolishes cell-to-cell movement is maintained to high levels in the viral RNA population by complementation. *Journal of Virology* **71**, 9157–9162.

Morse, S.S. (1994). Evolution of genetic exchange in RNA viruses. In S.S. Morse, ed., *The Evolutionary Biology of Viruses*, pp. 1–28. Raven Press, New York.

Morse, S.S. (1995). Factors in the emergence of infectious diseases. *Emerging Infectious Diseases* **1**, 7–15.

Moya, A., Elena, S.F., Miralles, R., and Barrio, E. (2000). The evolution of RNA viruses: a population genetics view. *Proceedings of the National Academy of Sciences USA* **97**, 6967–6973.

Moya, A., Holmes, E.C., and González-Candelas, F. (2004). The population genetics and evolutionary epidemiology of RNA viruses. *Nature Reviews Microbiology* **2**, 279–287.

Muller, H.J. (1964). The relation of recombination to mutational advance. *Mutation Research* **1**, 2–9.

Munster, V.J., Baas, C., Lexmond, P., Waldenström, J., Wallensten, A., Fransson, T., Rimmelzwaan, G.F., Beyer, W.E., Schutten, M., Olsen, B. *et al.* (2007). Spatial, temporal, and species variation in prevalence of influenza A viruses in wild migratory birds. *PLoS Pathogens* **3**, e61.

Nadin-Davis, S.A., Casey, G.A., and Wandeler, A. (1993). Identification of regional variants of the rabies virus within the Canadian province of Ontario. *Journal of General Virology* **74**, 829–837.

Nadin-Davis, S.A., Sampath, M.I., Casey, G.A., Tinline, R.R., and Wandeler, A. (1999). Phylogeographic patterns exhibited by Ontario rabies virus variants. *Epidemiology and Infection* **123**, 325–336.

Nagai, M., Sakoda, Y., Mori, M., Hayashi, M., Kida, H., and Akashi, H. (2003). Insertion of cellular sequence and RNA recombination in the structural protein coding region of cytopathogenic bovine viral diarrhoea virus. *Journal of General Virology* **84**, 447–452.

Navarro, B. and Flores, R. (1997). Chrysanthemum chlorotic mottle viroid: unusual structural properties of a subgroup of self-cleaving viroids with hammerhead ribozymes. *Proceedings of the National Academy of Sciences USA* **94**, 11262–11267.

Nee, S. (1987). The evolution of multicompartmental genomes in viruses. *Journal of Molecular Evolution* **25**, 277–281.

Nee, S., Mooers, A.Ø., and Harvey, P.H. (1992). The tempo and mode of evolution revealed from molecular phylogenies. *Proceedings of the National Academy of Sciences USA* **89**, 8322–8326.

Nee, S., Holmes, E.C., May, R.M., and Harvey, P.H. (1994a). Extinction rates can be estimated from molecular phylogenies. *Philosophical Transactions of the Royal Society of London Series B Biological Sciences* **344**, 77–82.

Nee, S., May, R.M., and Harvey, P.H. (1994b). The reconstructed evolutionary process. *Philosophical Transactions of the Royal Society of London Series B Biological Sciences* **344**, 305–311.

Nee, S., Holmes, E.C., Rambaut, A., and Harvey, P.H. (1995). Inferring population history from molecular phylogenies. *Philosophical Transactions of the Royal Society of London Series B Biological Sciences* **349**, 25–31.

Nei, M. and Gojobori, T. (1986). Simple methods for estimating the numbers of synonymous and nonsynonymous nucleotide substitutions. *Molecular Biology and Evolution* **3**, 418–426.

Nelson, M.I. and Holmes, E.C. (2007). The evolution of epidemic influenza. *Nature Reviews Genetics* **8**, 196–205.

Nelson, M.I., Simonsen, L., Viboud, C., Miller, M.A., Taylor, J., St. George, K., Griesemer, S.B., Ghedin, E., Sengamalay, N.A., Spiro, D.J. et al. (2006). Stochastic processes are key determinants of the short-term evolution of influenza A virus. *PLoS Pathogens* **2**, e125.

Nelson, M.I., Simonsen, L., Viboud, C., Miller, M.A., and Holmes, E.C. (2007). Phylogenetic analysis reveals the global migration of seasonal influenza A viruses. *PLoS Pathogens* **3**, e131.

Nelson, M.I., Edelman, L., Spiro, D.J., Boyne, A.R., Bera, J., Halpin, R., Ghedin, E., Miller, M.A., Simonsen, L., Viboud, C., and Holmes, E.C. (2008). Molecular epidemiology of A/H3N2 and A/H1N1 influenza virus during a single epidemic season in the United States. *PLoS Pathogens* **4**, e1000133.

Nerrienet, E., Santiago, M.L., Foupouapouognigni, Y., Bailes, E., Mundy, N.I., Njinku, B., Kfutwah, A., Muller-Trutwin, M.C., Barre-Sinoussi, F., Shaw, G.M. et al. (2005). Simian immunodeficiency virus infection in wild-caught chimpanzees from cameroon. *Journal of Virology* **79**, 1312–1319.

Neumann, A.U., Lam, N.P., Dahari, H., Gretch, D.R., Wiley, T.E., Layden, T.J., and Perelson, A.S. (1998). Hepatitis C viral dynamics *in vivo* and the antiviral efficacy of interferon-alpha therapy. *Science* **282**, 103–107.

Nielsen, R. and Yang, Z. (1998). Likelihood models for detecting positively selected amino acid sites and applications to the HIV-1 envelope gene. *Genetics* **148**, 929–936.

Nijhuis, M., Boucher, C.A.B., Schipper, P., Leitner, T., Schuurman, R., and Albert, J. (1998). Stochastic processes strongly influence protease-inhibitor therapy. *Proceedings of the National Academy of Sciences USA* **95**, 14441–14446.

Nisalak, A., Endy, T.P., Nimmannitya, S., Kalayanarooj, S., Thisayakorn, U., Scott, R.M., Burke, D.S., Hoke, C.H., Innis, B.L., and Vaughn, D.W. (2003). Serotype-specific dengue virus circulation and dengue disease in Bangkok, Thailand from 1973–1999. *American Journal of Tropical Medicine and Hygiene* **68**, 191–202.

Noppornpanth, S., Lien, T.X., Poovorawan, Y., Smits, S.L., Osterhaus, A.D.M.E., and Haagmans, B.L. (2006). Identification of a naturally occurring recombinant genotype 2/6 hepatitis C virus. *Journal of Virology* **80**, 7569–7577.

Nora, T., Charpentier, C., Tenaillon, O., Hoede, C., Clavel, F., and Hance, A.J. (2007). Contribution of recombination to the evolution of human immunodeficiency viruses expressing resistance to antiretroviral treatment. *Journal of Virology* **81**, 7620–7628.

Norja, P., Eis-Hübinger, A.M., Söderlund-Venermo, M., Hedman, K., and Simmonds, P. (2008). Rapid sequence change and geographical spread of human parvovirus B19; comparison of B19 evolution in acute and persistent infections. *Journal of Virology* **82**, 6427–6433.

Novella, I.S. (2003). Contributions of vesicular stomatitis virus to the understanding of RNA virus evolution. *Current Opinion in Microbiology* **6**, 399–405.

Novella, I.S. and Ebendick-Corpus, B.E. (2004). Molecular basis of fitness loss and fitness recovery in vesicular stomatitis virus. *Journal of Molecular Biology* **342**, 1423–1430.

Novella, I.S., Quer, J., Domingo, E., and Holland, J.J. (1999). Exponential fitness gains of RNA virus populations are limited by bottleneck effects. *Journal of Virology* **73**, 1668–1671.

Novella, I.S., Ball, L.A., and Wertz, G.W. (2004a). Fitness analyses of vesicular stomatitis strains with rearranged genomes reveal replicative disadvantages. *Journal of Virology* **78**, 9837–9841.

Novella, I.S., Zárate, S., Metzgar, D., and Ebendick-Corpus, B.E. (2004b). Positive selection of synonymous mutations in vesicular stomatitis virus. *Journal of Molecular Biology* **342**, 1415–1421.

Nowak, M.A. and May, R.M. (2000). *Virus Dynamics: Mathematical Principles of Immunology and Virology*. Oxford University Press, Oxford.

Nowak, M.A., Anderson, R.M., McLean, A.R., Wolfs, T.F.W., Goudsmit, J., and May, R.M. (1991). Antigenic diversity threshold and the development of AIDS. *Science* **254**, 963–969.

Obbard, D.J., Jiggins, F.M., Halligan, D.L., and Little, T.J. (2006). Natural selection drives extremely rapid evolution in antiviral RNAi genes. *Current Biology* **16**, 580–585.

Obenauer, J.C., Denson, J., Mehta, P.K., Su, X., Mukatira, S., Finkelstein, D.B., Xu, X., Wang, J., Ma, J., Fan, Y. *et al*. (2006). Large-scale sequence analysis of avian influenza isolates. *Science* **311**, 1576–1580.

Ochman, H., Lawrence, J.G., and Groisman, E.A. (2000). Lateral gene transfer and the nature of bacterial innovation. *Nature* **405**, 299–304.

O'Fallon, B.D., Adler, F.R., and Proulx, S.R. (2007). Quasi-species evolution in subdivided populations favours maximally deleterious mutations. *Proceedings of the Royal Society of London Series B Biological Sciences* **274**, 3159–3164.

Ohshima, K., Tomitaka, Y., Wood, J.T., Minematsu, Y., Kajiyama, H., Tomimura, K and Gibbs, A.J. (2007). Patterns of recombination in turnip mosaic virus genomic sequences indicate hotspots of recombination. *Journal of General Virology* **88**, 298–315.

Ohta, T. (1992). The nearly neutral theory of molecular evolution. *Annual Review of Ecology and Systematics* **23**, 263–286.

Okamoto, H., Fukuda, M., Tawara, A., Nishizawa, T., Itoh, Y., Hayasaka, I., Tsuda, F., Tanaka, T., Miyakawa, Y., and Mayumi, M. (2000). Species-specific TT viruses and cross-species infection in nonhuman primates. *Journal of Virology* **74**, 1132–1139.

Oldstone, M.B.A. (2006). Viral persistence: parameters, mechanisms and future predictions. *Virology* **344**, 111–118.

Olsen, B., Munster, V.J., Wallensten, A., Waldenstrom, J., Osterhaus, A.D.M.E., and Fouchier, R.A.M. (2006). Global patterns of influenza A virus in wild birds. *Science* **312**, 384–388.

Orr, H.A. (2002). The population genetics of adaptation: the adaptation of DNA sequences. *Evolution* **56**, 1317–1330.

Ortmann, A.C., Wiedenheft, B., Douglas, T., and Young, M. (2006). Hot crenarchaeal viruses reveal deep evolutionary connections. *Nature Reviews Microbiology* **4**, 520–528.

Pal, C., Maciá, M.D., Oliver, A., Schachar, I., and Buckling, A. (2007). Coevolution with viruses drives the evolution of bacterial mutation rates. *Nature* **450**, 1079–1081.

Palese, P. and Shaw, M.L. (2007). *Orthomyxoviridae*: the virus and their replication. In D.M. Knipe, P.M. Howley, D.E. Griffin, R.A. Lamb, and M.A. Martin, eds, *Fields Virology*, 5th edn, pp. 1647–1689. Lippincott Williams and Wilkins, Philadelphia, PA.

Palmemberg, A.C and Sgro, J.-Y. (1997). Topological organisation of picornaviral genomes: Statistical prediction of RNA structural signals. *Seminars in Virology* **8**, 231–241.

Pan Amercian Health Organization. (2002). *Framework: New Generation of Dengue Prevention and Control Programs in the Americas.* Pan American Health Organization, Washington DC.

Panavas, T., Panaviene, Z., Pogany, J., and Nagy, P.D. (2003). Enhancement of RNA synthesis by promoter duplication in tombusviruses. *Virology* **310**, 118–129.

Pariente, N., Sierra, S., Lowenstein, P.R., and Domingo, E. (2001). Efficient virus extinction by combinations of a mutagen and antiviral inhibitors. *Journal of Virology* **75**, 9723–9730.

Parrish, C.R. (1990). Emergence, natural history, and variation of canine, mink, and feline parvoviruses. *Advances in Virus Research* **38**, 403–450.

Parrish, C.R., Holmes, E.C., Morens, D.M., Park, E.-C., Burke, D.S., Calisher, C.H., Laughlin, C.A., Saif, L.J., and Dazsak, P. (2008). Cross-species viral transmission and the emergence of new epidemic diseases. *Microbiology and Molecular Biology Reviews* **72**, 457–470.

Peiris, M., Yuen, K.Y., Leung, C.W., Chan, K.H., Ip, P.L., Lai, R.W., Orr, W.K., and Shortridge, K.F. (1999). Human infection with influenza H9N2. *Lancet* **354**, 916–917.

Pelletier, J. and Sonenberg, N. (1988). Internal initiation of translation of eukaryotic mRNA directed by a sequence derived from poliovirus RNA. *Nature* **334**, 320–325.

Peng, C.-W., Peremyslov, V.V., Mushegian, A.R., Dawson, W.O., and Dolja, V.V. (2001). Functional specialization and evolution of leader proteinases in the family *Closteroviridae*. *Journal of Virology* **75**, 12153–12160.

Pepin, K.M. and Wichman, H.A. (2007). Variable epistatic effects between mutations at host recognition sites in phiX174 bacteriophage. *Evolution* **61**, 1710–1724.

Pepin, K.M. and Wichman, H.A. (2008). Experimental evolution and genome sequencing reveal variation in levels of clonal interference in large populations of bacteriophage phiX174. *BMC Evolutionary Biology* **8**, 85.

Pepin, K.M., Samuel, M.A., and Wichman, H.A. (2006). Variable pleiotropic effects from mutations at the same locus hamper prediction of fitness from a fitness component. *Genetics* **172**, 2047–2056.

Perelson, A.S., Neumann, A.U., Markowitz, M., Leonard, J.M., and Ho, D.D. (1996). HIV-1 dynamics in vivo: virion clearance rate, infected cell life-span, and viral generation time. *Science* **271**, 1582–1586.

Pinheiro, M.D., Power, M.E., Butler, B.J., Dayeh, V.R., Slawson, R., Lee, L.E., Lynn, D.H., and Bols, N.C. (2007). Use of *Tetrahymena thermophila* to study the role of protozoa in inactivation of viruses in water. *Applied and Environmental Microbiology* **73**, 643–649.

Pita, J.S., De Miranda, J.R., Schneider, W.L., and Roossinck, M.J. (2007). Environment determines fidelity for an RNA virus replicase. *Journal of Virology* **81**, 9072–9077.

Plikat, U., Nieselt-Struwe, K., and Meyerhans, A. (1997). Genetic drift can dominate short-term human immunodeficiency virus type 1 *nef* quasispecies in vivo. *Journal of Virology* **71**, 4233–4240.

Plotkin, J.B., Robins, H., and Levine, A.J. (2004). Tissue-specific codon usage and the expression of human genes. *Proceedings of the National Academy of Sciences USA* **101**, 12588–12591.

Plyusnin, A. and Morzunov, S.P. (2001). Virus evolution and genetic diversity of hantaviruses and their rodent hosts. *Current Topics in Microbiology and Immunology* **256**, 47–75.

Poch, O., Sauvaget, I., Delarue, M., and Tordo, N. (1989). Identification of four conserved motifs among the RNA-dependent polymerase encoding elements. *EMBO Journal* **8**, 3867–3874.

Pomeroy, L.W., Bjørnstad, O.N., and Holmes, E.C. (2008). A molecular perspective on the demographic history of the *Paramyxoviridae*. *Journal of Molecular Evolution* **66**, 98–106.

Poon, A. and Chao, L. (2004). Drift increases the advantage of sex in RNA bacteriophage ϕ6. *Genetics* **166**, 19–24.

Prangishvili, D., Forterre, P., and Garrett, R.A. (2006). Viruses of the Archaea: a unifying view. *Nature Reviews Microbiology* **4**, 837–848.

Pressing, J. and Reanney, D.C. (1984). Divided genomes and intrinsic noise. *Journal of Molecular Evolution* **20**, 135–146.

Preston, R.M. (1994). *The Hot Zone*. Knopf Publishing Group, New York.

Price, D.A., Goulder, P.J.R., Klenerman, P., Sewell, A.K., Easterbrook, P.J., Troop, M., Bangham, C.R.M., and Phillips, R.E. (1997). Positive selection of HIV-1 cytotoxic T lymphocyte escape variants during primary infection. *Proceedings of the National Academy of Sciences USA* **94**, 1890–1895.

Proutski, V., Gaunt, M.W., Gould, E.A., and Holmes, E.C. (1997). Secondary structure of the 3'-untranslated region of Yellow Fever virus: implications for virulence, attenuation and vaccine development. *Journal of General Virology* **78**, 1543–1549.

Pugachev, K.V., Guirakhoo, F., Ocran, S.W., Mitchell, F., Parsons, M., Penal, C., Girakhoo, S., Pougatcheva, S.O., Arroyo, J., Trent, D.W., and Monath, T.P. (2004). High fidelity of yellow fever virus RNA polymerase. *Journal of Virology* **78**, 1032–1038.

Pybus, O.G., Rambaut, A., and Harvey, P.H. (2000). An integrated framework for the inference of viral population history from reconstructed genealogies. *Genetics* **155**, 1429–1437.

Pybus, O.G., Charleston, M.A., Gupta, S., Rambaut, A., Holmes, E.C., and Harvey, P.H. (2001). The epidemic behaviour of the hepatitis C virus. *Science* **292**, 2323–2325.

Pybus, O.G., Rambaut, A., Holmes, E.C., and Harvey, P.H. (2002). New inferences from tree shape: numbers of missing taxa and population growth rates. *Systematic Biology* **51**, 881–888.

Pybus, O.G., Cochrane, A., Holmes, E.C., and Simmonds, P. (2005). The hepatitis C virus epidemic among injecting drug users. *Infection, Genetics and Evolution* **5**, 131–139.

Pybus, O.G., Rambaut, A., Freckleton, R.P., Belshaw, R., Drummond, A.J., and Holmes, E.C. (2007). Phylogenetic evidence for deleterious mutation load in RNA viruses and its contribution to viral evolution. *Molecular Biology and Evolution* **24**, 845–852.

Qiao, X., Qiao, J., and Mindich, L. (1997). Stoichiometric packaging of the three genomic segments of double-stranded RNA bacteriophage ϕ6. *Proceedings of the National Academy of Sciences USA* **94**, 4074–4079.

Qin, H., Wu, W.B., Comeron, J.M., Kreitman, M., and Li, W.-H. (2004). Intragenic spatial patterns of codon usage bias in prokaryotic and eukaryotic genomes. *Genetics* **168**, 2245–2260.

Rabadan, R., Levine, A.J., and Robins, H. (2006). Comparison of avian and human influenza A viruses reveals a mutational bias on the viral genomes. *Journal of Virology* **80**, 11887–11891.

Rambaut, A., Robertson, D.L., Pybus, O.G., Peeters, M., and Holmes, E.C. (2001). Phylogeny and the origin of HIV-1. *Nature* **410**, 1047–1048.

Rambaut, A., Pybus, O.G., Nelson, M.I., Viboud, C., Taubenberger, J.K., and Holmes, E.C. (2008). The genomic and epidemiological dynamics of human influenza A virus. *Nature* **453**, 615–619.

Ramsden, C., Holmes, E.C., and Charleston, M.A. (2009). Hantavirus evolution in relation to its rodent and insectivore hosts: no evidence for co-divergence. *Molecular Biology and Evolution* **26**, 143–153.

Ramsden, C., Melo, F.L., Figueiredo, L.M., Holmes, E.C., Zanotto, P.M. de A., and the VGDN Consortium. (2008). High rates of molecular evolution in hantaviruses. *Molecular Biology and Evolution* **25**, 1488–1492.

Raney, J.L., Delongchamp, R.R., and Valentine, C.R. (2004). Spontaneous mutant frequency and mutation spectrum for gene a of phi x174 grown in *E. coli*. *Environmental and Molecular Mutagenesis* **44**, 119–127.

Raoult, D., Audic, S., Robert, C., Abergel, C., Renesto, P., Ogata, H., La Scola, B., Suzan, M., and Claverie, J.-M. (2004). The 1.2-megabase genome sequence of mimivirus. *Science* **306**, 1344–1350.

Ravkov, E.V., Smith, J.S., and Nichol, S.T. (1995). Rabies virus glycoprotein gene contains a long 3' noncoding region which lacks pseudogene properties. *Virology* **206**, 718–723.

Real, L.A., Henderson, J.C., Biek, R., Snaman, J., Jack, T.L., Childs, J.E., Stahl, E., Waller, L., Tinline, R., and Nadin-Davis, S. (2005). Unifying the spatial population dynamics and molecular evolution of epidemic rabies virus. *Proceedings of the National Academy of Sciences USA* **102**, 12107–12111.

Reanney, D.C. (1982). The evolution of RNA viruses. *Annual Review of Microbiology* **36**, 47–73.

Rector, A., Lemey, P., Tachezy, R., Mostmans, S., Ghim, S.J., Van Doorslaer, K., Roelke, M., Bush, M., Montali, R.J., Joslin, J. *et al.* (2007). Ancient papillomavirus-host co-speciation in *Felidae*. *Genome Biology* **8**:R57.

Regoes, R.R., Crotty, S., Antia, R., and Tanaka, M.M. (2005). Optimal replication of poliovirus within cells. *American Naturalist* **165**, 364–373.

Remold, S.K., Rambaut, A., and Turner, P.E. (2008). Evolutionary genomics of host adaptation in vesicular stomatitis virus (VSV). *Molecular Biology and Evolution* **25**, 1138–1147.

Rice, W.R. and Chippindale, A.K. (2001). Sexual recombination and the power of natural selection. *Science* **294**, 555–559.

Rico, P., Ivars, P., Elena, S.F., and Hernández, C. (2006). Insights into the selective pressures restricting *Pelargonium Flower Break Virus* genome variability: evidence for host adaptation. *Journal of Virology* **80**, 8124–8132.

Rico-Hesse, R. (2003). Microevolution and virulence of dengue viruses. *Advances in Virus Research* **59**, 315–341.

Rico-Hesse, R. (2007). Dengue virus evolution and virulence models. *Clinical Infectious Diseases* **44**, 1462–1466.

Rico-Hesse, R., Harrison, L.M., Salas, R.A., Tovar, D., Nisalak, A., Ramos, C., Boshell, J., de Mesa, M.T.R., Nogueira, R.M.R., and da Rosa, A.T. (1997). Origins of dengue type 2 viruses associated with increased pathogenicity in the Americas. *Virology* **230**, 244–251.

Robertson, M.P., Igel, H., Baertsch, R., Haussler, D., Ares, M., Jr, and Scott W.G. (2005). The structure of a rigorously conserved RNA element within the SARS virus genome. *PLoS Biology* **3**, e5.

Robinson, D.M., Jones, D.T., Kishino, H., Goldman, N., and Thorne, J.L. (2003). Protein evolution with dependence among codons due to tertiary structure. *Molecular Biology and Evolution* **20**, 1692–704.

Rogers, G.N. and Paulson, J.C. (1983). Receptor determinants of human and animal influenza virus isolates: differences in receptor specificity of the H3 hemagglutinin based on species of origin. *Virology* **127**, 361–373.

Rokyta, D.R., Joyce, P., Caudle, S.B., and Wichman, H.A. (2005). An empirical test of the mutational landscape model of adaptation using a single-stranded DNA virus. *Nature Genetics* **37**, 441–444.

Romano, C.M., Zanotto, P.M. de A., and Holmes, E.C. (2008). Bayesian coalescent analysis reveals a high rate of molecular evolution in GB virus C. *Journal of Molecular Evolution* **66**, 292–297.

Ronquist, F. and Huelsenbeck, J.P. (2003). MrBayes 3: Bayesian phylogenetic inference under mixed models. *Bioinformatics* **19**, 1572–1574.

Rossmann, M.G., Arnold, E., Erickson, J.W., Frankenberger, E.A., Griffith, J.P., Hecht, H.-J., Johnson, J.E., Kamer, G., Luo, M., Mosser, A.G. *et al.* (1985). Structure of a human common cold virus and functional relationship to other picornaviruses. *Nature* **317**, 145–153.

Rouzine, I.M. and Coffin, J.M. (1999). Linkage disequilibrium test implies a large effective population number for HIV *in vivo*. *Proceedings of the National Academy of Sciences USA* **96**, 10758–10763.

Rudnick, A. (1978). Ecology of dengue virus. *Asian Journal of Infectious Diseases* **2**, 156–160.

Ruiz-Jarabo, C.M., Arias, A., Baranowski, E., Escarmís, C., and Domingo, E. (2000). Memory in viral quasispecies. *Journal of Virology* **74**, 3543–3547.

Ruíz-Jarabo, C.M., Arias, A., Molina-París, C., Briones, C., Baranowski, E., Escarmís, C., and Domingo, E (2002). Duration and fitness dependence of quasispecies memory. *Journal of Molecular Biology* **315**, 285–296.

Ruiz-Jarabo, C.M., Ly, C., Domingo, E. and de la Torre, J.C. (2003a). Lethal mutagenesis of the prototypic arenavirus lymphocytic choriomeningitis virus (LCMV). *Virology* **308**, 37–47.

Ruiz-Jarabo, C.M., Miller, E., Gómez-Mariano, G., and Domingo, E. (2003b). Synchronous loss of quasispecies memory in parallel viral lineages: a deterministic feature of viral quasispecies. *Journal of Molecular Biology* **333**, 553–563.

Russell, C.A., Smith, D.L., Childs, J.E., and Real, L.A. (2005). Predictive spatial dynamics and strategic planning for raccoon rabies emergence in Ohio. *PLoS Biology* **3**, e88.

Russell, C.A., Jones, T.C., Barr, I.G., Cox, N.J., Garten, R.J., Gregory, V., Gust, I.D., Hampson, A.W., Hay, A.J., Hurt, A.C. *et al.* (2008). The global circulation of seasonal influenza A (H3N2) viruses. *Science* **320**, 340–346.

Saag, M.S., Hahn, B.H., Gibbons, J., Li, Y., Parks, E.S., Parks, W.P., and Shaw, G.M. (1988). Extensive variation of human immunodeficiency virus type-1 *in vivo*. *Nature* **334**, 440–447.

Sacristán, S., Malpica, J.M., Fraile, A., and García-Arenal, F. (2004). Estimation of population bottlenecks during systemic movement of tobacco mosaic virus in tobacco plants. *Journal of Virology* **77**, 9906–9911.

Sage, R.F. (2004). The evolution of C_4 photosynthesis. *The New Phytologist* **161**, 341–370.

Sala, M. and Wain-Hobson, S. (2000). Are RNA viruses adapting or merely changing? *Journal of Molecular Evolution* **51**, 12–20.

Salazar-Gonzalez, J.F., Bailes, E., Pham, K.T., Salazar, M.G., Guffey, M.B., Keele, B.F., Derdeyn, C.A., Farmer, P., Hunter, E., Allen, S. *et al.* (2008). Deciphering human immunodeficiency virus type 1 transmission and early envelope diversification by single-genome amplification and sequencing. *Journal of Virology* **82**, 3952–3970.

Salehi-Ashtiani, K., Lupták, A., Litovchick, A., and Szostak, J.W. (2006). A genomewide search for ribozymes reveals an HDV-like sequence in the human CPEB3 gene. *Science* **313**, 1788–1792.

Salemi, M., Lewis, M., Egan, J.F., Hall, W.W., Desmyter, J., and Vandamme, A.-M. (1999). Different population dynamics of human T cell lymphotropic virus type II in intravenous drug users compared with endemically infected tribes. *Proceedings of the National Academy of Sciences USA* **96**, 13253–13258.

Sanchez, A., Trappier, S.G., Mahy, B.W., Peters, C.J., and Nichol, S.T. (1996). The virion glycoproteins of Ebola viruses are encoded in two reading frames and are expressed through transcriptional editing. *Proceedings of the National Academy of Sciences USA* **93**, 3602–3607.

Sanjuán, R. (2006). Quantifying antagonistic epistasis in a multifunctional RNA secondary structure of the Rous sarcoma virus. *Journal of General Virology* **87**, 1595–1602.
Sanjuán, R. (2008). Quasispecies. In B.W.J. Mahy and M.H.V. Van Regenmortal, eds, *Encyclopedia of Virology*, vol. 4, 3rd edn, pp. 359–365. Elsevier, Oxford.
Sanjuán, R. and Elena, S.F. (2006). Epistasis correlates to genomic complexity. *Proceedings of the National Academy of Sciences USA* **103**, 14402–14405.
Sanjuán, R., Codoñer, F.M., Moya, A., and Elena, S.F. (2004a). Natural selection and the organ-specific differentiation of HIV-1 V3 hypervariable region. *Evolution* **58**, 1185–1194.
Sanjuán, R., Moya, A., and Elena, S.F. (2004b). The distribution of fitness effects caused by single-nucleotide substitutions in an RNA virus. *Proceedings of the National Academy of Sciences USA* **101**, 8396–8401.
Sanjuán, R., Moya, A., and Elena, S.F. (2004c). The contribution of epistasis to the architecture of fitness in an RNA virus. *Proceedings of the National Academy of Sciences USA* **101**, 15376–15379.
Sanjuán, R., Cuevas, J.M., Moya, A., and Elena, S.F. (2005). Epistasis and the adaptability of an RNA virus. *Genetics* **170**, 1001–1008.
Sanjuán, R., Forment, J., and Elena, S.F. (2006a). In silico predicted robustness of viroids RNA secondary structures. I. The effect of single mutations. *Molecular Biology and Evolution* **23**, 1427–1436.
Sanjuán, R., Forment, J., and Elena, S.F. (2006b). In silico predicted robustness of viroid RNA secondary structures. II. Interaction between mutation pairs. *Molecular Biology and Evolution* **23**, 2123–2130.
Sanjuán, R., Cuevas, J.M., Furió, V., Holmes, E.C., and Moya, A. (2007). Selection for robustness in mutagenesized RNA viruses. *PLoS Genetics* **3**, e93.
Sankoff, D. (2003). Rearrangements and chromosomal evolution. *Current Opinion in Genetics and Development* **13**, 583–587.
Santiago, M.L., Rodenburg, C.M., Kamenya, S., Bibollet-Ruche, F., Gao, F., Bailes, E., Meleth, S., Soong, S.J., Kilby, J.M., Moldoveanu, Z. et al. (2002). SIVcpz in wild chimpanzees. *Science* **295**, 465–465.
Santiago, M.L., Lukasik, M., Kamenya, S., Li, Y., Bibollet-Ruche, F., Bailes, E., Muller, M.N., Emery, M., Goldenberg, D.A., Lwanga, J.S. et al. (2003). Foci of endemic simian immunodeficiency virus infection in wild-living eastern chimpanzees (*Pan troglodytes schweinfurthii*). *Journal of Virology* **77**, 7545–7562.
Santiago, M.L., Range, F., Keele, B.F., Li, Y., Bailes, E., Bibollet-Ruche, F., Fruteau, C., Noe, R., Peeters, M., Brookfield, J.F. et al. (2005). Simian immunodeficiency virus infection in free-ranging sooty mangabeys (*Cercocebus atys atys*) from the Tai Forest, Cote d'Ivoire: implications for the origin of epidemic human immunodeficiency virus type 2. *Journal of Virology* **79**, 12515–12527.
Sanz, A.I., Fraile, A., Gallego, J.M., Malpica, J.M., and García-Arenal, F. (1999). Genetic variability of natural populations of cotton leaf curl geminivirus, a single-stranded DNA virus. *Journal of Molecular Evolution* **49**, 672–681.
Sawyer, S.L., Emerman, M., and Malik, H.S. (2004). Ancient adaptive evolution of the primate antiviral DNA-editing enzyme APOBEC3G. *PLoS Biology* **2**, e275.
Sawyer, S.L., Emerman, M., and Malik, H.S. (2007). Discordant evolution of the adjacent antiretroviral genes TRIM22 and TRIM5 in mammals. *PLoS Pathogens* **3**, e197.
Scarlatti, G. (2004). Mother-to-child transmission of HIV-1: advances and controversies of the twentieth centuries. *AIDS Reviews* **6**, 67–78.

Schindler, M., Munch, J., Kutsch, O., Li, H., Santiago, M.L., Bibollet-Ruche, F., Muller-Trutwin, M.C., Novembre, F.J., Peeters, M., Courgnaud, V. *et al.* (2006). Nef-mediated suppression of T cell activation was lost in a lentiviral lineage that gave rise to HIV-1. *Cell* **125**, 1055–1067.

Scholtissek, C. (1987). Molecular aspects of the epidemiology of virus disease. *Experientia* **43**, 1197–1201.

Scholtissek, C., von Hoyningen, V., and Rott, R. (1978). Genetic relatedness between the new 1977 epidemic strains (H1N1) of influenza and human influenza strains isolated between 1947 and 1957 (H1N1). *Virology* **89**, 613–617.

Schrag, S.J., Rota, P.A., and Bellini, W.J. (1999). Spontaneous mutation rate of measles virus: direct estimation based on mutations conferring monoclonal antibody resistance. *Journal of Virology* **73**, 51–54.

Schuster, P. and Stadler, P.F. (1999). Nature and evolution of early replicons. In E. Domingo, R. Webster, and J.J. Holland, eds, *Origin and Evolution of Viruses*, pp. 1–24. Academic Press, London.

Schweizer, M., Schleer, H., Pietrek, M., Liegibel, J., Falcone, V., and Neumann-Haefelin, D. (1999). Genetic stability of foamy viruses: long-term study in an African green monkey population. *Journal of Virology* **73**, 9256–9265.

Scott, T.W., Weaver, S.C., and Mallampalli, V.L. (1994). Evolution of mosquito-borne viruses. In S.S. Morse, ed., *The Evolutionary Biology of Viruses*, pp. 293–324. Raven Press, New York.

Sentandreu, V., Jiménez-Hernández, N., Torres-Puente, M., Bracho, M.A, Valero, A., Gosalbes, M.J., Ortega, E., Moya, A., and González-Candelas, F. (2008). Evidence of recombination in intrapatient populations of hepatitis C virus. *PLoS ONE* **3**, e3239.

Shackelton, L.A. and Holmes, E.C. (2004). The evolution of large DNA viruses: combining genomic information of viruses and their hosts. *Trends in Microbiology* **12**, 458–465.

Shackelton, L.A. and Holmes, E.C. (2006). Phylogenetic evidence for the rapid evolution of human B19 erythrovirus. *Journal of Virology* **80**, 3666–3669.

Shackelton, L.A. and Holmes, E.C. (2008). The role of alternative genetic codes in virus evolution and emergence. *Journal of Theoretical Biology* **254**, 128–134.

Shackelton, L.A., Parrish, C.R., Truyen, U., and Holmes, E.C. (2005). High rate of viral evolution associated with the emergence of canine parvoviruses. *Proceedings of the National Academy of Sciences USA* **102**, 379–384.

Shankarappa, R., Margolick, J.B., Gange, S.J., Rodrigo, A.G., Upchurch, D., Farzadegan, H., Gupta, P., Rinaldo, C.R., Learn, G.H., He, X. *et al.* (1999). Consistent viral evolutionary changes associated with the progression of human immunodeficiency virus type 1 infection. *Journal of Virology* **73**, 10489–10502.

Shapiro, B., Rambaut, A., Pybus, O.G., Drummond, A., and Holmes, E.C. (2006). A phylogenetic method for detecting positive epistasis in gene sequences and its application to RNA virus evolution. *Molecular Biology and Evolution* **23**, 1724–1730.

Sharp, P.M. and Matassi, G. (1994). Codon usage and genome evolution. *Current Opinion in Genetics and Development* **4**, 851–860.

Sharp, P.M., Bailes, E., Gao, F., Beer, B.E., Hirsch, V.M., and Hahn, B.H. (2000). Origins and evolution of AIDS viruses: estimating the time-scale. *Biochemical Society Transactions* **28**, 275–282.

Sheehy, A.M., Gaddis, N.C., Choi, J.D., and Malim, M.H. (2002). Isolation of a human gene that inhibits HIV-1 infection and is suppressed by the viral Vif protein. *Nature* **418**, 646–650.

Sheridan, I., Pybus, O.G., Holmes, E.C., and Klenerman, P. (2004). High resolution phylogenetic analysis of hepatitis C virus adaptation and its relationship to disease progression. *Journal of Virology* **78**, 3447–3454.

Shih, A.C., Hsiao, T.C., Ho, M.S., and Li, W.-H. (2007). Simultaneous amino acid substitutions at antigenic sites drive influenza A hemagglutinin evolution. *Proceedings of the National Academy of Sciences USA* **104**, 6283–6288.

Shin, C.G., Taddeo, B., Haseltine, W.A., and Farnet, C.M. (1994). Genetic analysis of the human immunodeficiency virus type 1 integrase protein. *Journal of Virology* **68**, 1633–1642.

Shinya, K., Ebina, M., Yamada, S., Ono, M., Kasai, N., and Kawaoka, Y. (2006). Avian flu: influenza virus receptors in the human airway. *Nature* **440**, 435–436.

Shriner, D., Shankarappa, R., Jensen, M.A., Nickle, D.C., Mittler, J.E., Margolick, J.B., and Mullins, J.I. (2004). Influence of random genetic drift on human immunodeficiency virus type 1 *env* evolution during chronic infection. *Genetics* **166**, 1155–1164.

Shurtleff, A.C., Beasley, D.W.C., Chen, J.J.Y., Ni, H., Suderman, M.T., Wang, H., Xu, R., Wand, E., Weaver, S.C., Watts, D.M. *et al.* (2001). Genetic variation in the 3' non-coding region of dengue viruses. *Virology* **281**, 75–87.

Sierra, S., Dávila, M., Lowenstein, P.R., and Domingo, E. (2000). Response of foot-and-mouth disease virus to increased mutagenesis. Influence of viral load and fitness in loss of infectivity. *Journal of Virology* **74**, 8316–8323.

Silander, O.K., Weinreich, D.M., Wright, K.M., O'Keefe, K.J., Rang, C.U., Turner, P.E., and Chao, L. (2005). Widespread genetic exchange among terrestrial bacteriophages. *Proceedings of the National Academy of Sciences USA* **102**, 19009–19014.

Silvestri, G., Sodora, D.L., Koup, R.A., Paiardini, M., O'Neil, S.P., McClure, H.M., Staprans, S.I., and Feinberg, M.B. (2003). Nonpathogenic SIV infection of sooty mangabeys is characterized by limited bystander immunopathology despite chronic high-level viremia. *Immunity* **18**, 441–452.

Simmonds, P. (2004). Genetic diversity and evolution of hepatitis C virus – 15 years on. *Journal of General Virology* **85**, 3173–3188.

Simmonds, P. and Smith, D.B. (1999). Structural constraints on RNA virus evolution. *Journal of Virology* **73**, 5787–5794.

Simmonds, P., Tuplin, A., and Evans, D.J. (2004). Detection of genome-scale ordered RNA structure (GORS) in genomes of positive-stranded RNA viruses: implications for virus evolution and host persistence. *RNA* **10**, 1337–1351.

Simmons, H.E., Holmes, E.C., and Stephenson, A.G. (2008). Rapid evolutionary dynamics of zucchini yellow mosaic virus. *Journal of General Virology* **89**, 1081–1085.

Simon, J.H., Southerling, T.E., Peterson, J.C., Meyer, B.E., and Malim, M.H. (1995). Complementation of *vif*-defective human immunodeficiency virus type 1 by primate, but not nonprimate, lentivirus *vif* genes. *Journal of Virology* **69**, 4166–4172.

Simonsen, L., Viboud, C., Grenfell, B.T., Dushoff, J., Jennings, L., Smit, M., Macken, C., Hata, M., Gog, J., Miller, M.A., and Holmes, E.C. (2007). The rapid global spread of reassortant human influenza A/H3N2 viruses conferring adamantane resistance. *Molecular Biology and Evolution* **24**, 1811–1820.

Sironen, T., Kallio, E.R., Vaheri, A., Lundkvist, A., and Plyusnin, A. (2008). Quasispecies dynamics and fixation of a synonymous mutation in hantavirus transmission. *Journal of General Virology* **89**, 1309–1313.

Sittisombut, N., Sistayanarain, A., Cardosa, M.J., Salminen, M., Damrongdachakul, S., Kalayanarooj, S., Rojanasuphot, S., Supawadee, J., and Maneekarn, N. (1997). Possible occurrence of a genetic bottleneck in dengue serotype 2 viruses between the 1980 and 1987 epidemic seasons in Bangkok, Thailand. *American Journal of Tropical Medicine and Hygiene* **57**, 100–108.

Slemons, R.D., Johnson, D.C., Osborn, J.S., and Hayes, F. (1974). Type-A influenza viruses isolated from wild free-flying ducks in California. *Avian Diseases* **18**, 119–124.

Smallwood, S., Cevik, B., and Moyer, S.A. (2002). Intragenic complementation and oligomerization of the L subunit of the sendai virus RNA polymerase. *Virology* **304**, 235–245.

Smith, D.B., McAllister, J., Casino, C., and Simmonds, P. (1997). Virus "quasispecies": making a mountain out of a molehill? *Journal of General Virology* **78**, 1511–1519.

Smith, D.J., Lapedes, A.S., de Jong, J.C., Bestebroer, T.M., Rimmelzwaan, G.F., Osterhaus, A.D.M.E., and Fouchier, R.A.M. (2004). Mapping the antigenic and genetic evolution of influenza virus. *Science* **305**, 371–376.

Smith, D.R., Adams, A.P., Kenney, J.L., Wang, E., and Weaver, S.C. (2008). Venezuelan equine encephalitis virus in the mosquito vector *Aedes taeniorhynchus*: infection initiated by a small number of susceptible epithelial cells and a population bottleneck. *Virology* **372**, 176–186.

Sniegowski, P.D., Gerrish, P.J., Johnson, T., and Shaver, A. (2000). The evolution of mutation rates: separating causes from consequences. *BioEssays* **22**, 1057–1066.

Snijder, E.J., Bredenbeek, P.J., Dobbe, J.C., Thiel, V., Ziebuhr, J., Poon, L.L., Guan, Y., Rozanov, M., Spaan, W.J., and Gorbalenya, A.E. (2003). Unique and conserved features of genome and proteome of SARS-coronavirus, an early split-off from the coronavirus group 2 lineage. *Journal of Molecular Biology* **331**, 991–1004.

Sonoda, S., Li, H.C., Cartier, L., Nunez, L., and Tajima, K. (2000). Ancient HTLV type 1 provirus DNA of Andean mummy. *AIDS Research and Human Retroviruses* **16**, 1753–1756.

Soto-Ramirez, L.E., Renjifo, B., McLane, M.F., Marlink, R., O'Hara, C., Sutthent, R., Wasi, C., Vithayasai, P., Vithayasai, V., Apichartpiyakul, C. *et al.* (1996). HIV-1 Langerhans' cell tropism associated with heterosexual transmission of HIV. *Science* **271**, 1291–1293.

Spann, K.M., Collins, P.L., and Teng, M.N. (2003). Genetic recombination during coinfection of two mutants of human respiratory syncytial virus. *Journal of Virology* **77**, 11201–11211.

Stallknecht, D.E. and Shane, S.M. (1988). Host range of avian influenza virus in free-living birds. *Veterinary Research Communications* **12**, 125–141.

Stanhope, M.J., Brown, J.R., and Amrine-Madsen, H. (2004). Evidence from the evolutionary analysis of nucleotide sequences for a recombinant history of SARS-CoV. *Infection, Genetics and Evolution* **4**, 15–19.

Starkman, S., MacDonald, D.M., Lewis, J.C.M., Holmes, E.C., and Simmonds, P. (2003). Geographic and species association of hepatitis B virus genotypes in non-human primates. *Virology* **314**, 381–393.

Stavrinides, J. and Guttman, D.S. (2004). Mosaic evolution of the Severe Acute Respiratory Syndrome coronavirus. *Journal of Virology* **78**, 76–82.

Steele, J.H. and Fernandez, P.J. (1991). History of rabies and global aspects, In G.M. Baer, ed., *The Natural History of Rabies*, 2nd edn, pp. 1–26. CRC Press, Boca Raton, FL.

Steinhauer, D.A., de la Torre, J.C., and Holland, J.J. (1989). High nucleotide substitution error frequencies in clonal pools of vesicular stomatitis virus. *Journal of Virology* **63**, 2063–2071.

Stevens, J., Blixt, O., Tumpey, T.M., Taubenberger, J.K., Paulson, J.C., and Wilson, I.A. (2006). Structure and receptor specificity of the hemagglutinin from an H5N1 influenza virus. *Science* **312**, 404–410.

Strassman, B.I. and Dunbar, R.I.M. (1999). Human evolution and disease: putting the stone age in perspective. In S.C. Stearns, ed., *Evolution in Health and Disease*, pp. 91–101. Oxford University Press, Oxford.

Stremlau, M., Owens, C.M., Perron, M.J., Kiessling, M., Autissier, P., and Sodroski, J. (2004). The cytoplasmic body component TRIM5alpha restricts HIV-1 infection in Old World monkeys. *Nature* **427**, 848–853.

Summers, J. and Litwin, S. (2006). Examining the theory of error catastrophe. *Journal of Virology* **80**, 20–26.

Sun, F.J. and Caetano-Anollés, G. (2008). Evolutionary patterns in the sequence and structure of transfer RNA: early origins of archaea and viruses. *PLoS Computational Biology* **4**, e1000018.

Suzuki, Y. (2006a). Ancient positive selection on CD155 as a possible cause for susceptibility to poliovirus infection in simians. *Gene* **373**, 16–22.

Suzuki, Y. (2006b). Natural selection on the influenza virus genome. *Molecular Biology and Evolution* **23**, 1902–1911.

Suzuki, Y. and Nei, M. (2002a). Simulation study of the reliability and robustness of the statistical methods for detecting positive selection at single amino acid sites. *Molecular Biology and Evolution* **19**, 1865–1869.

Suzuki, Y. and Nei, M. (2002b). Origin and evolution of influenza virus hemagglutinin genes. *Molecular Biology and Evolution* **19**, 501–509.

Suzuki, Y., Katayama, K., Fukushi, S., Kageyama, T., Oya, A., Okamura, H., Tanaka, Y., Mizokami, Y., and Gojobori, T. (1999). Slow evolutionary rate of GB virus C/hepatitis G virus. *Journal of Molecular Evolution* **48**, 383–389.

Swetina, J. and Schuster, P. (1982). A model for polynucleotide replication. *Biophysical Chemistry* **16**, 329–345.

Switzer, W.M., Salemi, M., Shanmugam, V., Gao, F., Cong, M.-E., Kuiken, C., Bhullar, V., Beer, B., Vallet, D., Gautier-Hion, A. *et al.* (2005). Ancient co-speciation of simian foamy viruses and primates. *Nature* **434**, 376–380.

Swofford, D.L. (2003). *PAUP*. Phylogenetic analysis using parsimony (*and other methods)*, version 4. Sinauer Associates, Sunderland, MA.

Szathmáry, E. (2005). Life: in search of the simplest cell. *Nature* **433**, 469–470.

Szathmáry, E. and Demeter, L. (1987). Group selection of early replicators and the origin of life. *Journal of Theoretical Biology* **128**, 463–486.

Taber, S.W. and Pease, C.M. (1990). Paramyxovirus phylogeny: tissue tropism evolves slower than host specificity. *Evolution* **44**, 435–438.

Taddei, F., Radman, M., Maynard Smith, J., Toupance, B., Gouyon, P.H., and Godelle, B. (1997). Role of mutator alleles in adaptive evolution. *Nature* **387**, 700–702.

Tam, P.E. and Messner, R.P. (1999). Molecular mechanisms of coxsackievirus persistence in chronic inflammatory myopathy: viral RNA persists through formation of a double-stranded complex without associated genomic mutations or evolution. *Journal of Virology* **73**, 10113–10121.

Tanne, E. and Sela, I. (2005). Occurrence of a DNA sequence of a non-retro RNA virus in a host plant genome and its expression: evidence for recombination between viral and host RNAs. *Virology* **332**, 614–622.

Tarlinton, R.E., Meers, J., and Young, P.R. (2006). Retroviral invasion of the koala genome. *Nature* **442**, 79–81.

Taubenberger, J.K., Reid, A.H., Frafft, A.E., Bijwaard, K.E., and Fanning, TG. (1997). Initial genetic characterization of the 1918 "Spanish" influenza virus. *Science* **275**, 1793–1796.

Taubenberger, J.K., Reid, A.H., Janczewksi, T.A., and Fanning, T.G. (2001). Integrating historical, clinical and molecular genetic data in order to explain the origin and virulence of the 1918 Spanish influenza virus. *Philosophical Transactions of the Royal Society of London Series B Biological Sciences* **356**, 1857–1859.

Taubenberger, J.K., Reid, A.H., Lourens, R.M., Wang, R., Jin, G., and Fanning, T.G. (2005). Characterization of the 1918 influenza virus polymerase genes. *Nature* **437**, 889–893.

Teycheney, P.Y., Laboureau, N., Iskra-Caruana, M.L., and Candresse, T. (2005). High genetic variability and evidence for plant-to-plant transfer of Banana mild mosaic virus. *Journal of General Virology* **86**, 3179–3187.

Thorne, J.L. (2007). Protein evolution constraints and model-based techniques to study them. *Current Opinion in Structural Biology* **17**, 337–341.

Thu, H.M., Lowry, K., Myint, T.T., Shwe, T.N., Han, A.M., Khin, K.K., Thant, K.Z., Thein, S., and Aaskov, J. (2004). Myanmar dengue outbreak associated with displacement of serotypes 2, 3, and 4 by dengue 1. *Emerging Infectious Diseases* **10**, 593–597.

Thurner, C., Witwer, C., Hofacker, I.L., and Stadler, P.F. (2004). Conserved RNA secondary structures in *Flaviviridae* genomes. *Journal of General Virology* **85**, 1113–1124.

Trachtenberg, E., Korber, B., Sollars, C., Kepler, T.B., Hraber, P.T., Hayes, E., Funkhouser, R., Fugate, M., Theiler, J., Hsu, Y.S. *et al.* (2003). Advantage of rare HLA supertype in HIV disease progression. *Nature Medicine* **9**, 928–935.

Tristem, M., Marshall, C., Karpas, A., Petrik, J., and Hill, F. (1990). Origin of *vpx* in lentiviruses. *Nature* **347**, 341–342.

Troyer, J.L., Vandewoude, S., Pecon-Slattery, J., McIntosh, C., Franklin, S., Antunes, A., Johnson, W., and O'Brien, S.J. (2008). FIV cross-species transmission: an evolutionary prospective. *Veterinary Immunology and Immunopathology* **123**, 159–166.

Truyen, U., Evermann, J.F., Vieler, E., and Parrish, C.R. (1996). Evolution of canine parvovirus involved loss and gain of feline host range. *Virology* **215**, 186–189.

Tsetsarkin, K.A., Vanlandingham, D.L., McGee, C.E., and Higgs, S. (2007). A single mutation in chikungunya virus affects vector specificity and epidemic potential. *PLoS Pathogens* **3**, e201.

Turner, P.E. and Chao, L. (1998). Sex and the evolution of intrahost competition in RNA virus φ6. *Genetics* **150**, 523–532.

Turner, P.E. and Chao, L. (1999). Prisoner's dilemma in an RNA virus. *Nature* **398**, 441–443.

Twiddy, S.S., Farrar, J.F., Chau, N.V., Wills, B., Gould, E.A., Gritsun, T., Lloyd, G., and Holmes, E.C. (2002). Phylogenetic relationships and differential selection pressures among genotypes of dengue-2 virus. *Virology* **298**, 63–72.

Twiddy, S.S., Holmes, E.C., and Rambaut, A. (2003). Inferring the rate and time-scale of dengue virus evolution. *Molecular Biology and Evolution* **20**, 122–129.

Tzeng, W.P. and Frey, T.K. (2003). Complementation of a deletion in the rubella virus p150 nonstructural protein by the viral capsid protein. *Journal of Virology* **77**, 9502–9510.

Umemura, T., Tanaka, Y., Kiyosawa, K., Alter, H.J., and Shih, J.W.-K. (2002). Observation of positive selection within hypervariable regions of a newly identified DNA virus (SEN virus). *FEBS Letters* **510**, 171–174.

UNAIDS (2007). *2007 AIDS Epidemic Update*. www.unaids.org/en/KnowledgeCentre/HIVData/EpiUpdate/EpiUpdArchive/2007default.asp

Valli, A., López-Moya, J.J., and García, J.A. (2007). Recombination and gene duplication in the evolutionary diversification of P1 proteins in the family *Potyviridae*. *Journal of General Virology* **88**, 1016–1028.

Vandamme, A.-M., Bertazzoni, U., and Salemi, M. (2000). Evolutionary strategies of human T-cell lymphotropic virus type II. *Gene* **261**, 171–180.

van Gils, J.A., Munster, V.J., Radersma, R., Liefhebber, D., Fouchier, R.A.M., and Klaassen, M. (2007). Hampered foraging and migratory performance in swans infected with low-pathogenic avian influenza A virus. *PLoS ONE* **2**, e184.

Van Heuverswyn, F., Li, Y., Neel, C., Bailes, E., Keele, B.F., Liu, W., Loul, S., Butel, C., Liegeois, F., Bienvenue, Y. *et al*. (2006). Human immunodeficiency viruses: SIV infection in wild gorillas. *Nature* **444**, 164.

van Hulten, M.C., Goldbach, R.W., and Vlak, J.M. (2000). Three functionally diverged major structural proteins of white spot syndrome virus evolved by gene duplication. *Journal of General Virology* **81**, 2525–2529.

van Riel, D., Munster, V.J., de Wit, E., Rimmelzwaan, G.F., Fouchier, R.A.M., Osterhaus, A.D.M.E., and Kuiken, T. (2006). H5N1 virus attachment to lower respiratory tract. *Science* **312**, 399.

Vartanian, J.-P., Meyerhans, A., Asjo, B., and Wain-Hobson, S. (1991). Selection, recombination, and G-A hypermutation of human immunodeficiency virus type 1 genomes. *Journal of Virology* **65**, 1779–1788.

Vasilakis, N., Holmes, E.C., Fokam, E.B., Faye, O., Diallo, M., Sall, A.A., and Weaver, S.C. (2007a). Evolutionary processes among sylvatic dengue-2 viruses. *Journal of Virology* **81**, 9591–9595.

Vasilakis, N., Shell, E.J., Fokam, E.B., Mason, P.W., Hanley, K.A., Estes, D.M., and Weaver, S.C. (2007b). Potential of ancestral sylvatic dengue-2 viruses to re-emerge. *Virology* **358**, 402–412.

Vasilakis, N., Tesh, R.B., and Weaver, S.C. (2008). Sylvatic dengue virus type 2 activity in humans, Nigeria, 1966. *Emerging Infectious Diseases* **14**, 502–504.

Vaughn, D.W., Green, S., Kalayanarooj, S., Innis, B.L., Nimmannitya, S., Suntayakorn, S., Endy, T.P., Raengsakulrach, B., Rothman, A.L., Ennis, F.A., and Nisalak, A. (2000). Dengue viremia titer, antibody response pattern, and virus serotype correlate with disease severity. *Journal of Infectious Diseases* **181**, 2–9.

Viboud, C., Alonso, W.J., and Simonsen, L. (2006a). Influenza in tropical regions. *PLoS Medicine* **3**, e89.

Viboud, C., Bjørnstad, O.N., Smith, D.L., Simonsen, L., Miller, M.A., and Grenfell, B.T. (2006b). Synchrony, waves, and spatial hierarchies in the spread of influenza. *Science* **312**, 447–451.

Vidal, N., Mulanga-Kabeya, C., Nzilambi, N., Robertson, D.L., Ilunga, W., Sema, H., Tshimanga, K., Bongo, B., Delaporte, E., and Peeters, M. (2000). Unprecedented degree of HIV-1 group M genetic diversity in the Democratic Republic of Congo suggests that the HIV-1 pandemic originated in Central Africa. *Journal of Virology* **74**, 10498–10507.

Vieth, S., Torda, A.E., Asper, M., Schmitz, H., and Günther, S. (2004). Sequence analysis of L RNA of Lassa virus. *Virology* **318**, 153–168.

Vignuzzi, M., Stone, J.K., Arnold, J.J., Cameron, C.E., and Andino, R. (2005). Quasispecies diversity determines pathogenesis through cooperative interactions in a viral population. *Nature* **439**, 344–348.

Vignuzzi, M., Wendt, E., and Andino, R. (2008). Engineering attenuated virus vaccines by controlling replication fidelity. *Nature Medicine* **14**, 154–161.

Wain, L.V., Bailes, E., Bibollet-Ruche, F., Decker, J.M., Keele, B.F., Van Heuverswyn, F., Li, Y., Takehisa, J., Ngole, E.M., Shaw, G.M. *et al*. (2007). Adaptation of HIV-1 to its human host. *Molecular Biology and Evolution* **4**, 1853–1860.

Walker, P.J., Byrne, K.A., Riding, G.A., Cowley, J.A., Wang, Y., and McWilliam, S. (1992). The genome of bovine ephemeral fever rhabdovirus contains two related glycoprotein genes. *Virology* **191**, 49–61.

Walsh, C.P. and Xu, G.L. (2006). Cytosine methylation and DNA repair. *Current Topics in Microbiology and Immunology* **301**, 283–315.

Walsh, P.D., Biek, R., and Real, L.A. (2005). Wave-like spread of Ebola Zaire. *PLoS Biology* **3**, e371.

Wang, E., Ni, H., Xu, R., Barrett, A.D.T., Watowich, S.J., Gubler, D.J., and Weaver, S.C. (2000). Evolutionary relationships of endemic/epidemic and sylvatic dengue viruses. *Journal of Virology* **74**, 3227–3234.

Wang, W.-K., Sung, T.-L., Lee, C.-N., Lin, T.-Y., and King, C.-C. (2002). Dengue type 3 virus in plasma is a population of closely related genomes: quasispecies. *Journal of Virology* **76**, 4662–4665.

Wang, X.-H., Aliyari, R., Li, W.-X., Li, H.-W., Kim, K., Carthew, R., Atkinson, P., and Ding, S.-W. (2006). RNA interference directs innate immunity against viruses in adult *Drosophila*. *Science* **312**, 452–454.

Wang, Y. and Walker, P.J. (1993). Adelaide river rhabdovirus expresses consecutive glycoprotein genes as polycistronic mRNAs: new evidence of gene duplication as an evolutionary process. *Virology* **195**, 719–731.

Wearing, H.J. and Rohani, P. (2006). Ecological and immunological determinants of dengue epidemics. *Proceedings of the National Academy of Sciences USA* **103**, 11802–11807.

Weaver, S.C. and Barrett, A.D.T. (2004). Transmission cycles, host range, evolution and emergence of arboviral disease. *Nature Reviews Microbiology* **2**, 789–801.

Weaver, S.C., Brault, A.C., Kang, W., and Holland, J.J. (1999). Genetic and fitness changes accompanying adaptation of an arbovirus to vertebrate and invertebrate cells. *Journal of Virology* **73**, 4316–4326.

Webby, R.J. and Webster, R.G. (2001). Emergence of influenza A viruses. *Philosophical Transactions of the Royal Society of London Series B Biological Sciences* **356**, 1817–1828.

Webby, R.J., Hoffman, E., and Webster, R.G. (2004). Molecular constraints to interspecies transmission of viral pathogens. *Nature Medicine* **10**, S77-S81.

Webster, R.G., Yakhno, M., Hinshaw, V.S., Bean, W.J., and Murti, K.G. (1978). Intestinal influenza: replication and characterization of influenza viruses in ducks. *Virology* **84**, 268–278.

Webster, R.G., Bean, W.J., Gorman, O.T., Chambers, T.M., and Kawaoka, Y. (1992). Evolution and ecology of influenza A viruses. *Microbiological Reviews* **56**, 152–179.

Wei, X., Decker, J.M., Wang, S., Hui, H., Kappes, J.C., Wu, X., Salazar-Gonzalez, J.F., Salazar, M.G., Kilby, J.M., Saag, M.S. *et al.* (2003). Antibody neutralization and escape by HIV-1. *Nature* **422**, 307–312.

Weiss, R.A. (2001). The Leeuwenshoek lecture 2001. Animal origins of human infectious disease. *Philosophical Transactions of the Royal Society of London Series B Biological Sciences* **356**, 957–977.

Wertheim, J.O. and Worobey, M. (2007). A challenge to the ancient origin of SIVagm based on African Green monkey mitochondrial genomes. *PLoS Pathogens* **3**, e95.

Wichman, H.A., Badgett, M.R., Scott, L.A., Boulianne, C.M., and Bull, J.J. (1999). Different trajectories of parallel evolution during viral adaptation. *Science* **285**, 422–424.

Wiehe, T. (1997). Model dependency of error thresholds: the role of fitness functions and contrasts between the finite and infinite sites models. *Genetical Research* **69**, 127–136.

Wilke, C.O. (2003). Probability of fixation of an advantageous mutant in a viral quasispecies. *Genetics* **163**, 467–474.

Wilke, C.O. (2005). Quasispecies theory in the context of population genetics. *BMC Evolutionary Biology* **5**, 44.

Wilke, C.O. and Adami, C. (2001). Interaction between directional epistasis and average mutational effects. *Proceedings of the Royal Society of London Series B Biological Sciences* **268**, 1469–1474.

Wilke, C.O. and Adami, C. (2002). The biology of digital organisms. *Trends in Ecology and Evolution* **17**, 528–532.

Wilke, C.O. and Novella, I.S. (2003). Phenotypic mixing and hiding may contribute to memory in viral quasispecies. *BMC Microbiology* **3**, 11.

Wilke, C.O., Wang, J.L., Ofria, C., Lenski, R.E., and Adami, C. (2001). Evolution of digital organisms at high mutation rates leads to survival of the flattest. *Nature* **412**, 331–333.

Wilke, C.O., Lenski, R.E., and Adami, C. (2003). Compensatory mutations cause excess of antagonistic epistasis in RNA secondary structure folding. *BMC Evolutionary Biology* **3**, 3.

Wilkins, C., Dishongh, R., Moore, S.C., Whitt, M.A., Chow, M., and Machaca, K. (2005). RNA interference is an antiviral defence mechanism in *Caenorhabditis elegans*. *Nature* **436**, 1044–1047.

Williamson, S. (2003). Adaptation in the *env* gene of HIV-1 and evolutionary theories of disease progression. *Molecular Biology and Evolution* **20**, 1318–1325.

Wittke, V., Robb, T.E., Thu, H.M., Nimmannitya, S., Kalayanrooj, S., Vaughn, D.W., Endy, T.P., Holmes, E.C., and Aaskov, J.G. (2002). Extinction and rapid emergence of strains of dengue 3 virus during an interepidemic period. *Virology* **301**, 148–156.

Wittmann, T.J., Biek, R., Hassanin, A., Rouquet, P., Reed, P., Yaba, P., Pourrut, X., Real, L.A., Gonzalez, J.P., and Leroy, E.M. (2007). Isolates of Zaire ebolavirus from wild apes reveal genetic lineage and recombinants. *Proceedings of the National Academy of Sciences USA* **104**, 17123–17127.

Woelk, C.H. and Holmes, E.C. (2002). Reduced positive selection in vector-borne RNA viruses. *Molecular Biology and Evolution* **19**, 2333–2336.

Woelk, C.H., Jin, L., Holmes, E.C., and Brown, D.W.G. (2001). Immune and artificial selection in the hemagglutin (H) glycoprotein of measles virus. *Journal of General Virology* **82**, 2463–2474.

Woelk, C.H., Pybus, O.G., Jin, L., Brown, D.W.G., and Holmes, E.C. (2002). Increased positive selection pressure in persistent (SSPE) versus acute measles virus infections. *Journal of General Virology* **83**, 1419–1430.

Wolf, Y.I., Viboud, C., Holmes, E.C., Koonin, E.V., and Lipman, D.J. (2006). Long intervals of stasis punctuated by bursts of positive selection in the seasonal evolution of influenza A virus. *Biology Direct* **1**, 34.

Wolfe, N.D., Kilbourn, A.M., Karesh, W.B., Rahman, H.A., Bosi, E.J., Cropp, B.E., Andau, M., Spielman, A., and Gubler, D.J. (2001). Sylvatic transmission of arboviruses among Bornean orangutans. *American Journal of Tropical Medicine and Hygiene* **64**, 310–316.

Wolfe, N.D., Switzer, W.M., Carr, J.K., Bhullar, V.B., Shanmugam, V., Tamoufe, U., Prosser, A.T., Torimiro, J.N., Wright, A., Mpoudi-Ngole, E. *et al.* (2004). Naturally acquired simian retrovirus infections in central African hunters. *Lancet* **363**, 932–937.

Wolfe, N.D., Daszak, P., Kilpatrick, A.M., and Burke, D.S. (2005). Bushmeat hunting, deforestation, and prediction of zoonoses emergence. *Emerging Infectious Diseases* **11**, 1822–1827.

Wolfe, N.D., Dunavan, C.P., and Diamond, J. (2007). Origins of major human infectious diseases. *Nature* **447**, 279–283.

Woolhouse, M.E.J. (2002). Population biology of emerging and re-emerging pathogens. *Trends in Microbiology* **10**, S3–S7.

Woolhouse, M.E.J., Taylor, L.H., and Haydon, D.T. (2001). Population biology of multihost pathogens. *Science* **292**, 1109–1112.

Woolhouse, M.E.J., Haydon, D.T., and Antia, R. (2005). Emerging pathogens: The epidemiology and evolution of species jumps. *Trends in Ecology and Evolution* **20**, 238–244.

World Health Organization (2003). *Fact Sheet no. 211. Influenza.* www.who.int/mediacentre/factsheets/fs211/.

Worobey, M. and Holmes, E.C. (1999). Evolutionary aspects of recombination in RNA viruses. *Journal of General Virology* **80**, 2535–2544.

Worobey, M., Rambaut, A., Pybus, O.P., and Robertson, D.L. (2002). Questioning the evidence for genetic recombination in the 1918 "Spanish flu" virus. *Science* **296**, 211.

Worobey, M., Gemmel, M., Teuwen, D.E., Haselkorn, T., Kunstman, K., Bunce, M., Muyembe, J.-J., Kabongo, J.-M.M., Kalengayi, R.M., Van Mark, E. *et al.* (2008). Direct evidence of extensive diversity of HIV-1 in Kinshasa by 1960. *Nature* **455**, 661–664.

Xia, Y., Bjørnstad, O.N, and Grenfell, B.T. (2004). Measles metapopulation dynamics: a gravity model for epidemiological coupling and dynamics. *American Naturalist* **164**, 267–281.

Xiong, Y. and Eickbush, T.H. (1990). Origin and evolution of retroelements based upon their reverse transcriptase sequences. *EMBO Journal* **9**, 3353–3362.

Yamada, S,. Suzuki, Y., Suzuki, T., Le, M.Q., Nidom, C.A., Sakai-Tagawa, Y., Muramoto, Y., Ito, M., Kiso, M., Horimoto, T. *et al.* (2006). Haemagglutinin mutations responsible for the binding of H5N1 influenza A viruses to human-type receptors. *Nature* **444**, 378–382.

Yamashita, M., Krystal, M., Fitch, M., and Palese, P. (1988). Influenza B virus evolution: co-circulating lineages and comparison of evolutionary pattern with those of influenza A and C viruses. *Virology* **163**, 112–122.

Yang, Y., Yi, M., Evans, D.J., Simmonds, P., and Lemon, S.M. (2008). Identification of a Conserved RNA Replication Element (cre) within the 3Dpol-coding sequence of Hepatoviruses. *Journal of Virology* **82**, 10118–10128.

Yang, Z. (1996). Among-site variation and its impact on phylogenetic analyses. *Trends in Ecology and Evolution* **11**, 367–371.

Yang, Z. and Bielawski, J.P. (2000). Statistical methods for detecting molecular adaptation. *Trends in Ecology and Evolution* **15**, 496–502.

Yeh, S.-H., Wang, H.-Y., Tsai, C.-Y., Kao, C.-L., Yang, J.-Y., Liu, H.-W., Su, I.-J., Tsai, S.-F., Chen, D.-S., Chen, P.-J., and the National Taiwan University SARS Research Team (2004). Characterization of severe acute respiratory syndrome coronavirus genomes in Taiwan: molecular epidemiology and genome evolution. *Proceedings of the National Academy of Sciences USA* **101**, 2542–2547.

Yohn, C.T., Jiang, Z., McGrath, S.D., Hayden, K.E., Khaitovich, P., Johnson, M.E., Eichler, M.Y., McPherson, J.D., Zhao, S., Pääbo, S., and Eichler, E.E. (2005). Lineage-specific expansions of retroviral insertions within the genomes of African great apes but not humans and orangutans. *PLoS Biology* **3**, e110.

Yuste, E., Sánchez-Palomino, S., Casado, C., Domingo, E., and López-Galíndez, C. (1999). Drastic fitness loss in human immunodeficiency virus type 1 upon serial bottleneck events. *Journal of Virology* **73**, 2745–2751.

Zampino, R., Pickering, J., Iqbal, M., Gaud, U., Thomas, H.C., and Karayiannis, P. (1999). Hepatitis G virus/GBV-C persistence: absence of hypervariable E2 region and genetic analysis of viral quasispecies in serum and lymphocytes. *Journal of Viral Hepatitis* **6**, 209–218.

Zanotto, P.M. de A., Gibbs, M.J., Gould, E.A., and Holmes, E.C. (1996a). A reassessment of the higher taxonomy of viruses based on RNA polymerases. *Journal of Virology* **70**, 6083–6096.

Zanotto, P.M. de A., Gould, E.A., Gao, G.F., Harvey, P.H., and Holmes, E.C. (1996b). Population dynamics of flaviviruses revealed by molecular phylogenies. *Proceedings of the National Academy of Sciences USA* **93**, 548–553.

Zanotto, P.M. de A., Kallas, E.G., de Souza, R.F., and Holmes, E.C. (1999). Genealogical evidence for positive selection in the *nef* gene of HIV-1. *Genetics* **153**, 1077–1089.

Zeldovich, K.B., Chen, P., and Shakhnovich, E.I. (2007). Protein stability imposes limits on organism complexity and speed of molecular evolution. *Proceedings of the National Academy of Sciences USA* **104**, 16152–16157.

Zhang, C., Mammen, M.P., Jr, Chinnawirotpisan, P., Klungthong, C., Rodpradit, P., Monkongdee, P., Nimmannitya, S., Kalayanarooj, S., and Holmes, E.C. (2005). Clade replacements in dengue virus serotypes 1 and 3 are associated with changing serotype prevalence. *Journal of Virology* **79**, 15123–15130.

Zhang, L.Q., MacKenzie, P., Cleland, A., Holmes, E.C., Leigh Brown, A.J., and Simmonds, P. (1993). Selection for specific sequences in the external envelope protein of HIV-1 upon primary infection. *Journal of Virology* **67**, 3345–3356.

Zhang, T., Breitbart, M., Lee, W.H., Run, J.Q., Wei, C.L., Soh, S.W., Hibberd, M.L., Liu, E.T., Rohwer, F., and Ruan, Y. (2006). RNA viral community in human feces: Prevalence of plant pathogenic viruses. *PLoS Biology* **4**, e3.

Zhou, Y. and Holmes, E.C. (2007). Bayesian estimates of the evolutionary rate and age of hepatitis B virus. *Journal of Molecular Evolution* **65**, 197–205.

Zhou, Y., Mammen, M.P., Jr, Chinnawirotpisan, P., Klungthong, C., Vaughn, D.W., Nimmannitya, S., Kalayanarooj, S., Holmes, E.C., and Zhang, C. (2006). Comparative analysis reveals no consistent association between the secondary structure of the 3'UTR of dengue viruses and disease syndrome. *Journal of General Virology* **87**, 2595–2603.

Zhu, T.F., Mo, H.M., Wang, N., Nam, D.S., Cao, Y.Z., Koup, R.A., and Ho, D.D. (1993). Genotypic and phenotypic characterization of HIV-1 in patients with primary infection. *Science* **261**, 1179–1181.

Zlateva, K.T., Vijgen, L., Dekeersmaeker, N., Naranjo, C., and Van Ranst, M. (2007). Subgroup prevalence and genotype circulation patterns of human respiratory syncytial virus in Belgium during ten successive epidemic seasons. *Journal of Clinical Microbiology* **45**, 3022–3030.

Index

acute infection 52, 56, 60, 83–84, 91, 103, 135–138, 154–155, 199
adaptation
   constraints to 7–8, 10, 16, 28, 40, 46, 48, 50, 55, 60–61, 65, 82, 105–106, 112, 114, 123, 139, 159, 165, 181
   off-the-shelf 144
   tailor-made 144
agriculture 2, 154
allele fixation 59–61
ambisense genome 11
ancient DNA 44
ancient virus world 26–28
APOBEC3G 39, 65–66, 69
arbovirus 140, 181
archaebacteria (*Archaea*) 9, 19, 24–27
arenavirus 11, 13, 33, 50
Arteriviridae 29, 107–108
asymptomatic infection 43, 157, 190

bacteria 1, 7, 9, 14, 16–18, 24–25, 38, 47, 73, 80, 113, 117
   Buchnera sp. 17–18
   Escherichia coli 1
   Mycobacterium tuberculosis 155
   plasmid 19, 27, 31–35
bacteriophage 3, 9, 14, 18, 29, 36, 68, 113, 119, 199
   Leviviridae 113
   φ6 94
   φX174 38
   Qβ 37–38, 87, 93
baculovirus 24
basic reproductive number (R) 124, 133, 136–137, 139, 142, 154, 172
Bayesian skyline plot 57, 134
Beijerinck, Martinus 9
birnavirus 26
birth-death process 124–130, 133, 168, 176, 186–187

Black, Francis 137
BLAST algorithm 32–33
blood-borne transmission 151, 168
Bunyaviridae 25, 33, 43, 118
Burch, Christina 94
bushmeat 155

capsid protein 11, 17, 19, 26, 29, 49, 60–61, 114, 116–117, 119–120, 124
   encapsidation 67, 112
   jelly-roll capsid (JRC) 19, 26, 28
caulimovirus 11, 26, 31, 120
cell fusing agent virus 118
cell lysis 42, 138
cell receptor 60, 124, 140–146, 168
   angiotensin-converting enzyme 2 (ACE2) 144
   CD4 168, 173
   chemokine 168
   sialic acid 145
Chao, Lin 94
chimpanzee (*Pan troglodytes*) 130, 140, 144, 148, 168–169, 171–176
circovirus 10, 45, 105
clade replacement 186–188
closterovirus 44, 118
clonal evolution 61, 159
clonal interference 53, 93
coalescent 56, 132–135, 177
co-divergence 24, 40, 42–45, 135–141, 175
codon usage bias 67–70
consensus sequence 52, 73, 76–77, 83–84, 100
convergent evolution 19, 28, 82, 144, 199
complementation 7, 74–78, 86, 94, 111, 114
compensatory mutation 23, 54, 61, 77, 80–81
Coronaviridae 1, 3, 10, 29, 48, 50, 104, 106–113, 118–119, 144, 148
covarion substitution model 176
critical community size (CCS) 136–137, 139, 153–154, 190

cross-immunity 59, 191
Cystoviridae 113

deamination 45, 65, 107
defective interfering (DI) particle 74–75, 112, 138
deforestation 153, 155
deleterious mutation 6–7, 18, 23, 41, 45–48, 53–55, 58, 61, 63, 65, 70–76, 78, 80, 84, 94, 99–101, 107, 109, 111, 114–116, 146, 161, 177, 180
dendritic cells 187
dengue
  antibody-dependent enhancement (ADE) 186
  fever (DF) 181
  haemorrhagic fever (DHF) 181, 186
  human virus (DENV) 51–52, 59, 73–74, 76–77, 84, 96–97, 114, 124–127, 133, 140, 144, 146, 176, 180–191
  shock syndrome (DSS) 181, 186
  sylvatic virus 139, 144, 181–184, 188, 190
d'Herelle, Felix 9
digital organism 91, 93
DNA polymerase 11
  DNA polymerase II 20, 22
Domingo, Esteban 87, 93
Drake, Jan 38, 42, 45
Drosophila 1, 66, 68
drug resistance 8, 53, 55, 97, 104, 177–178
  adamantane 59, 62, 85–86
  zidovudine (AZT) 7, 59–60, 98, 168

Ebola virus (EBOV) 43, 50, 52, 141, 153
  Ebola virus Zaire (EBOV-Zaire) 43
ecological transition 153–155
effective population size (N_e) 54, 56–59, 60, 67–68, 92, 97, 134, 177–178
Eigen, Manfred 21–23, 47–48, 87, 91–92
  Eigen's paradox 22–24, 48
Elena, Santiago 71, 78
emergence 7–8, 45, 65, 109, 121–123, 131, 135–148, 155–158, 160, 167–168, 173, 188, 191
  cross-species transmission 7–8, 39, 63, 69, 121, 123, 128, 135–148, 154–155, 157–158, 167, 169, 175
  spill-over transmission 7, 24, 180, 184, 188, 194
epidemiological dynamics 132–134, 188, 199
envelope protein 11–13, 19, 26, 49, 76, 120, 124, 142,

epistasis 37, 53–54, 67, 77–82, 100
  antagonistic 54, 78–80
  synergistic 78–80, 111
episodic evolution 130, 162
error catastrophe 100–103
error threshold 22–23, 99–101
  relaxed error threshold 23
ExoN domain (3'-to-5' exoribonuclease) 48, 107–108, 119
extinction threshold 100–103
evolvability 94
evolutionary dynamics 28, 37, 43–44, 52, 59, 83, 87, 89, 93, 97–98, 100, 126, 128–129, 165, 167, 176–178, 186, 195, 199–200
evolutionary stasis 159
evolutionary trade-off 46–47, 65, 116, 123

feline calicivirus (FCV) 138
feline immunodeficiency virus (FIV) 137
Flexiviridae 84
fitness 71, 79, 90, 95, 143, 147, 160
Flaviviridae 70, 118, 121, 122, 127, 180–191
5-fluorouracil (FU) 95, 102
foot-and-mouth-disease virus (FMDV) 1, 9, 73, 99, 101–102
fungi 13, 24, 35, 167
  Pneumocystis carinii 167

gamma distribution (of among-site rate variation) 176
GB virus
  A (GBV-A) 43
  B (GBV-B) 84
  C (GBV-C) 43, 73, 138, 140
genetic code 35–36
gene duplication 6, 80, 109, 117–118
gene expression 10, 55, 66, 109, 112–114, 116
generation time 42–43, 60, 66
genetic drift 2, 37, 53, 55–70, 89, 92–93, 97, 116, 177, 199
genetic redundancy 80
genome evolution 34, 116–120
  gene content 18, 34
  gene order 31, 34, 114, 116
  genome orientation 34, 110, 113–114
  genome size 6–7, 10, 18, 23–24, 34, 47–48, 78, 82, 104–109, 112, 116–118
gibbon ape leukaemia virus (GALV) 128
greater spot-nosed monkeys (*Cercopithecus nictitans*) 175

Grenfell, Bryan 83
group selection 92

haemagglutinin esterase (HE) gene 118
hallmark gene 25
hantavirus 43–44, 73, 139
Harvey, Paul 133
helicase protein 26, 106, 108
hepatitis B virus (HBV) 22, 115, 140
hepatitis C virus (HCV) 53, 57–58, 83, 101, 134–135, 138, 151–152
hepatitis D (delta) agent (virus) (HDV) 10, 19, 21–22
herd immunity 188
herpesvirus 24, 26, 40, 137
Himalayan palm civet (*Paguma larvata*) 144, 147–148
hitch-hiking 62
Holland, John 93
human immunodeficiency virus (HIV)
    AIDS 1, 138, 155, 167–180
    circulating recombinant forms (CRFs) 169–170, 172
    groups 169, 171, 179
    HIV-1 38, 43, 46, 50, 57–59, 66, 98, 132, 134, 137, 140–144, 154, 167–180
    HIV-2 137, 140, 168–169, 171, 173, 175
    subtypes 154, 169–174
human leukocyte antigen (HLA) 43
    HLA-B27 60, 178, 180
human papillomavirus (HPV) 40
human T-cell leukaemia virus (HTLV) 15, 42, 138, 140
hunter-gatherer 154
hypermutation 65
hyperthermophiles 24

immune escape 46, 62, 138, 144, 177
    cytotoxic T-lymphocyte (CTL) escape 43 60–61, 178, 180
    epitope 60–61, 161, 191
immune-mediated selection 188
immune response
    adaptive 44, 46, 66, 142, 153, 177
    innate 46, 66, 70, 82, 177
    intrinsic 46, 65–66
    T cell 60, 66, 144, 173, 177
immunological landscape 123
influenza A virus 156–167
    antigenic cartography 162–166
    antigenic drift 161–167

antigenic shift 161
avian influenza virus (AIV) 15, 135, 145–148, 157–160
haemagglutinin (HA) 12, 57, 83, 98, 145, 148, 157–164-167
haemagglutinin inhibition (HI) assay 164
H1N1 subtype 57, 135, 157–159, 161–163, 165–166
H3N2 subtype 57, 62, 135, 158–159, 161–166
H5N1 subtype 142, 145, 157, 159–160
neuraminidase (NA) protein 12, 57, 148, 157–161, 165
'Spanish influenza' (1918) 15, 52, 156–157
sink-source model 149–151, 167
internal ribosome entry site (IRES) 80, 112, 120
International Committee on Taxonomy of Viruses (ICTV) 29
intra-host variation 41–45, 52–53, 57, 59–60, 73–74, 83–86, 91, 96–98, 103, 123, 132, 138–139, 146, 175–178, 180, 199
intron 22, 67, 116
    group II intron 26–27, 31–32
Ivanovsky, Dimitri 9

Japanese encephalitis virus (JEV) 181, 191
Jenner, Edward 8

Kamiti River virus 118
koala retrovirus (KoRV) 128–129
Koonin, Eugene 25–28

ladder-like phylogeny 98, 161–162, 178, 197
latent infection 42–43, 54
lateral gene transfer (LGT) 6–7, 9, 19, 24, 117–120
Levine, Arnold 9
life-history 48
lineage-through-time plot 133
linkage disequilibrium 53, 55, 60
lipid membrane 11
long terminal repeat (LTR) 128
Lynch, Michael 116
Lyssavirus 192–193

macroevolution 8, 37, 104–130, 176, 200
major histocompatibility complex (MHC) 141
May, Robert 88
Mayer, Adolf 9
Maynard Smith, John 153

measles virus 40, 127, 132, 138–139, 152
  subacute sclerosing panencephalitis (SSPE) 138–139
microevolution 37, 120, 186
mimicry 146–147
mimivirus 10, 14, 18, 25, 105, 200
mitochondrial DNA 10, 35, 40
  mtDNA plasmid 31, 35
mitovirus 10–11, 35, 104
mixed infection 49, 85–86, 111
modular evolution 118–120
molecular clock 127, 154, 184, 192, 195
  relaxed 133
  strict 66, 133, 183
molecular epidemiology 6, 83, 121, 131, 149, 191, 199
monocistronic RNA 112
Mononegavirales 29, 31, 33, 114
monopartite genome 13
mosquito 59, 63, 70, 76, 84, 118, 181, 186–187, 190
  Aedes aegypti 144, 182, 187–188, 190
mRNA molecule 11–12, 18–19, 21, 112–114
Muller's ratchet 53–54
multicomponent virus 13, 48–49, 59, 76, 109, 111–112, 120
multigene family 117
multipartite genome 13
multiplicity of infection (MOI) 75–76, 111
mutation rate 7, 24, 28, 37–39, 42, 44–48, 53–55, 70–71, 73, 75, 80, 87–94, 99–101, 103, 107, 115, 117
  universal mutation rate 38, 45
  deleterious mutation rate per genome replication (U) 53–54, 70
mutation-selection balance 87–88, 92, 103
mutational deterministic hypothesis 53–55
mutational meltdown 23, 47, 107
mutational robustness 23, 28, 47, 76, 78–80, 84, 89–90, 94–95, 100, 102
mutator strains 7, 47, 71
myxomavirus 9

Narnaviridae 35, 104
natural selection 1, 2, 9, 21, 37, 39, 46–47, 55–73, 84, 88–89, 91–92, 96, 98–99, 101, 103, 111, 116, 122–123, 134, 139, 159–161, 167, 176–178, 180, 187, 199
  immune selection 44, 98, 175 (also see immune escape)
  positive 44, 59–60, 63, 65–66, 96, 98, 109, 115, 144, 146, 161, 177
  purifying 43, 54, 63–64, 70, 73, 84, 177
  selective sweep 56, 177–178
  'survival of the fittest' 89, 93–94, 96
  'survival of the flattest' 80, 89–90, 94–95
Nee, Sean 133
neo-Darwinism 6
Neurospora Varkud satellite motif 21
neutral theory of molecular evolution 66, 98
  neutral mutation 67
  neutral network 165
  neutral space 71, 78
Nidovirales 29, 107–108
non-human primates 24, 139–140, 153, 155, 168–169, 173, 182–184, 188
non-structural protein 109, 157
nonsynonymous site 16, 67, 70
nonsynonymous substitution 40, 44, 175
Nowak, Martin 88
nucleocapsid protein 11, 13, 114, 157
nucleotide (base) composition 63, 68–70, 157
nucleotide substitution rate 39–46, 107–108, 128, 133, 159, 175, 178, 182, 195
  multiple substitution 83, 124

open reading frame (ORF) 10, 45, 107, 109–110, 112–117
  overlapping reading frame 10, 78, 83, 114–116
Orgel, Leslie 9
Orthomyxoviridae 33, 156

packaging 67, 106, 112
Paramyxoviridae 29
parallel evolution 28, 82–83
Partitiviridae 13
parvovirus 45, 142
  canine (CPV) 45, 142
  feline (FPV) 45, 142
  human 45
Pasteur, Louis 191
persistent infection 18, 43, 82–83, 136–139, 150, 155
phenotypic mixing 49–50
phylodynamics 131–135
phylogenetic incongruence 62, 165
phylogeography 148–153, 192, 195, 199
  gravity model 149–151
  source-sink model 149–150, 167
  transmission wave 149, 153, 191, 197
Picornavirales 29
pleiotropy 65, 67, 78, 82–83, 115

poliovirus 9, 38, 42, 50, 57, 69–70, 75, 94, 96, 98, 155
polycistronic RNA 113
polymerase chain reaction (PCR) 52, 83–84, 96
polyprotein 10, 31, 109–110, 112, 116, 181
population bottleneck 44, 54, 58–59, 153, 176–177, 186, 199
population subdivision 132, 134, 153–154, 178, 194
postweaning multisystemic wasting syndrome (PMWS) 45
potyvirus 29, 50, 119–120
primary infection 176–177
primate lentiviruses 128, 137, 148, 171, 173–176
proofreading 10–11, 39, 45, 107
protein structure 16, 20, 25–28, 34, 198–199
 α-helix 25–26
 β-sheet 25–26
pseudogene 67, 116
pyrosequencing 3, 199

quasispecies 7, 38, 46, 80, 87–103
 co-operation within 95–96
 lethal mutagenesis 47, 100–103
 master sequence 88–89
 memory 99–100
 mutational coupling 88–89, 91–92, 103
 phenotypic quasispecies 92
 quasispecies effect 94–95, 103

rabbit haemorrhagic disease virus (RHDV) 148
rabies virus (RABV) 153, 191–197
raccoon (*Procyon lotor*) 153, 191, 196
raccoon dog (*Nyctereutes procyonoides*) 195–197
reassortment 7, 16, 48–55, 62, 75–76, 86, 111, 118, 120, 147–148, 157–162, 165–166
recombination 6–7, 16, 37, 48–55, 63, 67, 75–76, 78, 98–99, 111, 117–118, 120, 125, 134, 147–148, 167, 169, 172–173
red fox (*Vulpes vulpes*) 153, 195–197
Reoviridae 25
repair 10–11, 39, 45, 53, 107–108
replication
 copy-choice 48–50
 geometric 39, 46
 stamping-machine 39, 46
retroelement 26, 32, 118
retrotransposon 19, 31, 32
retrovirus 4–5, 11–12, 18–20, 24–27, 31, 38–39, 42–43, 50, 54, 66 93, 99, 109, 118, 168, 175
endogenous retrovirus 43, 66, 128–130, 137, 168
reverse transcriptase (RT) 4–5, 11–12, 19, 25–26, 31–32, 39, 42, 50, 60, 98, 118, 128
Rhabdoviridae 25, 29, 192
ribavirin 47, 101
ribosomal frameshift 109–110, 113
ribozyme 20–24, 27
 CPEB3 ribozyme 22
 HDV 21–22
RNA-dependent RNA polymerase (RdRp)
 palm subdomain 19, 25–26, 28–29
RNA interference (RNAi) 66
RNA secondary structure 13, 21–23, 25, 34, 67, 77–78, 80–82, 84, 187
RNA world 7, 17, 19–24, 27, 103
Roniviridae 29, 48, 106–109, 121
rotavirus 1, 11, 83
Rush, Benjamin 181

Sanjuán, Raphael 78, 88
SARS coronavirus (SARS-CoV) 1, 108–110, 141, 144, 147–148
satellite RNA 10
segmented genome 4–5, 10, 12–13, 29, 33–34, 48–50, 54–55, 57, 61, 67, 104–114, 120, 156–162, 165–166
sexual reproduction 48–49, 53, 55, 67, 111
sexual transmission 8, 58, 136, 139, 168, 172, 177, 179–180
simian foamy virus (SFV) 24, 42, 136, 153, 175
simian immunodeficiency virus (SIV) 61, 144, 155, 168–171, 173–175
simian T-cell leukaemia virus (STLV) 15
Simmonds, Peter 81–82
slave trade 154, 182–183, 188, 190
sooty mangabey monkey (*Cercocebus torquatus atys*) 140, 168, 173
speciation 137, 191
 allopatric 121–124
 sympatric 121–124
stop codon 35, 54, 73, 76–77, 109, 114
subgenomic RNA 10, 30, 76, 109–114
Sulfolobus turreted icosahedral virus 26
superspreaders 58, 180
synonymous site 40, 63, 67–68, 80–81, 84
Szathmáry, Eörs 153

telomerase 26, 31–32
Tetrahymena thermophila 36
tobacco mosaic virus (TMV) 9, 44, 57
Tobamovirus 120
transcription 12, 13, 20, 67, 109–114
translation 9, 12–13, 35, 68–69, 80, 109–114
transmission network 135, 142, 145, 151
tropism 124
tRNA 19
Twort, Frederick 9
tymovirus 29, 114

untranslated region (3' and 5') 81, 117
urbanization 153–154, 190

vaccine 1, 8, 40, 95, 161, 167, 173, 181, 191
   vaccine failure 40, 55, 164–165
vector-borne transmission
   aphid 59, 63
   mosquito 59, 63, 70, 76, 84, 118, 144, 181–182, 186–188, 190
   tick 63, 70, 181, 184
Venezuelan equine encephalitis (VEE) virus 142
vertical transmission 128, 136, 139, 177

vesicular stomatitis virus (VSV) 46, 60, 70–71, 82–83, 89, 94–95, 113–114, 192
viraemia 173, 187, 190
viral supergroups 29–34, 120
viroid 4, 10, 20–22, 78, 94, 104
virion 11–12, 50, 57, 59, 70, 106, 124, 192
   icosahedral 25, 26, 28, 35
virophage 14, 200
virosphere 9, 13–14, 200
virulence 58, 81, 136, 148, 158, 172, 176, 180, 184, 186–187, 190
virusoid 10
virus origins 15–26

West Nile virus (WNV) 73, 76, 84, 139, 181
wild-type 6, 89
Wilke, Claus 88, 92, 103
Woolhouse, Mark 144

yellow fever virus (YFV) 38, 140, 154, 181–182, 188–191
   Philadelphia epidemic 181, 190